JOHN LESLIE

# TREATISES
## ON VARIOUS SUBJECTS OF
# NATURAL AND CHEMICAL PHILOSOPHY

Elibron Classics
www.elibron.com

Elibron Classics series.



ISBN 1-4021-6434-3 (paperback)
ISBN 1-4021-0538-X (hardcover)

This Elibron Classics Replica Edition is an unabridged facsimile of the edition published in 1838 by Adam and Charles Black, Edinburgh.

# TREATISES

## ON VARIOUS SUBJECTS OF

# NATURAL AND CHEMICAL PHILOSOPHY.

BY

SIR JOHN LESLIE,

PROFESSOR OF NATURAL PHILOSOPHY IN THE UNIVERSITY OF EDINBURGH.

WITH

A BIOGRAPHICAL MEMOIR.

REPUBLISHED FROM THE ENCYCLOPÆDIA BRITANNICA.

EDINBURGH:

ADAM AND CHARLES BLACK,

NORTH BRIDGE,

BOOKSELLERS TO HER MAJESTY FOR SCOTLAND.

MDCCCXXXVIII.

# ADVERTISEMENT.

This Volume contains all the more important Articles and Treatises, not already republished separately, which were contributed by the late Sir John Leslie to the Supplement to the former editions of the Encyclopædia Britannica, and which have been incorporated in the Seventh, now nearly completed. His Discourse on the "Progress of Mathematical and Physical Science during the Eighteenth Century," stands apart from his other contributions, and forms one of that series of similar Discourses which constitutes the first volume of the Work. With the exception of the Article on "Achromatic Glasses," which contains an historical review of those optical controversies and discoveries that terminated in the construction of the Achromatic Telescope, the rest may be said to belong chiefly to Pneumatics—one of the most extensive and interesting departments of Natural Philosophy. They are thus described in the work in which they originally appeared :—"The extensive branch of Pneumatics has been elucidated by a number of articles containing many new facts and views. Under the article Acoustics are examined the later experiments and opinions as to the theory of sound. The subject of Balloons has been resumed, in order to trace the progress of discovery with greater precision, and to explain the calculation of the ascent and stability of these machines. This has been done in the article Aeronautics, where also some remarkable aerial voyages are described. With a view to explain the successive steps which led to the discovery of atmospheric pressure, and to state and examine the various formulas that have been proposed for determining heights by the Barometer, there are two articles on that instrument, and its uses in measuring elevations. The connected subjects of Climate, Cold, Dew, and Meteorology, rank under Pneumatics, in as far as they can be considered as belonging to Natural Philosophy; but the phenomena to which they relate depend upon principles which place them partly within the province of Chemistry. In the very valuable series of Articles here given upon these subjects, the author has accordingly drawn his reasonings and illustrations from either department. The last of them contains a copious explanation of the theory and applications of the different instruments capable of being employed in Meteorological observations;—a portion of it calculated to be eminently useful to those engaged in such observations, or in directing the proper preparations for making them."

The Memoir of Sir John Leslie prefixed to these Pieces was written for the seventh edition of the Encyclopædia by its Editor, who, in addition to long personal knowledge of their Author, had the advantage, in writing it, of all the information possessed by his family. Though necessarily brief and circumscribed, to suit it to the plan of the Work for which it was composed, it has at least the merit of presenting an authentic and tolerably comprehensive sketch of the life, labours, and character of that distinguished Philosopher.

# CONTENTS.

# MEMOIR

OF

# SIR JOHN LESLIE.

BY

MACVEY NAPIER, ESQ. F.R.S. LOND. AND EDINB.

# MEMOIR.

John Leslie, Professor of Natural Philosophy in the University of Edinburgh, was born at the village of Largo, in the county of Fife, on the 16th of April 1760. He was the son of humble, but, in their line of life, highly respectable parents, who lived to enjoy the celebrity of their son, and for whom he ever cherished that affection which formed a marked feature of his character in regard to all the members of his family. His father, originally from the neighbourhood of St Andrews, lived for some time at Anstruther, but ultimately settled at Largo, as a joiner and cabinet-maker. His wife, Anne Carstairs, was a native of that place; and the subject of this Memoir was the youngest child of their marriage. Though he attained in manhood a robust frame, he was, in early youth, of a very feeble constitution; so much so, that when sent, at four years of age, to a sort of school, kept by an old woman, who plied her wheel whilst teaching the alphabet, he was indulged with a separate stool near the fire-place, which the dame set apart for the feeblest of her juvenile pupils. As long as he was permitted to monopolize this seat of honour, he seems to have been tolerably pleased with his situation; but being at length superseded by a younger or more favoured pupil, he eloped from the school, hid himself for a day in some obscure corner, and, when obliged to come forth, obstinately refused to return to the tutelage of the ancient spinster. He was in consequence placed in another school, where he remained six months, and was taught writing and arithmetic; his father and his eldest brother, who appear to have possessed some knowledge of the elementary parts of mathematics, giving him, at the same time, his first lessons in that science. He was afterwards entered at a school in the neighbouring town of Leven, where Latin was taught; but his intense dislike to that language, and his inability, from weakness, to walk and return daily a distance of three miles, induced his parents to discontinue his attendance, after a short trial of six weeks. Such was the brief and meagre *curriculum* which formed the whole training of the future Mathematician and Philosopher, previous to his being entered a student at the University of St Andrews! But we must not too hastily dispatch this early period of his life, when his genius, working with its own inward resources, already began to attract observation.

The first person of any sort of distinction who noticed his precocious attainments, was Mr Oliphant, who became Minister of Largo about the time when the boy had reached his eleventh or twelfth year. Struck with the knowledge in mathematical and physical science which he displayed, the Reverend gentleman kindly lent him some scientific books, with which he was but poorly provided; and he also strongly urged him to study Latin—telling him, by way of showing the necessity of an immediate commencement, that it had cost himself seven years to acquire that language. This was the worst argument that could be urged to the young philosopher, who unhesitatingly declared that he never would bestow half that time upon any language, and that he particularly disliked Latin. In this state of his knowledge and taste, he was, in his thirteenth year, sent to the University of St Andrews to study mathematics under Professor Vilant. On examination by the Professor, he was found already qualified for the second or senior class; and, at the close of the session, he obtained a prize. It is remembered, as a characteristic particular, that having previously discovered, in some of those antiquarian researches to which he was from early youth addicted, that it was not indispensable for students of the first year to wear a gown, he steadily refused, during this year, to exhibit himself in the accustomed academical habiliment.

This session proved a decisive one as to the course of his future studies. The Earl of Kinnoull, then Chancellor of the University, having been informed of his remarkable abilities, sent for his father, and proposed to him to defray the expense of his son's education, provided the father would agree to maintain him at college, with the view of qualifying him for the Church. The proposal was readily embraced; and the repugnance of the youth to apply himself to Latin was at length overcome, by making the permission to attend the Natural Philosophy class of next session—the great object of his desire—conditional on his agreeing to qualify himself, during the vacation, for attending also the Latin class. With this lure before him, he applied assiduously to his lessons, under the direction of a private teacher, and succeeded in fitting himself for admission into that class. No one could discover, in his after life, any traces of this early and vehement dislike to Latin; for though he ever held* that the learned languages were suffered to engross too much attention in our system of education, and was by no means sparing of reprehension upon this subject, his own scholarship had become considerable; and indeed his writings manifest a more than ordinary degree of fondness for embellishing the conclusions of science with illustrations from the Greek and Latin Classics. He continued to the last to read them occasionally, particularly Lucretius, whose bold and imaginative philosophy, and splendid descriptions, were peculiarly adapted to his taste. He came also to be well versed in the theory of grammar; and his observations upon language often evinced learning as well as ingenuity.

His health, at the period above alluded to, was still so delicate, that it became necessary to moderate and regulate his studious habits; but he succeeded, during his second session, in acquiring additional honours; and also in attracting in a more remarkable degree the flattering attentions of the Chancellor,

* "The great error in modern education consists in the undue attention paid to the dead languages, which consumes the precious time that should be devoted, during the freshness of youth, to the higher intellectual pursuits."—Preface to *Rudiments of Geometry*, published in 1828.

who kindly invited him to Dupplin Castle, where, notwithstanding the bashfulness of his manners, he contrived to impress the other visitors with a high opinion of his powers. About the same time, he became known to Mr Playfair, on occasion of a visit which the latter, alongst with the Reverend Dr Small, made to St Andrews. Dr Small's son was a fellow-student and companion of young Leslie, and hence his introduction to both these mathematicians. Mr Playfair was at this time parish Minister of Liff, in Forfarshire. Here he was afterwards visited by his young acquaintance, neither of them then dreaming of that lot which was to place both, in succession, in two conspicuous Chairs of the University of Edinburgh, destined to derive from their talents additional lustre and recommendations. His visits to Mr Playfair were continued after the latter, in 1782, resigned his clerical charge, in order to superintend the education of Robert Ferguson, Esq. of Raith, and his brother, General Sir Ronald Ferguson. It was in this way, we believe, that he first became known to these excellent and distinguished men, then youths of his own age; and the acquaintance, in after years, ripened into a warm and lasting friendship, alike honourable to both parties, and which formed one of the chief and most valued solaces of his life.

In 1783 or 1784, he quitted St Andrews, and proceeded to Edinburgh, with the intention of entering himself as a student of Divinity in the Metropolitan University. He was accompanied, we believe, by another young Mathematician, destined, like himself, to obtain a distinguished niche in the Temple of Fame —James Ivory; and they lived together for some time. He never had any liking for the Church as a profession; and though he was formally entered at the Divinity Hall, he contrived to devote his first session to the sciences, particularly to Chemistry. In fact, he seems early to have relinquished all thoughts of the Church;—a resolution perhaps hastened by the death of his patron, the Earl of Kinnoull, which took place soon after his removal to Edinburgh. He continued to study at this University till the close of the session of 1787; and, as is customary with students of greater ability and industry than means, devoted part of his time to private tuition. One of the young men whose studies he assisted was nearly related to, and became the heir of Dr Adam Smith;—a circumstance which he was accustomed to recollect with pleasure, as having made him known to that illustrious Philosopher, who treated him kindly, and occasionally favoured him with directions as to his own pursuits. His first essay as an author must have been composed about the time of his leaving the University. Its subject was *The Resolution of Indeterminate Problems;* and it was read to the Royal Society of Edinburgh, by Mr Playfair, in 1788, and afterwards published in its *Transactions*.*

In this year he was prevailed upon by two students of the name of Randolph, and of the distinguished American family of that name, to accompany them, in capacity of tutor, to Virginia; and he accordingly left Scotland alongst with them. They arrived at the place of their destination in the beginning of November; and his time afterwards seems to

* See vol. ii, p. 193.

have passed both agreeably and usefully. He was all his life fond of visiting other countries, and perhaps a little disposed to underrate his own. Thus he was wont to say, that the thunder-storms of Virginia took away all feelings of awe at those of Scotland, just as the Alps of Switzerland left him nothing to admire in the Scottish mountains. His stay in the New World did not much exceed a year; owing to the breaking up of the family establishment by the death of the father of his young friends. After visiting New York and Philadelphia, he returned to Britain, and to his native place, towards the close of the year 1789.

From some letters to his family, written about the time of his leaving America, his thoughts seem to have been anxiously directed to his future means of employment and support; and one of his schemes appears to have been to try his fortune in India,—probably as a Civil Engineer. This notion recurred afterwards, but without leading to any results. His next field of adventure was London, whither he proceeded in January 1790, carrying with him various letters of recommendation; one of which was written by Dr Adam Smith, then drawing near the close of his memorable career. It was on this occasion, if our recollection does not mislead us, that Dr Smith exhorted his young friend never to approach any author whose favour he might wish to win, without first reading his works, lest the conversation should happen to turn that way. One of Mr Leslie's objects in visiting the capital, was to ascertain what success he might expect from a course of Lectures of Natural Philosophy; and the information he received soon satisfied him, as he says in a letter to one of his brothers, that "rational lectures would not succeed." He therefore employed himself for some time in writting for the *Monthly Review*, and in executing literary jobs delegated to him by his countryman Dr William Thomson, author of the continuation of Watson's *History of Philip the Third*, and of many other works now forgotten; and who was much in the habit of lending his versatile pen, as well as his name, to those who required the assistance or recommendation of either. But a more eligible and suitable connection was ere long opened to him, by the invitation of the younger Wedgwoods, who had been his fellow-students at Edinburgh, to reside with them, and superintend their studies. He readily acceded to this proposal; and proceeded, in April 1790, to their residence at Etruria, in Staffordshire, where he remained till the close of 1792, in the enjoyment of a liberal salary, and of society at once agreeable and intelligent. This was, in every sense, a happy and prosperous period of his life. The time not devoted to tuition was assiduously employed in experimental investigations, and in completing a translation of Buffon's *Natural History of Birds*, which he had previously undertaken for a London bookseller. It was published in 1793, after he left Etruria, in nine volumes octavo. Though executed with fidelity and vigour, it was valued by himself, at least after he became otherwise eminent, only as having, by the sum which it procured him, laid the foundation of that pecuniary competency which he early foresaw to be necessary to the independent prosecution of his favourite studies; and which his industrious and prudent habits enabled him in no long time, in a moderate degree, to attain. The preface, which is written in that nervous, but strained and ornate style, displayed in all his after writings, nevertheless alludes to this "first attempt" with considerable anxiety; and endeavours to bespeak favour for it as executed "at an early period of life, and in the retirement of the country." His first contribution to Natural Philosophy, entitled *Observations on Electrical Theories*, was also written at Etruria; and was, in 1791, transmitted to the Royal Society of Edinburgh, for insertion in its *Transactions.* This was promised; but performance having, as he thought, been long and unhandsomely deferred, he indignantly recalled his paper, and laid it aside for many years. It did not, indeed, see the light till the year 1824, when he was induced to publish it in the *Edinburgh Philosophical Journal*,* conducted by his distinguished friend Professor Jameson.

About the period of the close of his agreeable engagement at Etruria, which was occasioned, we believe, by the ill health of Mr Thomas Wedgwood, and the marriage of his younger brother, a Cornish gentleman with whom he had become acquainted at that place, proposed to engage him as a companion in an extensive tour through Greece, Egypt, and the Holy Land. Ever fond, as he was, of visiting foreign countries, particularly those hallowed by memorable events or classical associations, the proposal could not but prove agreeable to him; and it was accordingly embraced with alacrity. The tour was to commence early in 1793; but a change of plan having been decided upon without his concurrence, and by which he

* See vol. xi. of that work.

found that his observations were to be transferred to another, he immediately relinquished his engagement. His intention to visit these interesting countries, when a fit opportunity should occur, remained with him through life; and, at a period nearly forty years subsequent to this,—in his last years, indeed,—he actually made some preparations for a year's sojourn in Egypt and Palestine, from which he was diverted only by engagements and avocations of which he found it difficult to disencumber himself.

After leaving Etruria, he passed some months in Holland; a country which, like Descartes, he seems to have thought peculiarly suited to secluded study, and where he at this time acquired the German language. Thereafter, he returned to Largo, where he remained for about two years, devoted to experimental researches; in the course of which he invented and perfected his Differential Thermometer,—the parent, if we may so speak, of that beautiful family of Philosophical Instruments* with which he enriched the Treasury of Science, and amplified and variegated the means of physical inquiry. His ingenuity had been early exercised in some attempts to construct an accurate Hygrometer, and these ultimately suggested

* These instruments, viz. the Differential Thermometer, Hygrometer, Hygroscope, Photometer, Pyroscope, Æthrioscope, and Atmometer, are all described, and their principles explained, by himself, in the articles Climate and Meteorology contributed to the *Encyclopædia Britannica*, and reprinted in this collection. Properly speaking, the Hygrometer was the *parent* instrument,—the Differential Thermometer having been invented in the course of his endeavours to improve it; but as the Hygrometer, in its latest form, is only a modification of the other, it may, properly enough, be represented as derived from it.

to him the well-known contrivance above named—a contrivance happily adapted to the measurement of the smallest variations of temperature, and which richly rewarded his inventive powers by its ministry to the achievement of his subsequent discoveries. It has generally, indeed, been allowed to be one of the most useful as well as elegant inventions that inductive genius ever applied to the investigation of chemical changes. At a later period, when his name had attained a high degree of celebrity, he, like most of the other sons of genius, was made to feel that fame brings with it pains as well as pleasures; for it was now rumoured that the Differential Thermometer, instead of being an invention of his own—perfected, as he has himself recorded, in the course of a series of experiments on the evaporation of ice, which the severe winter of 1794–5 afforded him an opportunity of performing*—was in reality a plagiarism, if not from Van Helmont, who died in 1644, at any rate from John Christopher Sturmius, who died some sixty years later. Such was one attempt to impeach the originality of this eminently inventive experimentalist. An instrument, of which all the world remained ignorant, till he, by means of it, told the world the better part of all it yet knows concerning the phenomena of radiated heat, was discovered to have been furtively purloined from one or other of these antiquated worthies! Neither the authors nor the abettors of this allegation pretended that either Van Helmont or Sturmius ever dreamt of such applications, or derived such results from their supposed invention, as were re-

* See the article Meteorology referred to in the preceding Note.

served, by some caprice of chance, no doubt, for its luckier plagiarist! and it is, we believe, generally admitted, that this is just one of those cases of partial anticipation so frequently to be met with in the history of science, and where the ultimate inventor shews, by his skilful and successful employment of the disputed invention, how much he surpassed, and how little he needed the help of those whom he is ungenerously supposed to have robbed of their legitimate honours.

In the Spring of 1796, Mr Leslie received an invitation from his friend, Mr Thomas Wedgwood, to accompany him in a tour through the north of Germany and Switzerland. To this proposal, which was every way agreeable, he immediately acceded. They arrived at Hamburg about the first of May, and employed that and the next four months in their tour, part of which was performed on foot. He alludes to some of their observations in one of the notes to his work on Heat; but of this, as of all his subsequent tours on the Continent, he kept regular Journals, which are still preserved, and which shew that he was no less observant of the social, moral, and economical condition of the countries he visited, than of their geological, meteorological, and physical aspects.

For some years after the period just mentioned, he seems to have employed himself in experimental pursuits, and to have divided his time chiefly between London and Largo, to which place, and to the society of his family, he ever was fondly attached. His earnings by literary labour, and his allowances, had raised his humble fortune to what a philosopher might view as an independence; and he accordingly employed or amused himself as his own inclinations dictated. Early in the Summer of 1799, he set out with an old college acquaintance on a tour to the northern kingdoms of Europe, in the capitals of which the latter, who had for some time been settled in Spain as a merchant, had business to transact. After traversing Denmark, part of Norway, and Sweden, they proceeded through Brandenburg to Berlin, and from thence returned to England about the end of November. From his Journal of this tour, which is more detailed than any of those he subsequently kept, the Swedish mines appear to have formed particular objects of his attention. One of his entries records his having, before quitting Hamburg, written an account of his Hygrometer, for insertion in Voght's *Magazine*, and in the *Annales de Chimie.*

In the following year, he published the paper just mentioned, but with some alterations, in Nicholson's *Philosophical Journal.* It is entitled a *Description of an Hygrometer and Photometer*, and was followed with *Additional Remarks* on these instruments. In the same year he also published, in that Journal, a Paper *On the Absorbent Powers of the Different Earths;* and other two, containing *Observations and Experiments on Light and Heat, with Remarks on the Inquiries of Dr Herschel on these objects.** These small pieces are very valuable, as shewing the progress of his researches and discoveries in that field of inquiry which their titles indicate.

---

* See Nicholson's Journal for 1800, vol. iii. p. 461–518, and vol. iv. p. 196, 344, 416.

The results of his more extended investigations were, ere long, to appear before the world in a different shape. Having collected at Largo all the necessary apparatus, he prosecuted with ardour a series of experiments, which enabled him, in the years 1801 and 1802, to compose the bulk of his celebrated work on Heat. In the latter year, "the gleam of peace," as he tells us in his usual ornate style, "tempted him to indulge in a temporary suspension, and to repair to the famed capital where the treasures of art and science are so profusely displayed. In that vortex of pleasure, and centre of information, I spent," he adds, "several months very agreeably; but the work I had undertaken recalled my thoughts, and I hastened again to my retreat."* His *Experimental Inquiry into the Nature and Properties of Heat* was here at length completed. It was published at London in the Spring of 1804, in an octavo volume, with a dedication, couched in terms of strong and affectionate friendship, to Mr Thomas Wedgwood, the companion of his studies at Etruria, and of his first Continental tour. The early death of that ingenious and excellent person, whose delicate health is here feelingly alluded to, was always mentioned by him as a public as well as private misfortune. The originality and boldness of the peculiar doctrines of the "Inquiry," and the number of new and important facts disclosed by its singularly ingenious experimental combinations, conspired to render it an object of extraordinary interest throughout the scientific world; and, indeed, it must ever be viewed as constituting an era in the

* Preface to the Inquiry on Heat.

history of that recondite branch of Chemical Science which forms its subject. The Council of the Royal Society of London unanimously adjudged to its author the Rumford Medals appropriated as the reward of discoveries regarding the nature of Heat. As a philosophical disquisition, it is far, however, from perfection. Its hypotheses are not warranted by the sober maxims of inductive logic, and its method and style are alike liable to serious criticism. But it would be difficult to name any work in the whole range of modern science more strongly indicative of vigorous and inventive genius; and it must be allowed by all, that its skilful experiments, and its large stock of new observations, more than atone for its questionable doctrines; and for that desultory arrangement, and those ambitious modes of expression, which so often mar its reasonings and obscure its sense. More than ten years before the appearance of this work, Mr Leslie wrote an Essay on *Heat and Climate*, which contains some of its theoretical opinions, as well as the germs of its discoveries. It was read at two successive meetings of the Royal Society of London in the Spring of 1793, but it was not admitted into the *Transactions* of that body. Its author was not of a disposition to be checked in the career of inquiry by this repulse; and he did not shrink from bringing his paper forth, in 1819, from its long oblivion of twenty-six years,—he having then published it, for the first time, in Dr Thomson's *Annals of Philosophy*.*

---

* See vol. xiv. p. 5–27, and vol. xvi. p. 7, for some just observations by the able Editor.

Twice before that period of his life at which we are now arrived, Mr Leslie had appeared as a candidate for an Academical Chair;—first in the University of St Andrews, afterwards in that of Glasgow, and on both occasions without success. Early in the year 1805, a vacancy occurred in the Mathematical Chair of the University of Edinburgh, owing to the removal of Professor Playfair, on the death of Dr John Robison, to that of Natural Philosophy. This afforded a new opening to his academical ambition; and he was particularly desirous to occupy a Chair upon which the names of the Gregories, of Maclaurin, of Matthew Stewart, and of Playfair, had shed so much lustre. He accordingly presented himself as a candidate; and, with his now high reputation as a discoverer and original thinker, and his acknowledged eminence in mathematical science, it was not to be expected that he could have any formidable competition to surmount. Nor did there occur any, in as far as fame or talents were concerned. His principal competitor, though a man of amiable character, and respectable attainments, was wholly unknown in the scientific world, and as inferior to him in abilities as in renown. But he was one of the Ministers of the City, and supported with all the influence of that active body, then pretty generally suspected of a wish to secure a monopoly of the philosophical chairs of the University. It soon, however, became known, that the Patrons were determined to decide upon a comparison of claims, and that Mr Leslie must triumph. In this state of things, and in an evil hour for themselves, the City Clergy were induced to raise an objection to his eligibility, on the serious ground of his having, in one of the Notes appended to his work on Heat, approved of a doctrine directly leading to atheism. It is far from being agreeable to notice this discreditable proceeding; but it forms too memorable an occurrence in Mr Leslie's life to allow it to be passed without some animadversion. In the Note alluded to, the author, though no metaphysician, and in general rather a contemner of metaphysical science, was naturally enough led to illustrate what he had said in the text, with reference to the unphilosophical opinion that impulse is necessary to motion, by some remarks on the doctrine of Causation. He prefaces these remarks by observing, that Mr Hume was the first who treated this subject "in a truly philosophical manner;" and that "the unsophisticated notions of mankind are in perfect unison with the deductions of logic, and imply nothing more at bottom, in the relation of cause and effect, than a consequent and invariable sequence."* Founding upon these observations, the Ministers of Edinburgh accused him of having "laid a foundation for rejecting all the argument that is derived from the works of God to prove either his Being or Attributes."† This heavy charge was preferred in a formal Protest, tendered by them to the Patrons of the University; in which they alleged that, in the election of Professors, the former were, by the Charter of erection, bound to act "with the advice of the City Ministers." That advice was, in the present instance, given with sufficient emphasis; but

---

* *Inquiry into the Nature of Heat*, p. 135, and Note 16, p. 522.

† See Professor Stewart's *Short Statement of Facts relative to the Election of a Mathematical Professor in the University of Edinburgh*, p. 44.

the Patrons, much to their honour, treated it as it deserved; and Mr Leslie, to the great joy of all liberal minds, was, in March 1805, elected to the Mathematical Chair. The efforts of the disappointed junto did not, however, cease with this rebuff; nor did they desist from their ill-starred opposition, till a decision of the General Assembly of the National Church, pronounced on the 23d of May, after a memorable debate of two days,* satisfied them that persecution had now exhausted its resources, and that its hopes must, however sorrowfully, be relinquished.

Dismissing every supposition of interested designs, and even allowing that Mr Leslie's expressed opinion as to Causation, if taken apart from the subject-matter of his book, or left unexplained, was calculated to occasion some alarm in the minds of the pious, still, impartial history ever must brand the proceedings of his opponents as alike uncharitable, unfair, and arrogant; and, as deeply injurious to the character of that Church with whose name and authority they audaciously clothed themselves. It was on all hands admitted, that if Mr Leslie had, by a single word, limited his remarks to *physical* Causation, they would have been wholly free from objection; and, surely, it was a most perverse and intolerant construction, to extend to any but *physical* causes, his observations on those theories which resort to certain invisible *intermedia*, in order to account for the connection of physical sequences. But this was not all. Mr Leslie, on being informed

* See *Report of the Debate*, published in an octavo volume, at Edinburgh, in October 1805.

of the charge, immediately declared, in a very pointed Letter laid before the junto, that his reasonings "*referred entirely* to the relation between cause and effect, considered as an object of *physical* examination."* Yet was this prompt explanation disregarded —nay suppressed; whilst his persecutors—owing to an ignorant blunder in their statement of what they conceived to be the true notion of Causation—were themselves obliged to have recourse to explanation, in order to show that their own doctrine was not identical with that of the Fatalists and Spinoza! †

Mr Leslie commenced and prosecuted his official duties with great ardour. He entertained lofty ideas of the dignity and utility of the professorial character, and was thus disposed to make all the exertions necessary to success. Though the bent of his genius lay more to Physics than to Pure Mathematics, he had cultivated the study of Geometry with kindred relish; and with an admiration, in particular, of the analytical

---

* See Professor Stewart's *Short Statement*, p. 36, and *Report of the Debate in the General Assembly*, p. 16.

† See *Short Statement*, p. 77-94, and *Report of the Debate, passim.* This remarkable controversy gave rise to a number of other publications; but none of them, with the exception of these two, an admirable *Letter* to the author of a Reply to the former, by Professor Playfair, and Dr Thomas Brown's *Observations on the nature and tendency of the doctrine of Mr Hume concerning cause and effect*, with other two pieces by that most acute metaphysician, have the least chance of interesting posterity. The two pieces alluded to were, *A short criticism of the terms of the charge against Mr Leslie in the Statement of the Ministers of Edinburgh*, and *An examination of some remarks in Dr Inglis's Reply to Professor Playfair.* The Reply by Dr Inglis was the ablest production on the clerical side, and certainly evinced a good deal of controversial ability.

investigations of the ancient Geometers, which led to his happiest essays in that science. As a teacher, he not only laboured to promote the study, but to procure for it a larger share of attention than our academical system usually assigns to it. His instructions were better suited, perhaps, to youths of superior ability than to ordinary students; but great reputation and intellectual power produced their usual results; and secured for him an attendance as numerous as could be expected by any teacher, during the whole of the fifteen years that he occupied the Mathematical Chair. Soon after his election, he resolved to compose and to publish, at successive intervals, a complete "Course of Mathematics," digested and arranged according to his own ideas of what was wanted towards promoting a purer taste in the cultivation of the science. Of this Course, the first volume, comprising *Elements of Geometry, Geometrical Analysis, and Plane Trigonometry*, was published in 1809; and it has gone through several extensive editions. The first part of it has generally been regarded as the least perfect and useful; but his own favourite portion, on *Geometrical Analysis*, was extolled even by the most unsparing critics of the former, as "a great acquisition to Elementary Geometry; and as calculated to keep alive the knowledge of a most beautiful and interesting branch of the mathematics, which has been too much overlooked during the improvement of the more general and powerful methods of algebraic investigation."* Abroad, it seems to have been viewed in a light equally favourable; as it was speedily translated into

* *Edinburgh Review*, vol. xx. p. 98.

the French and German languages. He reproduced it, with considerable emendations, in the second volume of his "Mathematical Course," which, besides, contains the *Geometry of Curve Lines.* This volume was not published till 1821—an interval of twelve years having thus elapsed from the appearance of the first. A third volume, on *Descriptive Geometry, and the Theory of Solids*, was still wanting to complete his original design; but his removal to another Chair, and other circumstances, called his attention to different objects, and his "Mathematical Course" was thus left unfinished. He was induced, at a late period of life, to recast the first volume in a greatly abridged form. His object in doing so was to accommodate it to the use of those who, in riper years, become desirous to supply the defects of early education, and to qualify themselves for obtaining some knowledge of Natural Philosophy. This abridgment appeared in 1828, under the title of *Rudiments of Plane Geometry, including Geometrical Analysis, and Plane Trigonometry.* In connection with his mathematical works, though forming no part of his "Course," we may here mention the profound and learned treatise on the *Philosophy of Arithmetic*, which he published in 1817. It was a republication, with considerable alterations and additions, of one of the numerous articles contributed by him to the *Supplement* to the Encyclopædia Britannica.

It was not to be expected that the labours of the Mathematical Chair would wean Professor Leslie from his favourite experimental inquiries. His fine instruments were always at hand, and always in use, in connection with some ingenious conception, requiring to be prosecuted by means of their ministry. Early in the summer of 1810, he determined to proceed with a set of experiments previously suggested in the course of his researches with his Hygrometer, but which had for some time been suspended; and they now, on being resumed, conducted him to the discovery of that beautiful process of Artificial Congelation, by which he was enabled to produce ice, and even to freeze mercury, at pleasure. The discovery was achieved by means of a happily conceived combination of the powers of rarefaction and absorption,—effected by placing a very strong absorbent under the receiver of an air-pump. It was in the month of June of that year that the discovery was consummated.* We happened to witness this consummation,—at least the performance of the first successful repetition of the process,—and we never can forget the joy and elation which beamed on the face of the discoverer, as, with his characteristic good nature, he patiently explained its principles, and the steps by which he had been led to it. We could not but feel, on looking at, and listening to him, how noble and elevating must be the satisfaction derived from thus acquiring a mastery over the powers of Nature, and enabling man, weak and finite as he is, to reproduce at pleasure her wondrous works! Proportioned to the admiration which such achievements are calculated to inspire, ought to be the disapprobation of any unfair endeavours to lower or depreciate them. We have already alluded to an attempt to divest this illustrious experimentalist of the honours con-

* See the article COLD in this Collection. The successive steps of the discovery are there recited by himself, in those *verba ardentia* which the bent of his genius so strongly prompted him to employ.

nected with his Differential Thermometer; and we have now to add, that several years after his discovery of the process of Artificial Congelation, some similar endeavours were exhibited to transfer the merit of that discovery to a person of the name of Nairne. The claim for him was founded on a Paper published in 1777, in the *Transactions* of the Royal Society of London; from which he appears to have been acquainted with the facts, that evaporation produces cold, and that sulphuric acid, the absorbent employed by Mr Leslie, imbibes moisture. But, in order to decorate Mr Nairne with the laurels of the latter, his depreciators ought to have been able to shew, that the former had actually combined the properties alluded to, in a manageable process for the production of ice *ad libitum*. In such an attempt they must have failed ignominiously; for there is not in the whole history of science any more triumphant reply to a charge of plagiarism, than is furnished by the admitted facts, that with Mr Nairne's Paper before it for a long course of years, the scientific world remained utterly ignorant of the existence of any such process till the date of Mr Leslie's discovery;—nay, that with his description of that process in their hands, the most distinguished experimentalists of the capital failed in their attempts to repeat it, till it was exhibited there by himself, in the ensuing summer! *

* Some equally complete and curious legal evidence of these facts was adduced by Professor Leslie, in 1822, in a prosecution which he was advised to institute against the publisher of a well-known *Magazine*, for a series of libels inserted in that work; in one of which he was accused of having stolen the discovery alluded to from Mr Nairne. Amongst other witnesses, two very distinguished Chemists were ex-

In a letter from London to one of his friends, written in June 1811, he says—"My package has at last arrived, and I shall proceed without delay to make my *debut.*" It was only now that the experiment was first successfully performed in the capital. This took place in presence of some members of the Royal Society. The discovery was announced in the same year in the Memoirs of the French Institute, and the process itself was afterwards exhibited before that body by M. Pictet and M. Gay-Lussac.* He did not himself publish any detailed account of the experi-

---

amined on that occasion—the late Dr Marcet, and Dr Thomson, Professor of Chemistry in the University of Glasgow. The evidence of both was equally honourable to Professor Leslie. We shall extract a small portion of that of Dr Marcet, to whom the Professor was personally but little known:—"*Q.* Is it your opinion that Mr Leslie is to be considered as having borrowed or stolen this discovery, or do you consider his discovery to be original?—*A.* Some of the facts were known long before, but the process itself is perfectly original.—*Q.* Is the discovery of Mr Leslie analogous to other discoveries in the science of chemistry?—*A.* There is hardly any discovery of the least value that has been made in that science but from the known properties of bodies. It is by combining those properties, so as to produce certain effects, that a discovery is made.—*Q.* Then you mean to say that Mr Leslie has done what none before him ever accomplished? *A.* Certainly. *He has done what the whole philosophic world, with all the facts before them for a long period, had not been able to accomplish.*—*Q.* When and by whom was the experiment first successfully performed in London?—*A.* It was first successfully performed in London by Mr Leslie himself. My belief is, that no one succeeded in this experiment, in London, until Mr Leslie himself shewed the way.—*Q.* Do you know that Sir Humphrey Davy tried and failed? *A.* I cannot positively say; I believe he tried it, but without success."—(*Report of the Trial by Jury, Professor Leslie against William Blackwood*, p. 82–6.)

* See *Mémoires de la Classe des Sciences Mathématiques et Physiques*, t. xii. p. 80; t. xiv. p. 117–18.

ments connected with this discovery till 1813, when he illustrated them at considerable length in a small volume, entitled, *A Short Account of Experiments and Instruments depending on the Relations of Air to Heat and Moisture.* This publication, which was partly intended to promote the circulation of his philosophical instruments, would have been infinitely more useful, had its author superadded the powers of methodical exposition to his other high endowments. But, notwithstanding its defects in point of arrangement, it was greatly extolled by those competent to appreciate its value as a contribution to science.* Closely connected with the subject of this treatise, and which we may therefore notice here, though it did not appear till some years later, was an ingenious Paper, published in 1818, in the *Transactions* of the Royal Society of Edinburgh,—*On Certain Impressions of Cold transmitted from the Higher Atmosphere; with a Description of an Instrument adapted to measure them.* This instrument, to which he gave the name of Æthrioscope, is described, in another place, in the poetical language of its author, as "fitted to extend its sensation through indefinite space, and to reveal the condition of the remotest atmosphere!"

---

* See *Edinburgh Review*, vol. xxiv. p. 339–52, and Dr Thomson's *Annals of Philosophy*, vol. ii. p. 457–62, for a skilful analysis of the treatise. In the trial above alluded to, the party prosecuted took an *Issue* to shew, that the above article in the *Edinburgh Review*, in commendation of the work, was written by Mr Leslie himself! The attempt was not made; and it is hardly worth while to mention, with reference to such a charge, that the article was written by that very able chemist, the late Dr John Murray.

In the Autumn of 1814, Professor Leslie indulged himself with a tour of six weeks in France and the Netherlands. He never was satisfied if he allowed a vacation to pass without visiting some foreign scenes. One or two extracts from letters written by him on the present occasion may be here introduced, as either curious in themselves, or characteristic of the writer. Writing from Paris, on the 1st of August, he says,—"You know that it was not my intention at present to mix much with the *Scavans*. But I have been so well received, and even feasted by them, that I may perhaps depart a little from my original design. Humboldt has been very kind and attentive to me, and introduces me wherever I want. They are much better acquainted with what we are doing than I should have imagined. My book on Heat is better known than in England. I was even reminded of some passages in it which in England were considered as fanciful, but which the recent discoveries on the Polarity of Light have confirmed. Even Laplace has, in consequence of some observations of mine, silently omitted a passage in the last edition of his *Systeme du Monde*. I paid a visit the other day at Arcueil. Berthollet has a fine chateau seated on a bank amidst gardens, vineyards, &c.; and Laplace has another, little inferior, and adjoining to the grounds. I dine with Laplace next Sunday. Some person had informed him that I was the author of a critique in the *Edinburgh Review* on a paper of his, and he had sent an answer to me, which, however, I never received."* The following extract

* The paper of Laplace, here alluded to, was that *On the Motion of*

from another letter, written at Bordeaux in the beginning of September, gives a lively sketch of his journey to that place:—"My tour has furnished what I wanted—a number of images of the milder and hotter regions of the Continent. From Paris I proceeded to Macon, over a rich and well-cultivated country, covered with wheat and vineyards, the crop for the most part already gathered in. Thence I descended the Somme, a fine clear river, to Lyons, the banks covered with luxuriant vineyards, stretching to a range of hills, and the waved surface sprinkled with trees, intermixed with frequent villages and lively villas, all of white limestone. At Lyons I met with the celebrated Baron Zach, and was conducted by him in his carriage to the fountain of Vaucluse, Avignon, and thence to Marseilles. His society was particularly entertaining and instructive. We now passed through the country of the mulberry, the fig, and the olive, but I confess that Provence did not come up to my expectation. I have seen, what I had much longed to see, the awful fountain of Vaucluse, and the blue expanse of the Mediterranean; but the shores at Marseilles are terminated by gray, naked rocks, shooting fantastically to great heights. I staid some days at Marseilles, and spent most of my time at a country-house occupied by the old Duchess of Saxe-Gotha, with whom Baron Zach lives in quality of Chamberlain. I found her very kind and affable, and

*Light in Diaphanous Media*, published in the *Mémoires de la Societé d'Arcueil.* These Memoirs had been reviewed by Professor Leslie, in two articles, in the fifteenth volume of the *Edinburgh Review.* They formed his first contributions to that Journal.

extremely glad to hear any news of the Royal Family of England. From Marseilles I returned to Avignon, and then crossing the Rhine, proceeded to Nismes, over a country extremely fertile and interesting. This is almost a Roman town. There is a temple to Augustus, beautifully Corinthian, a fine temple of Diana, and an amphitheatre almost entire, capable of holding twenty thousand spectators. I now proceeded to Montpellier, and saw the majestic range of the Pyrenees stretching on our left, and covered with eternal snow. Montpellier is an interesting place—its Botanic Garden rich, and its promenades superb. I then proceeded to Toulouse. In leaving it, the carriage plunged into a hollow across the road; it was eleven at night, but the moon shone full, and lighted up a fine rich plain. I shuddered when told that we had just crossed 500 dead bodies, which had been thrust into a cut or trench of the road after the late battle. The road to Bordeaux runs near the course of the Garonne, through one of the finest and richest countries I have ever seen." From this place he returned to Paris, and, after a short stay, proceeded through the Netherlands to Rotterdam, where he took shipping for Scotland.

The publication of the *Supplement* to the fourth, fifth, and sixth editions of the *Encyclopædia Britannica* commenced towards the close of the year 1815, and was continued progressively till its completion early in 1824. To that work, which was undertaken upon a very comprehensive plan, and which aimed at procuring the highest attainable assistance, Mr Leslie was an extensive contributor. His contributions display all the powers and attainments for which he was remarkable. Nor was it by his writings alone that he aided that publication. His advice, his information—amazing alike for its minuteness and extent—and his exertions in procuring useful assistance, were always at the service of its Editor, whose acknowledgment of these various favours has long been before the public.* But it is due to his memory here to particularize what the work owes to his genius and knowledge. Ranged in alphabetical order, his contributions occur under the following heads:—Achromatic Glasses; Acoustics; Aeronautics; Andes; Angle; Angle, trisection of; Arithmetic, palpable and figurate; Atmometer; Barometer; Barometrical Measurements; Climate; Cold and Congelation; Dew; Interpolation; Meteorology.

Another work which at intervals enjoyed the benefit of his co-operation, was the *Edinburgh Review*. But his contributions, though commencing in the year 1809, were not numerous. They helped, however, and that in an eminent degree, to strengthen and diversify the scientific character of that Journal; whilst those in the interesting department of Voyages and Travels combined scientific observation with powers of writing of no ordinary description. Among his principal articles may be mentioned those on the Memoirs of the Society of Arcueil; on the History of the Barometer; on Delambre's Arithmetic of the Greeks; on Von Buch's Travels; on Humboldt's Physical View of the Equatorial Regions, and his Travels; on the attempts to discover a North-West Passage

---

* See Preface to the Supplement, p. 13.

to Asia. The picture in the last, of the revolving year, as observed within the arctic circle, is executed with a force of conception, and a splendour of colouring, that have not often been surpassed.

In the year 1819, a new field of professorial labour was opened to Mr Leslie by the vacancy in the Chair of Natural Philosophy, occasioned by the death of one of the greatest ornaments of the University, Mr Playfair. Mr Leslie was on his return from a Summer trip to the Continent when this lamented occurrence took place. The news met him on being put ashore at Largo, and is thus mentioned in a letter from that place, written on the 1st of August, in which he also alludes to an accident that had nearly, as he says, deprived the University of another Professor: "After having been detained for about a week in Holland, and after a tedious but agreeable passage of near ten days, I was at last put ashore here, from a sloop bound from Rotterdam to Grangemouth. Every thing looked joyous; but I had soon the tidings of poor Playfair's death, which was most unexpected and distressing. The loss to the University is severe. I suspect that you were nearly deprived of two Professors at once; for on Tuesday the 13th ultimo, I met with an alarming accident in Holland. I had passed the evening at a clergyman's with an intelligent merchant, late Provost of Aberdeen. On coming out of the house, at the end of the Boomtiges Street, or the street with rows of trees, the Maas running close, I stepped hastily forward to show him the Comet, and the quay being quite dark, I fell eight feet, and then plunged four feet in the water. This was the affair of an instant, and I felt that I was drowning; but I quickly recovered myself, and was helped out, with no damage but that of a bruise sustained in the fall." The eyes of the Patrons, and others interested in the welfare of the University, were now turned to Mr Leslie, as the person best fitted to fill the vacant chair. Three years before, when Mr Playfair was abroad, and an arrangement had failed for carrying on the Class, Mr Leslie, at the very commencement of the session, unhesitatingly undertook the task for his absent colleague;* but, independently of this circumstance, his eminence in mathematical and physical science was such, that no one else could reasonably be thought of; and he accordingly was, without difficulty, appointed to the Chair—being thus a second time, but in more melancholy circumstances than on the former occasion, nominated a successor to his early friend.

The Chair to which he was thus unexpectedly called was unquestionably that for which he was best suited; and had he happened to be placed in it at the commencement of his professorial career, Science in all probability would have derived greater benefits than she actually reaped from his powers. The time spent in his mathematical compositions would have been

---

* Mr Playfair expressed his gratitude, in very warm terms, for the promptness with which Mr Leslie, on this occasion, relieved him from what would have otherwise proved an unpleasant predicament; and it is but justice to the latter to mention the fact, as showing that the placability of his nature had entirely obliterated those feelings which the somewhat severe critique of his *Geometry* in the *Edinburgh Review*, known to have been written by Mr Playfair, would have left to rankle, and produce the ordinary results, in a mind more irritable, or less magnanimous.

more profitably employed in that wider and richer province which he was so peculiarly qualified to cultivate and improve. But let us not vainly repine. One of the first cares of his new situation was, the extension of the apparatus required in that greatly enlarged series of experiments which he thought necessary for the illustration of the Course. This, indeed, was an object of which, from the first to the last hour of his incumbency, he never lost sight; and it is due to him to state, that it was through his exertions that the means of experimental illustration, in the Natural Philosophy Class, were for the first time made worthy of the University. Viewing him merely as a lecturer, it must be admitted that he was not eminent. He was apt to forget, or rather did not perceive, that the connecting links between premises and conclusion, though familiar to the teacher, may, to the learner, be all unknown; that views quickly mastered by the acquired perceptions of the one, must be disclosed, step by step, to the yet imperfect vision of the other; and, that it is the imperative duty of a public instructor to bring down knowledge from its highest spheres, and place it on a level adapted to the powers of unpractised understandings. His ideas of the nature and dignity of science were grand and animating; and his strictures on the great discoveries which constitute the epochs of its history, sometimes swelled into lofty strains of eloquence; but, generally, in lecturing, as in writing, he wanted that consecution of thought, and that perspicuity of exposition, without which reasoning cannot be made intelligible, nor its conclusions satisfactory. Still, the attraction of his numerous experiments, the celebrity of his name, and the opinion entertained of his extraordinary powers, joined with great simplicity and affability of manner, concurred to secure the homage of the students, and to sustain the glory of the University. In 1823, he published, chiefly for the use of his class, a volume entitled *Elements of Natural Philosophy*; being the first of a Course intended to extend to three volumes, and to exhibit a comprehensive view of the principles of that congeries of sciences which we are accustomed to class under the above term. Here, as was the case with his Mathematical Course, his plan was not completed; for he published no part of it but the volume just mentioned, which includes only Mechanics and Hydrostatics. A second edition of this volume, corrected, and augmented with Notes, was published in 1829, three years before his death.

Mr Leslie had early determined to visit Italy at a fit opportunity; and he at last, in the summer of 1823, carried his design into execution. He set out with a mind still glowing with youthful enthusiasm, at the prospect of beholding the "eternal city;" but, whether from his entertaining too lofty conceptions, or from his view being too hurried and superficial, the tour ended in disappointment. The following letter, written at Innsbruck on the 11th of August, contains a brief outline of his journey: "I have thus reached the frontiers of Germany on my return from Italy. I have fortunately achieved the principal objects of my tour; and though I have travelled slowly, I have seen a great deal in a short time. From Geneva I proceeded by Lausanne up the Valais to Brieg—thence crossed the Simplon, and descended through the plains of Lombardy to Milan. From this place I advanced by Parma and Modena to Bologna—next crossed the Apennines to Florence—again crossed another part of that broad chain to Rome. For various reasons, I made the "eternal city" the limit of my journey. I therefore traversed the Apennines again to Ancona—skirted a considerable portion of the Adriatic—returned to Bologna, and thence proceeded to Mantua. I spent two days, about twenty miles from that place, at an old chateau, the residence of Acerbi, who accompanied me to Verona. Thence I went to Trent, and journeying through the Tyrol, have reached this spacious, interesting city. Italy has rather fallen below my expectations, whilst Switzerland has surpassed my early impressions. The passage of the Simplon alone is worth the journey; but the rout through the Tyrol, though not so sublime, is highly picturesque. Italy has every thing on a grand scale. The plains, mountains, rivers, works of art, are all majestic and noble. But there is no comfort. Rome itself stands in the midst of a desert. Its grandeur has always been artificial—the result of force or fraud. I have seen whatever is most interesting; and I am inclined to differ very widely from our ordinary travellers. I heard much of the *malaria*, but I escaped untouched. I suffered little from the heat, and bore it better than the natives. Conformably to the custom, however, I was a sort of prisoner during a great part of the day—the windows shut to exclude every ray of light; and I was only called at times to look out, by the babbling chant of the Monks, with their torches and crosses, carrying the dead to their graves, and followed by the Charitable Brethren, like ghosts, apparelled in white sheets, with only holes for their eyes. They seemed better fitted to terrify the living than to comfort the dying." Two years after this, he made another tour to the continent, in which he seems only to have gone over ground in France, the Netherlands, and Holland, which for the most part he had traversed before. This, we believe, was the last of his journeys abroad.

The only important production of Mr Leslie's latter years was that which formed his crowning benefaction to the Encyclopædia Britannica—his Discourse on the *Progress of Mathematical and Physical Science during the eighteenth century;* which, with others of a similar description, constitute its first volume. The opening tribute to Mr Playfair, of whose history of the earlier progress of these sciences this Discourse is a continuation, does honour alike to the writer's candour and taste. "The progress of mathematical and physical science during the brilliant period which closed with Newton and Leibnitz, has," he says, "been traced with fidelity and sustained interest by the hand of a master, whose calm judgment weighed impartially the different claims of discovery, whose powers of illustration could expand the fine results, and whose luminous eloquence was commensurate with the dignity of the subject." Nor is his observation on his own task less just; namely, that the more crowded field of discovery which it presents rendered it one of increased difficulty;—"its multifarious materials often lying scattered among the countless volumes of the Transactions of learned Societies." His arrangement of these materials, and his view of the whole subject, is comprehensive, vigorous, and spirited; and the greater ease and perspicuity of its style makes this the most agreeable perhaps of all his writings.

The volumes of the *Edinburgh Philosophical Journal*, published between 1824 and 1829, contain some small contributions, which may be here mentioned alongst with the other labours of his latter years. The first consists of *Remarks on the Light of the Moon and of the Planets*; the second, of an *Enumeration of the Instruments requisite for Meteorological Observations*; the third, of a *Letter on the Goniometer*; and the last, of *Observations on the Theory of Compression, applied to discover the internal constitution of our Earth.* The characteristic boldness and poetical dress of his speculations are abundantly displayed even in these brief productions. In the first of them, he endeavours to shew that the moon is a phosphorescent substance, like the Bolognian stone; and he anticipates a period when "she will no longer cheer our nights by her soft and silvery beams; when she will become dim and wane, and seem almost blotted from the blue vault of heaven. To our most distant posterity," he adds, "this prospect is indeed gloomy; but other changes will arise to renovate and embellish the spectacle of the universe." In the last, he arrives at the startling conclusion, that the crust of the planet on which we tread includes "an immense concavity, not dark and dreary, as poets have fabled, but containing *light* in its most concentrated state, shining with intense refulgence and overpowering splendour!"

Early in the year 1832, which unhappily proved to be that of his death, he was, on the recommendation, we believe, of Lord Brougham, then Lord High Chancellor, created a Knight of the Guelphic Order.

This honour was also conferred on several other distinguished men of science, about the same time. In a letter mentioning the occurrence, he says,—"My holiday title is now of course Sir John; but I shall always retain an affection for my old distinction of PROFESSOR." He had but few other titular distinctions of any sort; for he was far from setting any value on those arising from fellowship with Scientific Bodies. He was a member of the Royal Society of Edinburgh, but not of that of London. The only distinction of this kind that he in the least degree prized was his election as a Corresponding Member of the Royal Institute of France. This took place, with much honour, in 1820; the choice, if we recollect rightly, having fallen upon him, in preference to others then proposed, by a majority of thirty-three out of thirty-seven votes.

For a few years before the fatal one just mentioned, his occupations were agreeably diversified, by his attention to the improvement of a small estate, called Coates, situate near his native place, of which he had made a purchase. Here, the house and garden being every way commodious and suitable, and surrounded with scenes endeared by his earliest recollections, he loved to reside; and even those of his friends who most regretted that his precious time should be wasted on rural occupations, could not but sympathize with his feelings, and rejoice that the honourable labours of the Philosopher had enabled him to secure such a retreat. No one could enjoy more vigorous or constant health; and, though of a corpulent habit of body, he was exceedingly active, and fond of exercise. His strength, and the longevity of his family—a circumstance on which he himself always founded flattering hopes—alike gave promise of longer life; and it is melancholy to think that its close was but too probably hastened by one of his foibles—a contempt for medicine, and an unwillingness to think that he could be seriously ill. In the last days of October, whilst engaged in superintending some improvements on his grounds, he exposed himself to wet, and caught a severe cold. This was followed by erysipelas in one of his legs, which he neglected, and again imprudently exposed himself in the fields. He soon afterwards became dangerously ill, and expired at Coates, on the evening of Saturday the 3d of November 1832, in the sixty-seventh year of his age.

---

It has been well observed by Dr Johnson, that "of every great and eminent character, part breaks forth into public view, and part lies hid in domestic privacy. Those qualities which have been exerted in any known and lasting performances, may, at any distance of time, be traced and estimated; but those peculiarities which discriminate every man from all others, if they are not recorded by those whom personal knowledge enabled to observe them, are irrecoverably lost."* To prevent "this mutilation of character," as the same writer calls it, we shall close our narrative with a few details more particularly illustrative of the mind, opinions, and dispositions of this remarkable person.

* Johnson's *Life of Sir Thomas Browne.*

His discoveries, and the facts and controversies connected with them, may be discussed by many; and the full detail of them would require more space than can here be afforded; but, having enjoyed all the advantages for observation which long and intimate "personal knowledge" alone can supply, we think it right to endeavour, though briefly, to prevent some characteristic features from being "irrecoverably lost." The portrait may be imperfect; we cannot, indeed, complete it to our own satisfaction on our narrow canvass; but, in as far as the sketch extends, we can say that it is faithfully copied from nature. *

It would be impossible, we think, for any intelligent and well-constituted mind, thoroughly acquainted with the powers and attainments of Sir John Leslie, not to entertain a strong feeling of admiration for his vigorous and inventive genius, and of respect for that extensive and varied knowledge, which his active curiosity, his excursive reading, and his happy memory, had enabled him to amass. Some few of his contemporaries, in the same walks of science, may have excelled him in profundity of understanding, in philosophical caution, and in logical accuracy; but we doubt if any surpassed him, whilst he must be allowed to have surpassed most, in that creative faculty—one of the highest and rarest of nature's gifts—which leads to, and is necessary for discovery, though not all-

* The substance of some of the following observations appeared in the Newspapers immediately after the death of Sir John Leslie. They may, without impropriety, be used here, by the pen from which they originally proceeded.

sufficient of itself for the formation of safe conclusions; or in that subtilty and reach of discernment which seizes the finest and least obvious qualities and relations of things—which elicits the hidden secrets of nature, and ministers to new and unexpected combinations of her powers. "Discoveries in science," to use his own words, "are sometimes invidiously referred to mere fortuitous incidents. But the mixture of chance in this pursuit should not detract from the real merit of the invention. Such occurrences would pass unheeded by the bulk of men; and it is the eye of genius alone that can seize every casual glimpse, and discern the chain of consequences." With genius of this sort he was richly gifted. Results overlooked by others were by him perceived with a quickness approaching to intuition. To use a poetical expression of his own, they seemed "to blaze on his fancy." He possessed the inventive in a far higher degree of perfection than the inductive powers; and it thus sometimes happened, that his views and opinions were not only at variance with those of the majority of the learned, but inconsistent with one another. Notwithstanding the contrary testimony, explicitly recorded, by the founders of the English Experimental School, he denied all merit and influence to the immortal delineator of the Inductive Logic. He freely derided the supposed utility of Metaphysical Science, without perceiving that his own observations on Causation virtually expressed the important truth, that Physical is indebted to Mental Philosophy for the correct indication of its legitimate ends and boundaries. His writings are replete with bold suppositions; yet he laments the "ascendancy which the passion for hypotheses has obtained in the world." * His credulity in matters of ordinary life was, to say the least of it, as conspicuous as his tendency to scepticism in science. It has been profoundly remarked by Mr Dugald Stewart, that "though the mathematician may be prevented, in his own pursuits, from going far astray, by the absurdities to which his errors lead him, he is seldom apt to be revolted by absurd conclusions in other matters." "Thus, even in physics," he adds, "mathematicians have been led to acquiesce in conclusions which appear ludicrous to men of different habits." Something of this sort was observable in the mind of this distinguished mathematician. He was apt, too, to indulge in fanciful applications of mathematical reasoning to subjects altogether foreign to the science; as when he finds an analogy between circulating decimals and the lengthened cycles of the seasons! But when the worst has been said, it must be allowed that genius has struck its captivating impress over all his works. Whether his excursive speculations lead him to figure the earth as enclosing a stupendous concavity filled with light of overpowering splendour; or to predict the moon's arrival at an age when her "silvery beams" will become extinct; or to ascribe the phenomena of radiated heat to aerial pulsations; we every where perceive proofs of the originality of his conceptions, and the soaring character of his scientific aspirations. But this is not all. His hypothetical notions may be condemned and thrown aside; but

* See Introduction to *Elements of Natural Philosophy.*

his exquisite instruments, and his experimental devices, will ever attest the utility no less than the originality of his labours, and act as helps to farther discovery. We have already alluded to the extent and excursiveness of his reading. It is rare, indeed, to find a man of so much invention, and who valued the inventive far above all other powers, possessing so vast a store of information. Nor was it in the wide field of science alone that its amplitude was conspicuous. It was so in regard to every subject that books have touched upon. In Scottish history, in particular, his knowledge was alike extensive and minute; and he had, in acquiring it, gone deep into sources of information—such as parish records, family papers, and criminal trials—which ordinary scholars never think of exploring. The ingenious mathematician, the original thinker and discoverer, the rich depositary of every known fact in the progress of science, would have appeared to any one ignorant of his name and character, and who happened to hear him talk on this subject, as a plodding antiquary; or, at best, as a curious and indefatigable reader of history, whom nature had blessed with one strong faculty, that of memory. His conversation shewed none of that straining after "thoughts that breathe, and words that burn," so conspicuous in his writings. In point of expression, it was simple, unaffected, and remarkably correct. Though he did not shine in mixed society, and was latterly unfitted, by a considerable degree of deafness, for enjoying it, his conversation, when seated with one or two, was highly entertaining. It had no wit, little repartee, and no fine turns of any kind; but it had a strongly original and racy cast, and was replete with striking remarks and varied information.

Viewing his mind with reference to its moral attributes and habitudes, we must allow that it was not free from imperfections. He had prejudices, of which it would have been better to be rid; he was not over charitable in his views of human virtue; he was not so ready, on all occasions, to do justice to kindred merit as was to be expected in so ardent a worshipper of genius; and his care of his fortune went somewhat beyond what is seemly in a philosopher. But his infirmities were far more than compensated by his many good qualities;—by his equanimity, his cheerfulness, his simplicity of character, his straightforwardness, his perfect freedom from affectation, and his unconquerable good nature. He was, indeed, one of the most placable of human beings; and, notwithstanding his attention to his own interest, it is yet undeniable, that he was a warm and good friend, and a relation on whose affectionate assistance a firm reliance ever could be placed. He was fond of society, and greatly preferred and prized that of the intelligent and refined; but no man ever was more easily pleased; no fastidiousness ever interfered with his enjoyment of the passing hour; he could be happy, and never failed to converse in his usual way, though in the humblest company; and we have often known him pass an afternoon with mere boys, discoursing to them pleasantly upon all topics that presented themselves, just as if they had been his equals in age and attainments. He was thus greatly liked by many who knew nothing of his learning or science, except that he was famous for both.

He was never married. As to his person,—he was rather below the middle size, and corpulent, but well limbed; and though his face was large and florid, there was that about his eyes and forehead which seemed to shew that he was no ordinary man. There is a Bust of him by Joseph; a Portrait of the common size, taken a few years before his death, by Wilkie; and a Head, drawn at an earlier period, by Henning, which presents a very striking likeness.

# TREATISES

## ON VARIOUS SUBJECTS

OF

# NATURAL AND CHEMICAL

# PHILOSOPHY.

# ACHROMATIC GLASSES.

---

Achromatic Telescopes are telescopes contrived to remedy the aberrations in colours. The invention of the telescope, by which the powers of vision are extended to the utmost boundaries of space, forms an epoch in the history of science. The human intellect had at last emerged from the long night of error, and begun to shine with unclouded lustre. The age of erudition, which arose on the revival of letters, had been succeeded by the age of science and philosophy. The study of the ancient classics had infused some portion of taste and vigour. But men did not long remain passive admirers; they began to feel their native strength, and hastened to exert it. A new impulsion was given to the whole frame of society; the bolder spirits, bursting from the trammels of authority, ventured to question inveterate opinions, and to explore, with a fearless yet discerning eye, the wide fields of human knowledge. Copernicus had partly restored the true system of the world; Stevinus had extended the principles of mechanics; the fine genius of Galileo had detected and applied the laws of motion; the bold excursive imagination of Kepler had, by the aid of immense labour, nearly completed his discovery of the great laws which control the revolutions of the heavenly bodies; and our countryman Napier had just rendered himself immortal by the sublime discovery of logarithms. At this eventful period, amidst the fermentation of talents, the refracting telescope was produced by an obscure glass-grinder in Holland,—a country then fresh from the struggle against foreign oppression, and become the busy seat of commerce and of the useful arts. Yet the very name of that meritorious person, and the details connected with his invention, are involved in much obscurity. On a question of such peculiar interest we shall afterwards endeavour to throw some light, by comparing together such incidental notices as have been transmitted by contemporary writers. In the mean time, we may rest assured that the construction of the telescope was not, as certain authors would insinuate, the mere offspring of chance, but was, like other scientific discoveries, the fruit of close and patient observation of facts, directed with skill, and incited by an ardent curiosity. A new and perhaps incidental appearance, which would pass unheeded by the ordinary spectator, arrests the glance of genius, and sets all the powers of fancy to work. But the inventor of the telescope, we are informed, was acquainted besides with the elements of geometry, which enabled him to prosecute his views, and to combine the results with unerring success. No sooner was this fine discovery—admirable for the very simplicity of its principle—whispered abroad, than it fixed the attention of the chief mathematicians over Europe. Kepler, with his usual fertility of mind, produced a treatise on *Dioptrics*, in which he investigated at large the distinct effects of the combinations of different lenses.* Galileo, from some very obscure hints, not

* Kepler explained the construction of the astronomical telescope with two convex lenses; he likewise proposed a third glass to restore the inverted image. But Scheiner first employed the astronomical telescope, and described his observations with it in 1630. Father Rheita placed the third lens of Kepler near the primary focus, and thus enlarged the field of view. Such is the arrangement in the common spy-glass, which he gave in 1665.

only divined the composition of the telescope, but actually constructed one, with a concave eye-glass, which still bears his name. This telescope is shorter, but gives less light than another one proposed by Kepler, and called the astronomical telescope, which inverts the objects, and consists likewise of only two lenses, that next the eye being convex. With such an imperfect instrument—the same, indeed, though of rather higher magnifying power, with our modern opera-glass—did the Tuscan artist, as our great poet quaintly styles the philosopher, venture to explore the heavens.* He noticed the solar spots, surveyed the cavernous and rocky surface of the moon, observed the successive phases of the planet Venus, and discovered the more conspicuous of Jupiter's satellites. The truths thus revealed shook the inveterate prejudices of the learned, and furnished the most triumphant evidence to the true theory of the universe.

It is painful to remark, that the application of the first telescope in the country which had given it birth was directed to a very different purpose. The maker, after having finished one, judging it of singular use in the military profession, was naturally induced, by the hope of patronage, to present it to the younger Prince Maurice, whose bravery and conduct had so beneficially contributed to the independence of the United Provinces. But at this moment a bloody tragedy was acting in Holland. The chief of the republic, not content with that high station which the gratitude of his fellow-citizens had conferred upon him, sought

---

* ———— like the moon, whose orb
Through optic glass the Tuscan artist views
At evening from the top of Fesolé,
Or in Valdarno, to descry new lands,
Rivers, or mountains, on her spotty globe.
*Paradise Lost*, book i. 286—290.

to aggrandize his power by crushing all opposition. In the prosecution of his ambitious designs, he artfully gained the favour of the undiscerning populace, and joining his intrigues to the violence of the Calvinistic clergy, he succeeded in preferring the charge of a plot against the more strenuous supporters of the commonwealth, which involved them in ruin. Not only was the celebrated Grotius condemned to the gloom of perpetual imprisonment, but the aged senator Barneveldt, whose wise and upright counsels had guided the state amidst all the troubles of a long revolutionary conflict, was led to the scaffold, on the 14th of May 1619, while his persecutor, ashamed to approach the spectacle of his sufferings, beheld at a distance, with the coolness of a tyrant, from the windows of his palace, and *by help of a telescope,* the gesture and aspect of the venerable patriot, and all the melancholy circumstances attending the decollation.*

The skill and ingenuity of artists and mathematicians were now exerted in attempts to improve the construction of an instrument so fortunately contrived. The perfection

---

* The discovery of the telescope, from the mystery at first practised, is involved in considerable uncertainty. The most probable statement, however, ascribes the invention so early as 1590 to Zachary Jansen, an intelligent spectacle-maker at Middleburg. This intelligent person, led by accident to exercise his ingenuity on the subject, appears to have in private matured the execution of that wonderful though simple instrument. In a short time, however, the secret had transpired; and Laprey or Lippersheim, a townsman of the same profession, produced telescopes for sale between the years 1600 and 1610. But, in 1608, Jansen likewise constructed the compound microscope; and both instruments, by the activity of trade, were now spread quickly over Europe. The telescope was copied, and perhaps improved, by Adrian Metius, son of the celebrated mathematician. It was publicly sold at Frankfort in 1608, and in the following year the instruments were brought by Drebbel for sale to London.

of the telescope would require the union, as far as they are capable of being conjoined, of three different qualities,—distinctness of vision, depth of magnifying power, and extent of field. Of these requisites, the first two are evidently the most important, and to attain them was an object of persevering research. For the condition of amplitude and clearness, it was necessary that the principal image, or the one formed by the eye-glass, should be large, bright, and well defined. On the supposition then generally received, that, in the passage of light through the same media, the angle of incidence bears a constant ratio to the angle of refraction, which is very nearly true in the case of small angles, it followed, as a geometrical consequence, that the spherical figure would accurately collect all the rays into a focus. To obtain the desired improvement of the telescope, therefore, nothing seemed to be wanting but to enlarge sufficiently its aperture, or to employ for the eye-glass a more considerable segment of the sphere. On trial, however, the results appeared to be at variance with the hasty deductions of theory, and every sensible enlargement of aperture was found to occasion a corresponding glare and indistinctness of vision. But a discovery made soon afterwards in optics led to more accurate conclusions. Willebrord Snell, a very ingenious Dutch mathematician, who was snatched away at an early age, traced out by experiment, about the year 1629, the true law that connects the angles of incidence and of refraction; which the famous Descartes, who had about this time chosen Holland for his place of residence, published, in 1637, in his *Dioptrics*, under its simplest form; establishing, that the sines of those angles, and not the angles themselves, bore a constant ratio in the transit of light between the same diaphanous media. It hence followed, that the lateral rays of the light which enter a denser medium, bounded by a spherical surface, in the direction of the axis, will not meet this axis precisely in the same point, but will cross it somewhat nearer the surface. In short, the constant ratio or index of refraction will be that of the distances of the actual focus from the centre of the sphere, and from the point of external impact. Since an arc differs from its sine by a quantity nearly proportioned to its cube, the deviation of the extreme rays from the correct focus, or what is called the spherical aberration, must likewise proceed in that ratio, and consequently will increase with extreme rapidity, as the aperture of the telescope is enlarged. It was now attempted to modify the figure of the object glass, and to give it those curved surfaces which an intricate geometrical investigation marks out as fitted to procure a perfect concentration of all the refracted rays. Various contrivances were accordingly proposed for assisting the artist in working the lenses into a parabolic or spheroidal shape, and thus obtaining the exact surfaces generated by the revolution of the different conic sections. All those expedients and directions, however, were found utterly to fail in practice, and nature seemed, in this instance, to oppose insurmountable barriers to human curiosity and research. Philosophers began to despair of effecting any capital improvement in dioptrical instruments, and turned their views to the construction of those depending on the principles of catoptrics, or formed by certain combinations of reflecting specula. In 1663, the famous James Gregory, who in many respects may be regarded as the precursor, and in some things even the rival of Newton, published his *Optica Promota;* a work distinguished by its originality, and containing much ingenious research and fine speculation. In this treatise, a complete description is given of the reflecting telescope now almost universally adopted, consisting of a large perforated concave reflector combined with another very small and deep speculum placed before the principal focus. But such was still the low state of the mechanical arts in England, that no person was found capable of casting and polishing the metallic specula with any tolerable delicacy, and the great inventor never enjoyed the satisfaction and transport of witnessing the magic of his admirable contrivance. It was after the lapse of more than half a century, that Hadley—to whom we likewise owe another instrument scarcely less valuable, the quadrant, or sextant, known by his name—at last succeeded in executing the reflecting telescope. In the first attempt, silvered mirrors had been substituted for the specula; nor did the reflectors come to obtain much estimation, till, about the year 1733, the ingenious Mr Short distinguished himself by constructing them in a style of very superior excellence.

But though thus late in guiding the efforts of artists, the optical treatise of Gregory proved the harbinger of that bright day which soon arose to illume the recesses of physical science. The capacious mind of Newton, nursed in the calm of retirement and seclusion, was then teeming with philosophical projects. In 1665, when the tremendous visitation of the plague raged in London, and threatened Cambridge and other places communicating with the capital, this sublime genius withdrew from the routine of the university to his rural farm near Grantham, and devoted himself to the most profound meditation. Amidst his speculations in abstruse mathematics and theoretical astronomy, Newton was induced to examine the opinions entertained by the learned on the subject of light and colours. With this view he had recently procured from the Continent some prisms of glass, to exhibit the phenomena of refraction. Having placed the axis of the prism or glass wedge at right angles to a pencil of light from the sun, admitted through a small hole of the window-shutter in a darkened room, he contemplated the glowing image or spectrum now formed on the opposite wall or screen. This illuminated space was not round, however, as the young philosopher had been taught to expect, but appeared very much elongated, stretching out five times more than its breadth, and marked by a series of pure and brilliant colours. It was therefore obvious that the colours were not confined to the margin of the spectrum, nor could proceed from any varied intermixture of light and shade; and the conclusion seemed hence irresistible, that the white pencil, or solar beam, is really a collection of distinct rays, essentially coloured and differently refracted; that the ray, for instance, which gives us the sensation of the violet, is always more bent aside from its course by refraction than the ray which we term green,—and that this green ray again is more refracted than the red. When the spectrum was divided, by interposing partially a small screen, and each separate parcel of rays made to pass through a second prism, they still retained their peculiar colour and refractive property, but now emerged in parallel, and not in diverging lines as at first. The sun's light is thus decomposed by the action of the prism into a set of primary coloured rays; and these rays, if they be afterwards recombined in the same proportions, will always form a white pencil. It was hence easy to discern the real cause of the imperfection of dioptrical instruments, which is comparatively little influenced by the figure of the object-glass or spherical aberration, but proceeds mainly from the unequal refraction of light itself. The focal distance of the red ray being, in the most favourable case, about one fortieth part shorter than that of the violet ray, the principal image is necessarily affected with mistiness, and its margin always encircled by a coloured ring; for each point of the remote object from which the light arrives is not represented by a corresponding point in the image, but by a small circle composed of graduating colours, the centre being violet and the circumference red. This radical defect seemed at that time to be altogether irremediable. Newton had recourse, therefore, to the aid of catoptrics, and contrived his very simple though rather incommodious reflecting telescope, consisting of a concave speculum, with a small plane one placed obliquely before it, to throw the image towards the side of the tube. This instrument he actually constructed; and with all its rudeness, it promised essential advantages to astronomy. The Newtonian reflector, after having been long neglected, was lately revived by Dr Herschel; and from its great simplicity and moderate dissipation of light, it is perhaps on the whole not ill calculated for celestial observations.

These unexpected and very important discoveries, which entirely changed the face of optics, were soon communicated to the Royal Society, and published in the *Philosophical Transactions* for 1672. They were not received, however, by the learned with that admiration to which they were justly entitled, but gave occasion to so much ignorant opposition and obstinate controversy, that the illustrious author, thoroughly disgusted at such unmerited reception, henceforth pursuing his experimental researches in silence, made no disclosure of them to the world till more than thirty years afterwards, when his fame being mature, and his authority commanding respect, he suffered his *Treatise on Optics* to appear abroad. This celebrated production has long been regarded as a model of pure inductive science. The experiments which it relates appear ingeniously devised; the conclusions from them are drawn with acuteness, and pursued with exquisite skill; and the whole discourse proceeds in a style of measured and elegant simplicity. Though the researches were conducted by a process of strict analysis, the composition of the work itself is cast into the synthetical or didactic form, after the manner followed in the elementary treatises of the ancient mathematicians. But with all its beauty and undisputed excellence, it must be confessed that the treatise of optics is not exempt from faults, and even-material errors. We should betray the interests of science, if we ever yielded implicit confidence even to the highest master. It is the glory of Newton to have led the way in sublime discovery, and to have impressed whatever he touched with the stamp of profound and original genius. The philosopher paid the debt of human infirmity, by imbibing some tincture of the mystical spirit of the age, and taking a slight bias from the character of his studies. The difficult art of experimenting was still in its infancy, and inquirers had not attained that delicacy and circumspection which, in practice, are indispensable for obtaining accurate results. Most of the speculations in the second and third books of Newton's *Optics*, as we shall afterwards have occasion to observe, are built on mistaken or imperfect views of some facts, which the admixture of extraneous circumstances had accidentally disguised. The very ingenious, but hasty, and often untenable hypotheses, which are subjoined, under the modest and seemingly hesitating title of *Queries*, have, on the whole, been productive of real harm to the cause of science, by the splendid example thus held forth to tempt the rashness of loose experimenters, and of superficial reasoners. Even in the first book of *Optics*, some of the capital propositions are affected by hasty and imperfect statements. The term *refrangibility*, applied to the rays of light, is at least unguarded; it conveys an indistinct conception, and leads to inaccurate conclusions. The different refractions which the primary rays undergo are not absolute properties inherent in these rays themselves, but depend on the mutual relation subsisting between them and the particular diaphanous medium. When the medium is changed, the refraction of one set of rays cannot be safely inferred from that of another. Nay, in the passage among certain media, those rays which are designated as the most refrangible will sometimes be the least refracted. To ascertain correctly, therefore, the index of refraction, it becomes necessary, in each distinct case, to examine the bearing or disposition of the particular species of rays; since the principle, that the refraction of the extreme rays is always proportioned to that of the mean rays, involves a very false conclusion.

When Newton attempted to reckon up the rays of light decomposed by the prism, and ventured to assign the famous number *seven*, he was apparently influenced by some lurking disposition towards mysticism. If any unprejudiced person will fairly repeat the experiment, he must soon be convinced, that the various coloured spaces which paint the spectrum slide into each other by indefinite shadings; he may name four or five principal colours, but the subordinate divisions are evidently so multiplied as to be incapable of enumeration. The same illustrious mathematician, we can hardly doubt, was betrayed by a passion for analogy, when he imagined, that the primary colours are distributed over the spectrum after the proportions of the diatonic scale of music, since those intermediate spaces have really no precise and defined limits. Had prisms of a different kind of glass been used, the distribution of the coloured spaces would have been materially changed. The fact is, that all Newton's prisms, being manufactured abroad, consisted of plate or crown glass, formed by the combination of soda, or the mineral alkali, with silicious sand. The refined art of glass-making had only been lately introduced into England, and that beautiful variety called crystal, or flint-glass, which has so long distinguished this country, being produced by the union of a silicious material with the oxyde of lead, was then scarcely known. The original experimenter had not the advantage, therefore, of witnessing the varied effects occasioned by different prisms, which demonstrate that the power of refraction is not less a property of the peculiar medium than of the species of light itself. He mentions, indeed, prisms formed with water confined by plates of glass; but the few trials which he made with them had evidently been performed with no sufficient attention. In spite of his habitual circumspection, he could not always restrain the propensity so natural to genius, that of hastening to the result, and of trusting to general principles more than to any particular details. But the same indulgent apology will not be conceded to some later authors. It is truly astonishing that systematic writers on optics, in obvious contradiction to the most undoubted discoveries related by themselves, should yet repeat with complacency the fanciful idea of the harmonical composition of light.

Admitting the general conclusion which Newton conceived himself entitled to draw from analogy and concurring experiment, that "the sine of incidence of every ray considered apart, is to the sine of refraction in a given ratio;" it was strictly demonstrable, that no contrary refractions whatever, unless they absolutely restored the pencil to its first direction, could collect again the extreme rays, and produce, by their union, a white light. Thus, let the ratios of the sines of the angles of incidence and refraction of the violet rays in their transit from air to other two denser mediums, be expressed by $1 : M$ and $1 : m$; and the like ratios of the red rays under the same circumstances, by $1 : N$ and $1 : n$; where $M$, $m$ and $N$, $n$ respectively denote the refracting indices of those extreme rays. It is manifest that the refracting indices, corresponding to the passage of the violet and red rays from the first to the second medium, will be represented by $M-N$, and $m-n$. But by hypothesis, $M : m :: N : n$, and consequently $M : m :: M-N : m-n$; so that the extreme rays would be still separated and dispersed in proportion to the mean extent of the final refraction. The great philosopher appears to have contemplated with regret the result of his optical principle; and he had the penetration to remark, that if a different law had obtained, the proper combination of distinct refracting media would have corrected the spherical aberration. With this view, he would propose for the object-glass of a telescope, a compound lens, consisting of two exterior meniscuses of glass, their outsides being equally convex, and their insides of similar but greater concavity, and having the interior space filled with pure water, as in the figure annexed. He gives a rule, though without demonstration, and evidently disfigured or imperfect, for determining the curvature of the two surfaces: "And by this means," he subjoins, "might telescopes be brought to sufficient perfection, were it not for the different refrangibility of several sorts of rays. But, by reason of this different refrangibility, I do not see any other means of improving telescopes by refractions alone, than that of increasing their lengths."

These remarks appeared to preclude all attempts to improve the construction of the refracting telescope. Brightness and range of sight were sacrificed to distinctness. Instead of enlarging the aperture, recourse was had to the expedient of increasing the length of the focus. For nice astronomical observations, telescopes were sought of the highest magnifying powers, and their tubes had by degrees been extended to a most enormous and inconvenient size. But the famous Dutch mathematician Huygens contrived to supersede the use of these in certain cases, by a method which required, however, some address. Many years afterwards the reflecting, or rather catadioptric telescope, of the Gregorian construction, was executed with tolerable perfection. But a long period of languor succeeded the brilliant age of discovery. Not a single advance was made in the science of light and colours, till thirty years after the death of Newton.* His immortal *Principia* had not yet provoked discussion, and philosophers seemed inclined to regard the conclusions in the *Treatise of Optics* with silent and incurious acquiescence. This memorable fact not only evinces the danger of yielding, in matters of science, implicit confidence even to the highest authority, but shows, amidst all the apparent bustle o research, how very few original experiments are made, and how seldom these are repeated with the due care and attention.

The impossibility of correcting the colours in object-glasses of telescopes was therefore a principle generally adopted; though some vague hopes, grounded chiefly on the consideration of final causes, were still at times entertained of removing that defect. As the eye consists of two distinct humours, with a horny lens or cornea interposed, it was naturally imagined that such a perfect structure should be imitated in the composition of glasses. This inviting idea is concisely mentioned by David Gregory, the nephew of James, in his little tract on *Dioptrics*. It has also been stated that a country gentleman, Mr Hall of Chesterhall, in Worcestershire, discovered, about the year 1729, the proper composition of lenses by the united segments of crown and flint-glass, and caused a London artist, in 1733, to make a telescope under his directions, which was found on trial to answer extremely well. But whatever might be the fact, no notice was taken of it at the time, nor indeed till very long after, when circumstances had occurred to call forth public attention.

* The fine discovery of the apparent aberration of the fixed stars, made by our countryman Dr Bradley in 1729, cannot be justly deemed an exception to this remark. It belongs more to astronomy than to optics, and is indeed merely the result, however important, of the progressive motion of light, detected near sixty years before by the Danish philosopher Roemer, combined with the revolution of the earth in her orbit.

The Newtonian principle was first openly rejected, and a discussion excited, which eventually led to a most valuable discovery in optics, by a foreign mathematician of great celebrity and transcendent talents. Leonard Euler was one of those rare mortals who arise, at distant intervals, to shed unfading lustre on our species. Endowed with a penetrating genius and profound capacity, he was capable of pursuing his abstruse investigations with unremitting ardour and unwearied perseverance. To him the modern analysis stands chiefly indebted for its prodigious extension; and he continued to enrich it in all its departments with innumerable improvements and fine discoveries, during the whole course of a most active, laborious, and protracted life. Unfortunately the philosophical character of Euler did not correspond to his superlative eminence as a geometer. Bred in the school of Leibnitz, he had imbibed the specious but delusive metaphysics of the *sufficient reason,* and of the necessary and absolute constitution of the laws of nature. He was hence disposed in all cases to prefer the mode of investigating *a priori,* and never appeared to hold in due estimation the humbler yet only safe road to physical science, by the method of experiment and induction. Euler expressed the indices of refraction by the powers of a certain invariable root, and fancied that the exponents of those powers are proportional for the several rays in different media. Instead of making, in short, the numbers themselves proportional, as Newton had done, he assigned this property to their logarithms. In the *Berlin Memoirs* for 1747, he inserted a short paper, in which he deducted from his optical principle, by a clear analytical process conducted with his usual skill, the composition of a lens formed after certain proportions with glass and water, which should remove entirely all extraneous colours, whether occasioned by the unequal refraction of the several rays, or by spherical aberration; and in concluding, he remarked, with high satisfaction, the general conformity of his results with the wonderful structure of the eye.

But this paper met with opposition in a quarter where it could have been least expected. John Dolland, who had afterwards the honour of completing one of the finest and most valuable discoveries in the science of optics, was born in 1706, in Spitalfields, of French parents, whom the revocation of the edict of Nantes had compelled to take refuge in England, from the cruel persecution of a bigoted and tyrannical court. Following his father's occupation, that of a silk-weaver, he married at an early age; and being fond of reading, he dedicated his leisure moments to the acquisition of knowledge. By dint of solitary application, he made some progress in the learned languages; but he devoted his main attention to the study of geometry and algebra, and the more attractive parts of mixed or practical mathematics. He gave instructions in these branches to his son Peter, who, though bred to the hereditary profession, soon quitted that employment, and commenced the business of optician, in which he was afterwards joined by his father. About this time the volume of the *Berlin Memoirs*, containing Euler's paper, fell into the hands of the elder Dolland, who examined it with care, and repeated the calculations. His report was communicated by Mr Short to the Royal Society in 1752, and published in their Transactions for that year. Dolland, as might well be expected, could detect no mistake in the investigation itself, but strenuously contested the principle on which it was built, as differing from the one laid down by Newton, which he held to be irrefragable. "It is, therefore," says he, rather uncourteously, and certainly with little of the prophetic spirit, "it is, therefore, somewhat strange that any body now-a-days should attempt to do that which so long ago has been demonstrated impossible."

The great Euler replied with becoming temper, but persisted in maintaining that his optical principle was a true and necessary law of nature, though he frankly confessed that he had not been able to reduce it yet to practice. The dispute now began to provoke attention on the Continent. In 1754, Klingenstierna, an eminent Swedish geometer, demonstrated that the Newtonian principle is in some extreme cases incompatible with the phenomena, and therefore ought not to be received as an undoubted law of nature. Thus pressed on all sides, Dolland at length had recourse to that appeal which should have been made from the beginning,—to the test of actual experiment. He constructed a hollow wedge with two plates of glass, ground parallel, in which he laid inverted a common glass prism, and filled up the space with clear water, as in the annexed figure.

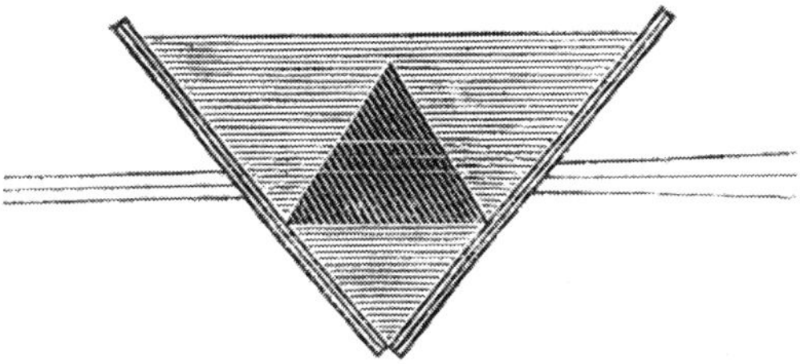

He now continued to enlarge the angle of the wedge, till the refraction produced by the water came to counterbalance exactly the opposite refraction of the glass, which must obtain whenever an object is seen through the compound prism, without change of direction, in its true place. But, contrary to what he so firmly expected, the external objects appeared glaringly bordered with coloured fringes; as much, indeed, as if they had been viewed through a glass prism with an angle of thirty degrees. It was therefore quite decisive that Newton had not performed his experiment with scrupulous accuracy, and had trusted rather too hastily to mere analogical inference. But to remove every shadow of doubt from the subject, Mr Dolland, finding that large angles were inconvenient for observation, ground a prism to the very acute angle of nine degrees, and adjusted, by careful trials, a wedge of water to the same precise measure of refraction. Combining the opposite refraction as before, he beheld, on looking through the apparatus (as here represented), their various objects real position, but distinctly marked with the prismatic colours. In these experiments, although the mean ray pursues the same undeviating course, the extreme rays which enter parallel with it emerge from the compound prism, spreading out on both sides.

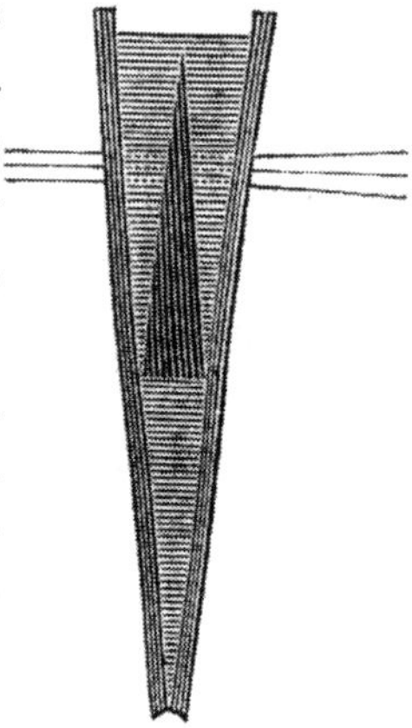

The capital point being completely ascertained, Dolland next tried so to adapt the opposite refractions as to destroy all extraneous colour. This effect he found to take place when the angle of the wedge had been further increased, till the refracting power of the water was to that of the glass in the ratio of five to four. His conclusive experiments were made in 1757, and he lost no time in applying their results to the improvement of the object-glasses of telescopes. Following the proportion just ascertained, he conjoined a very deep convex lens of water with a concave one of glass. In this way he succeeded in removing the colours occasioned by the unequal refraction of light; but the images formed in the foci of the telescopes so constructed, still wanted the distinctness which might have been expected. The defect now proceeded, it was evident, merely from spherical aberration; for the excess of refraction in the compound lens being very small, the surfaces were necessarily formed to a deep curvature.

But this partial success only stimulated the ingenious artist to make further trials. Having proved that the separation of the extreme rays, or what has been since termed the *dispersive power*, is not proportioned to the mean refraction in the case of glass and water, he might fairly presume that like discrepancies must exist among other diaphanous substances, and even among the different kinds of glass itself. The charm of uniformity being once dispelled, he was encouraged to proceed, with the confident hope of ultimately achieving his purpose. His new researches, however, were postponed for some time by the pressure of business. But on resuming the inquiry, he found the English crown-glass and the foreign yellow or straw-coloured, commonly called the Venice glass, to disperse the extreme rays almost alike, while the crystal, or white flint-glass, gave a much greater measure of dispersion. On this quarter, therefore, he centred his attention. A wedge of crown and another of flint-glass were ground till they refracted equally, which took place when their angles were respectively 29 and 25 degrees, or the indices of refraction were nearly as 22 to 19; but on being joined in an inverted position, they produced, without changing the general direction of the pencil, a very different divergence of the compound rays of light. He now reversed the experiment, and formed wedges of crown and flint-glass to such angles as might destroy all irregularity of colour by their opposite dispersions. When this condition was obtained, the refractive powers of those wedges of crown and flint-glass were nearly in the ratio of three to two, and consequently the sines of half their angles, or the angles them-

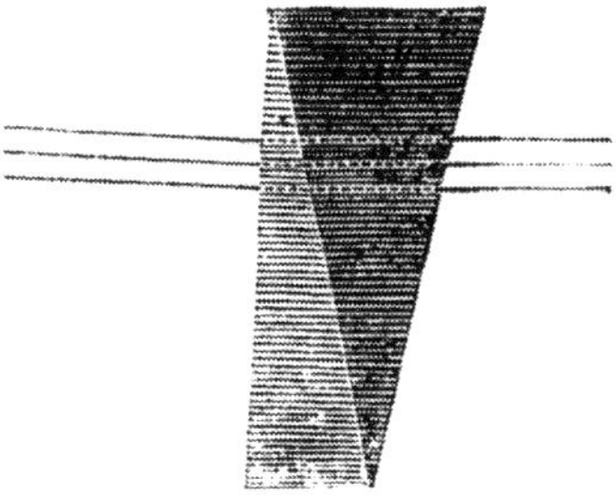

selves, if small, were as 33 to 19, or nearly as 7 to 4. The rays which enter parallel now escape likewise parallel, but all of them deflected equally from their course.

The appearance was rendered still more conspicuous by repeating the combination of the glass wedges as in the figure here adjoined. It will be perceived that the pencils of rays which enter at equal distances on both sides of the common junction, must nearly meet in the same point of the axis; for in small arcs the chords are almost proportional to the arcs themselves. This arrangement, indeed, with the projecting wedge of crown-glass in front, represents actually the composition of an object-glass formed of two distinct and opposing lenses, which would produce a similar effect. It was only required to apply a semi-convex lens of crown-glass before a semi-concave one of flint-glass, such that the curvature of the former be to that of the latter nearly as 7 to 4; but with some modifications in this ratio, according to the peculiar qualities of the glass. [The figure annexed represents this combination.] But the depth of the lenses might be diminished, by giving them curvature on both sides. Thus, if a double convex of crown-glass were substituted, of the same power, and consequently with only half the curvature on each side; the lens of flint-glass adapted to it having, therefore, their common surface of an equal concavity, would need, in order to produce the former quantity of refraction, and consequently to maintain the balance of opposite dispersions, a concavity eight times less than before on the other surface. Or if a double concave of flint-glass with half its

first depth were used, the front convexity of the lens of crown-glass would be five-sevenths of the former curvature, as here represented. The surface where the two lenses are united may hence have its curvature changed at pleasure; but every alteration of this must occasion corresponding changes in the exterior surfaces.

In all these cases, the refraction of the convex pieces being reduced to one-third by the contrary refraction of the concave piece, the focal distance of the compound glass must be triple of that which it would have had singly. But a most important advantage results from the facility of varying the adaptation of the lenses; for, by rightly proportioning the conspiring and counteracting curvatures, it was possible to remove almost entirely the errors arising from spherical aberration. This delicate problem Mr Dolland was the better prepared to encounter, as he had already, in 1753, improved the telescope materially, by introducing no fewer than six eye-glasses, disposed at proper distances, to divide the refraction. The research itself, and the execution of the compound lens, presented peculiar difficulties; but the ingenuity and toilsome exertions of the artist were at length, in 1758, rewarded with complete success. "Notwithstanding," says he, in concluding his paper, "so many difficulties as I have enumerated, I have, after numerous trials, and a resolute perseverance, brought the matter at last to such an issue, that I can construct refracting telescopes, with such apertures and magnifying powers, under limited lengths, as, in the opinion of the best and undeniable judges, who have experienced them, far exceed any thing that has been produced, as representing objects with great distinctness, and in their true colours."

The Royal Society voted to Mr Dolland, for his valuable discovery, the honour of the Copley medal. To this new construction of the telescope Dr Bevis gave the name of *Achromatic* (from *a* privative, and χρωμα, *colour,)* which was soon universally adopted, and is still retained. The inventor took out a patent, but did not live to reap the fruits of his ingenious labours. He died in the year 1761, leaving the prosecution of the business to his son and associate Peter Dolland, who realised a very large fortune by the exclusive manufacture, for many years, of achromatic glasses, less secured to him by the invidious and disputed provisions of legal monopoly, than by superior skill, experience, and sedulous attention. In 1765, the younger Dolland made another and final improvement, to which his father had before advanced some steps. To correct more effectually the spherical aberration, he formed the object-glass of three instead of two lenses, by dividing the convex piece; or he inclosed a concave lens of flint-glass between two convex lenses of crown-glass, as exactly represented in the figure here annexed. He showed a telescope of this improved construction, having a focal length of three feet and a half, with an aperture of three inches and three quarters, to the celebrated Mr Short, who tried it with a magnifying power of one hundred and fifty times, and who, superior to the jealousy of rivalship, and disposed to patronise rising merit, most warmly recommended it, and declared that he found "the image distinct, bright, and free from colours."

What were the curvatures of those distinct component lenses, Dolland has not mentioned, and perhaps he rather wished to conceal them. The Duke de Chaulnes was enabled, however, by means of a sort of micrometer, to ascertain the radii of the several surfaces, in the case of one object-glass of the best composition. He found these radii, beginning with the front lens, to be respectively $311\frac{1}{2}$, 392; 214, 294; 294 and $322\frac{1}{2}$, in French lines, which corresponded, in English inches, to 32·4, 40·8; 22·2, 30·6; 30·6 and 33·5. If these measures were correct, however, it would follow, that the middle lens of flint-glass was not perfectly adapted to the curvature of the lens of crown glass placed immediately before it. Similar admeasurements have been repeated by others, but the results differ considerably, and no general conclusion can be safely drawn. There is no doubt that the artist varied his practice, according to the nature of the glass which he was obliged to use. The more ordinary proportions for the curv..tures of the component would be represented by a truncated prism, formed with a double cluster of wedges, the outer ones having angles of 25° 53′, and 14° 27′, and consisting of crown-glass, and the inner one made of flint-glass, with an inverted angle of 27° 8′. These two wedges of crown-glass would produce the same refraction, it might be shown, as a single one having an angle of 40° 54′; wherefore this refraction will be diminished, by the opposite influence of the wedge of flint-glass, in the ratio of 49 to 16, or reduced to nearly one-third.

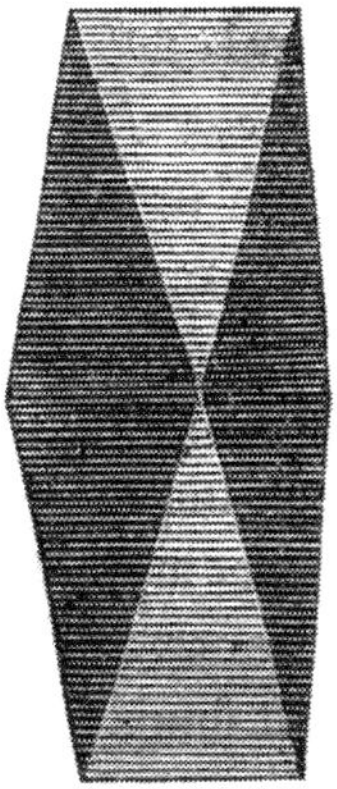

Thus was achieved, and fully carried into practical operation, the finest and most important detection made in optics since the great discovery of the unequal refraction of the several rays of light. It was drawn forth by a long series of trials, directed with judgment and ingenuity, but certainly very little aided by the power of calculation. Such a slow tentative procedure was perhaps the best suited, however, to the habits of an artist, and it had at least the advantage of leaving no doubt or hesitation behind it. On this occasion, we cannot help being struck with a remark, that most of those who have ever distinguished themselves in the philosophical arts by their original improvements, were seldom regularly bred to the profession. Both the Dollands, we have seen, began life by plying at the loom; Short had a liberal education, being designed for the Scottish Church, but, indulging a taste for practical optics, he afterwards followed it as a trade, in which he rose to pre-eminence; Ramsden, whose ingenuity and exquisite skill were quite unrivalled, was bred a clothier in Yorkshire; Tassie, who revived or created among us the nice art of casting gems, was originally a stone-mason at Glasgow; and Watt, who, by his very happy applications of mechanics, and his vast improvements on the steam-engine, has, more than any other individual perhaps, contributed to the great national advancement, was early an ivory-turner in that same city, and still found pleasure, in his declining years, with the amusement of the lathe. We might easily enlarge this catalogue; but enough has been said to prove the justness of the observation, and it suggests reflections which are not favourable to fixed and systematic plans of education.

The theory of achromatic telescopes, embraced in all its extent, opened a field of abstruse and difficult investigation. But the English mathematicians at that period, though they might appear to be especially invited to the discussion, very generally neglected so fine an opportunity for the exercise of their genius. They coldly suffered the artists to grope their devious way, without offering to guide their efforts by the lights of science. On the Continent the geometers of the first order were all eager to attempt the solution of problems at once so curious and important. For several years subsequent to 1758, the Transactions of the foreign academies were filled with memoirs on the combination of achromatic lenses, displaying the resources and refinements of the modern analysis, by Euler, Clairaut, and D'Alembert,—by Boscovich, Klingenstierna, Kæstner, and Hennert. On this, as on other occasions, however, we have to regret the want of close union between artists and men of science. Those profound investigations are generally too speculative for any real use; they often involve imperfect or inaccurate data; and the results appear wrapped in such comprehensive and intricate formulæ, as to deter the artist from endeavouring to reduce them into practice. We should have thought it preferable, on the whole, not to load the solution of the main problem with minute conditions, but to aim at a few general rules, which could afterwards be modified in their application according to circumstances. All this might have been accomplished without scarcely travelling beyond the limits of elementary geometry.

Euler and his adherents at Berlin were still not disposed to abandon his favourite optical hypothesis. It was even pretended that Dolland must have owed his success to a nice correction of spherical aberration, and not to any really superior dispersive power belonging to the flint-glass. But that candid philosopher afterwards yielded to the force of reason and testimony; and, collecting his various optical papers, he published, in the successive years 1769, 1770, and 1771, a complete treatise on *Dioptrics*, occupying three quarto volumes, which contain a store of ingenious and elegant disquisitions.

The last memoir which Clairaut ever wrote related to achromatic glasses. D'Alembert prosecuted the subject with diligence and ardour; and the volumes of his *Mathematical Opuscules*, published between the years 1761 and 1767, contain some elaborate dioptrical investigations. Among other conclusions which he deduced from his multiplied researches, he proposed a new composition for the object-glass of a telescope, to consist of three lenses, the outmost one being a meniscus of crown-glass, or having a convex and a concave surface, then a meniscus of flint-glass in the middle, and adapted to this, on the inside, a double convex of crown-glass. Of all the continental works, however, which treat of achromatic combinations, the tracts of Boscovich, who possessed a very fine taste for geometry, may be held as the simplest and clearest. We cannot help noticing, by the way, a curious theorem of his concerning the form and arrangement of eye-glasses, which would be free from irregular colours. It is, that the correction will be produced by means of two lenses of the same kind of glass, if separated from each other by an interval equal to half the sum of their focal distances. This principle furnishes a very simple construction for the common astronomical telescope, through which the objects are seen inverted. In the annexed figure, the

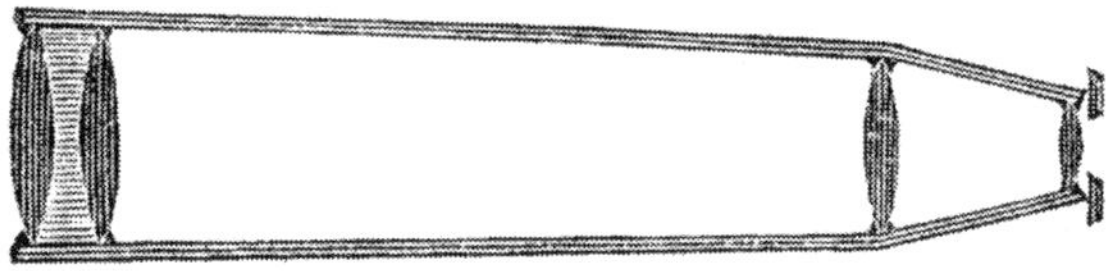

object-glass, as usual, is achromatic, being composed of two convex lenses of crown-glass, with a concave one of flint-glass fitted between them; but the eye-glass consists of two distinct lenses of crown-glass, both of them convex, and exactly similar, the first having every dimension triple that of the other, and their mutual distance double the focal length of the smaller.

Supposing, however, that the errors occasioned by spherical aberration were completely removed, the principle of achromatic combination is yet far from being so perfect as it has often been represented. Although the opposite dispersions of the flint and of the crown-glass should bring together the extreme rays, we are not, from this coincidence, warranted to infer that the several intermediate rays would likewise be accurately blended. In fact, a wedge of flint-glass not only separates all the rays much more than a similar one of crown-glass, but divides the coloured spaces after different proportions. While the combined lenses formed of those two kinds of glass give an image entirely free from the red and violet borders, they may still introduce secondary shades of green or yellow, sufficient to cause a certain degree of indistinctness. The mode of correcting this defect would be, to produce a counterbalance of colours, by conjoining several media endued with different refractive and dispersive powers. In these qualities, crown-glass itself admits of some variation, owing to the measure of saline ingredient; but flint-glass differs widely with regard to its optical properties, owing chiefly to the diversified proportion of minium or oxyde of lead which enters into its composition, and partly to the variable admixture of manganese employed to discharge the yellow tint occasioned by the lead. Manifest advantages, therefore, would result from a choice combination of three or more varieties of glass, since both the primary and the secondary deviations of colour would be corrected. Without pretending to any theoretical perfection, every thing really wanted in practice would be thus attained. A series of nice experiments on the optical relations of glass could not fail, by their results, to reward the assiduity of the ingenious artist. He would trace and determine the separate influence exerted on the refractive and dispersive powers by soda in the crown-glass, and by minium and manganese in the flint-glass. It is highly probable, that with perseverance he might discover a vitreous composition better adapted than any yet known for achromatic purposes. It is very generally believed, that the achromatic telescopes now manufactured in London are not of the same excellence with those first made by Peter Dolland. This declension of such a beautiful art has frequently been imputed to the baneful operation of a severe and oppressive system of excise. Whether the new mode of charging the duty on glass at the annealing arch has produced any beneficial effects, we are still to learn.

An extensive and ingenious set of experiments on the dispersive powers of different liquids, was undertaken, about the year 1787, and successfully prosecuted for some time afterwards, by Dr Robert Blair, for whom there had been recently created, under royal patronage, the chair of practical astronomy in the University of Edinburgh,—one of the very few professorships in that distinguished seminary which have been suffered to remain inefficient and merely nominal. Of these experiments, a judicious account was, in 1790, communicated by their author to the Royal Society of Edinburgh, in a paper drawn up with evident ability, but rather too diffuse, and unnecessarily digressive. Dr Blair had a very small brass prism perforated with a hole, which he filled with a few drops of the liquid to be examined, and confined each end by a plate of glass with parallel surfaces. He then applied, inverted to the prism in succession, a number of glass wedges which he had provided of different angles, and observed, when the bars of the window, seen through this compound prism, appeared colourless, the angle of the wedge now expressed the relative dispersive power of the liquid. This way of experimenting was sufficiently simple, but a more accurate and expeditious method might have easily been devised. For instance, if the prism, furnished with a graduated arch, had remained fixed, and a single glass wedge made to turn upon it, and present successive inclinations to the observer, the refracting angle at which the irregular colours were united could be deduced by an easy calculation. Dr Blair found, by his trials, that muriatic acid, in all its combinations, but particularly with antimony and mercury, shows a very great dispersive power. The essential oils stood the next with regard to that property, though differing considerably among themselves. In Dr Blair's first attempts to improve the achromatic telescope, he conjoined two compound lenses, the one formed with a double concave of crown-glass and a semi-convex of essential oil, and the other composed of a double convex filled with essential oil, of great dispersive power, and of a semi-concave, likewise containing essential oil, but less apt for dispersion. This very complex arrangement seemed, however, to produce the desired effect, not only discharging from the image the extreme fringes of red and violet, but excluding also the intermediate shades of green or yellow. A simpler combination was afterwards used, requiring merely one liquid, composed of muriatic acid joined with antimony, or the triple salt of that acid united in certain proportions to ammonia and mercury. This liquid, being accurately prepared, was inclosed between two thin glass shells, to form a double convex lens: on the front was applied a semi-convex of crown-glass, and a meniscus of the same material behind, the whole being secured by a glass ring. An object-glass so constructed seemed to perform its office with great perfection, effectually correcting both the primary and the secondary admixture of colours. This kind of eye-glass Dr Blair proposed to denominate *aplanatic*, (from *α* privative, and πλαναω, to *err* or *wander*,) and he obtained a patent for his invention. The late George Adams, optician in Fleet Street, was intrusted with the fabrication and sale of the telescopes thus constructed. Some of them were said to answer extremely well; but, whether from want of activity on the part of the tradesman, or from defect of temper in the patentee, these instruments never acquired much circulation.

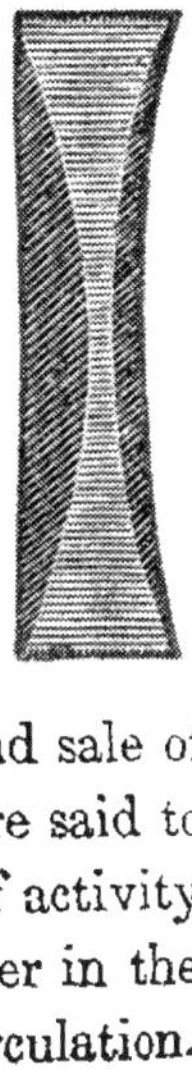

It was alleged that the liquid by degrees lost its transparency. Indeed, we suspect that there is no combination in which liquids are concerned, which can be judged sufficiently permanent for optical purposes. It seems hardly possible to preclude absolutely the impression of the external air; the liquid must, therefore, have a tendency both to evaporate and to crystallize; and, in the course of time, it will probably, by its activity, corrode the surfaces of the glass.

The manufacture of achromatic telescopes in England furnished, for a long period, a very profitable article of exportation. Even after the introduction of those instruments was prohibited by several foreign governments, the object-glasses themselves, in a more compendious form, were smuggled abroad to a large amount. In fact, no flint-glass of a good quality was then made on the Continent. A very material alteration, however, in that respect, has taken place, at least in France, where the stimulus impressed by the revolution has worked so many changes, and where ingenuity and science, in most of the mechanical arts, have so visibly supplied the scantiness of capital. The French now construct achromatic telescopes, equal, if not superior, to any that are made in England.

# ACOUSTICS.

THE doctrine of sound is unquestionably the most subtle and abstruse in the whole range of physical science. It has given occasion, in recent times, to much controversy and discussion, and has eventually called forth all the mighty resources of a refined and elaborate calculus. Yet an evident obscurity still remains to overcloud the subject. The discrepancies between theory and observation have been made entirely to disappear from astronomy, which has at last attained a degree of perfection befitting the sublimity of the science. But some latent suspicions pervade the structure of acoustics, sufficient to disturb that feeling of confidence which is calculated to invigorate our pursuits.

The impression of sound is conveyed by means of a certain tremor or internal agitation, which shoots, with more or less celerity and force, through any substance, whether solid or fluid. Nor is it requisite that the conducting medium should belong to the class of bodies which are commonly denominated elastic. In fact, all bodies whatever, in the minute and sudden alterations of their form, exert a perfect elasticity, and only seem to want this energy when they undergo such great changes that their component particles take a new *set* or arrangement, which prevents the full effect of re-action.

It is not every kind of tremulous motion, however, that will excite the sensation of sound. A certain degree of force and frequency in the pulsations appears always necessary to affect our sense of hearing. Yet the impression of sound is not confined to the mere external organ: the auditory nerves have a considerable expansion, and sympathize with those of taste and of smell. The only inlet of vision is by that very narrow aperture, the pupil of the eye; but the reception of sound partakes more of the character of the general sense of feeling, which, though most vivid at the extremities of the fingers, is likewise diffused over the whole surface of the body. The intimation of the ear is accordingly assisted by the consent of the palate, the teeth, and the nostrils. Fishes hear very acutely under water, though the organ itself lies so concealed in the head as to have long escaped the diligence of anatomists.

It was formerly supposed, that the transmission of impulse through a solid body is perfectly instantaneous. This formed, indeed, one of the Cartesian tenets, which Newton himself has tacitly admitted. But accurate observations have since proved, that motion is always really progressive, and propagated in succession. Professor Leslie has shown that the darting of impact through any substance, whether hard or soft, is accomplished by the agency of the same interior mechanism as that of sound, and has furnished the method of calculating, in some of the more difficult cases, the celerity of transmission.

All bodies may be considered as composed of physical points, without any sensible magnitude, but connected together by a system of mutual attraction and repulsion. When those integrant particles are compressed by external violence, a repulsive force is exerted to regain their first position; or if they be dilated, a corresponding attraction now draws them back to their neutral site of equilibrium. We may further presume, that in solids these constituent forces are confined to the proximate particles only, but that in the case of liquids or other fluids they embrace the particles in their near vicinity, and include a sphere of action varying in its extent. Hence, the former suffer disruption, without bending or giving way to powerful pressure; while the latter, acting by a sympathetic union, gently recede and take a new arrangement. In fact, the attribute of hardness applied to body is only a relative, and not an absolute quality; in the inferior degrees it relapses into softness, and softness again passes through interminable shades to the most yielding fluidity. The application of heat, by enlarging the system of internal connexion, generally promotes softness, and heightens the degree of fluidity itself. The effect is conspicuous in the increased flow from a capillary syphon, when kept warm. But even liquids, when struck with a blow so rapid and sudden as to preclude the sympathy of their adjacent molecules, will assume all the character of the hardest substances. This fact has a familiar illustration in the play of duck and drake; but it is beautifully exemplified in the successive rebounds made by cannon-shot from the surface of the sea.

In confirmation of the remark, we may quote a very singular and curious circumstance, mentioned by travellers, relative to the method of catching fish, which is successfully practised in some of the more northern countries. The hardy peasant, when the smaller lakes and rivers of Lapland or Siberia are completely frozen over, as soon as he observes, through the clear ice, a fish, perhaps at a considerable depth, but lying close to the bottom, strikes a smart blow against the firm surface, and the impulse sent through the vertical column of water instantly stuns or kills his prey, which he draws up by a large hook let down through the hole just made in the ice.

If we conceive a conducting substance to be struck at one

extremity, the proximate particles, yielding at first to the impulsion, will again expand themselves, like the recoil of a spring, and press against the next particles in the chain. The vibratory commotion will thus be conveyed, by a successive transfer of impressions, along the whole series of physical points. Analogous also to the oscillations of a spring or a pendulum, this multitude of concatenated internal pulses, whatever be the force or extent of agitation, will constantly be performed in the same instants of time. The celerity of transmission must depend on the elasticity of the medium compared with its gravity. This estimate is most readily obtained by determining what may be called the *modulus* of elasticity, or the height of a column of the same density as the conducting substance, whose weight would measure that elasticity; or, to speak more precisely, that the thousandth part of such a column, for instance, should be equivalent to the repulsive force corresponding to a condensation of one thousandth part in the vibrating body. It may be demonstrated from the principles of dynamics, that the celerity of the transmission of impulse or sound through any medium, is equal to what a falling body would acquire in falling through half the height of the modulus of elasticity. Hence this celerity for each second will be expressed in English feet by multiplying the square root of half the modulus by 8, or by extracting the square root of the modulus multiplied by 32.

Mr Leslie has pointed out a very simple method for ascertaining the modulus of elasticity in the case of solid rods or planks, by observing, when they are laid in a horizontal position, with their ends resting against two props, the *swag* or curvature which they take. By an experiment of this kind, he found that Memel fir had a modulus equal to 671,625 feet. Wherefore, an impulse would shoot through the substance of a deal-board with the velocity of 4,636 feet each second, or about four times the rapidity of sound. Professor Chladni, who has thrown so much curious light on the convoluted curves formed by vibrations spreading along the surface of solid bodies, inferred, from a very different procedure,—from the musical note which a bar of the substance emits when struck,—the celerity of the transmission of sound through iron and glass, which he reckoned for both at 17,500 feet, or above three miles each second, more than fifteen times swifter than the ordinary communication through the atmosphere.

The rate with which the tremor of sound is transmitted through cast-iron, was very lately ascertained, from actual experiment, by the ingenious M. Biot. This philosopher availed himself of the opportunity of the laying of a system of iron pipes, to convey water to Paris. These pipes were about eight feet each in length, connected together by narrow leaden rings. A bell being suspended within the cavity, at one end of the train of pipes, on striking the clapper at the same instant against the side of the bell and against the internal surface of the pipe, two distinct sounds were successively heard by an observer stationed at the other extremity. In these observations M. Biot was often assisted by the late M. Malus, who has, too soon for the progress of science, been hurried away by death, after having opened the delicate discovery of the *polarization of light.* With a train of iron pipes of 2550 feet, or nearly half a mile in length, the interval between the two sounds was found, from a mean of two hundred trials, to be 2.79 seconds. But the transmission of sound through the internal column of air would have taken 2.5 seconds; which leaves .29″ for the rapidity of the tremor conducted through the cast-iron. From other more direct trials, it was concluded that the exact interval of time during which the sound performed its passage through the substance of the train of pipes, amounted only to 26-100th parts of a second; being ten or twelve times less than the ordinary transmission through the atmosphere.

Except the observations of M. Hassenfratz, in the famous subterranean quarries which extend under almost the whole of Paris, we are not acquainted with any attempts that have been made to measure the elasticity of stone or brick. Yet sound is conveyed through these materials with great effect. The rattling of a carriage on the street spreads a very sensible tremor along the most solid buildings and the stateliest edifices. If a large stone be rubbed against the outside of the wall of a house, it will occasion within doors a strange rumbling noise. A miner will strike his pick against the side of a long gallery, when he wishes to give intimation to his companion, who listens at the other extremity. But stones or bricks, without being directly excited, may yet form a part of the chain which transmits sound, by receiving the tremulous impressions from the air on the one side, and delivering them again to that fluid on the other. We all know how easily the voice is heard through a thin partition. The mode of obstructing the passage of sound is, either to employ very thick masonry, or to interrupt the facility of communication and transfer, by means of sub-divisions opposed. Hence another distinct use of lath and plaster. Experiments on the elasticity of stones and other articles of building are not only curious, but of real importance; for, in many cases, their efficient strength must depend on their fitness to resist incidental impressions. This consideration is peculiarly necessary in selecting and combining the materials employed in the construction of bridges.

Respecting the elasticity of water and other liquids, our information is more satisfactory and complete. It was long held as an axiom, that the substance of water is absolutely incompressible. Yet the experiments on which this belief was grounded would, if weighed attentively, point to an opposite inference. On such a subject it were idle to cite Lord Bacon, whose credulity and ignorance of mathematical science betrayed him so often into false or shallow conclusions. The philosophers of the Florentine academy *del Cimento* tried the compression of water in three different ways, which are described in the account of the experiments printed in 1661. 1. Having provided two glass tubes terminated by hollow balls, they filled the one partly and the other to excess with pure water, and joined the tubes hermetically, so as to form one piece. Then applying heat to the first ball till the water boiled, they forced its vapour to press against the column in the other stem. But no contraction of the fluid took place, though a copper ball was afterwards substituted; and when the action of the heat was still farther urged, the tube at last burst with violence. 2. Into a glass tube, immediately above six pounds of water, they introduced eighty pounds of quicksilver, without causing any diminution of volume. 3. Their most noted experiment was, having filled a hollow silver ball with water by a small hole, afterwards soldered accurately, to give it a few smart blows with a hammer; when, far from suffering compression, the water was seen to ooze or spirt from the pores, as they imagined, of the silver.

Mr Boyle, whose practice it was generally to repeat the more striking experiments made on the Continent, had a round tin or pewter vessel filled carefully with water, and tightly plugged: the blow of a wooden mallet beat it flat, but on piercing the tin with the point of a small nail, the confined water instantly sprung to the height of two or three feet. About the year 1752, Dr Peter Shaw, who read public lectures in London, exhibited a stout copper ball of four inches in diameter, and filled with water by a small orifice, into which a screw was fitted, and forced to enter by turning an iron arm or lever: the globe was partly opened by this enormous squeeze, and the water spouted from the crevice as from a fountain.

These experiments all concur to show that water is capable of sustaining an immense pressure without undergoing any very sensible contraction; but they prove, at the same time, the actual existing of such a contraction, since the projecting of the water, after a crack has once begun in the vessel that confines it, could only proceed from the evolution of an internal repulsive force. Divers, accordingly, at considerable depths under water, hear distinctly the collision of two stones, or the remote ringing of a bell. Authentic instances are mentioned of sounds being transmitted audibly more than two miles through that fluid.

The compressibility of water was first demonstrated by the ingenious Mr Canton in 1762, by a very simple and conclusive experiment. To a glass ball of rather more than an inch and half in diameter, he joined hermetically a tube about four inches long, and having a bore equal to the hundredth part of an inch. The relative capacity of this ball and of the stem he ascertained by introducing mercury, and weighing nicely its separate portions. The stem was then marked by the edge of a file into divisions, corresponding each to the hundred thousandth part of the whole capacity of the ball. This instrument was now filled with distilled water, carefully purged of its adhering air, and placed under the receiver of a pneumatic machine; on producing an exhaustion, the water appeared constantly to swell, rising four divisions and three-fifths in the stem, or a space nearly equal to the mercurial expansion corresponding to half a degree of heat on Fahrenheit's scale. In a condensing engine, the water sunk just as much, for each additional pressure of an atmosphere,—the bulb remaining always at the same temperature, or at the fiftieth degree of Fahrenheit. Since the stem was left open, the pressure exerted by the air, both on the inside and the outside of the instrument, must in all cases have been precisely the same; and consequently, the glass had no disposition to alter its figure, and modify the results. The contraction or expansion produced was, therefore, confined wholly to the body of water and to the thin shell of glass, of which indeed the influence might be rejected as insignificant. It was hence decided that the purest water suffers a visible concentration, or a diminution of its volume, under a powerful compression. But, in the course of his experiments, Mr Canton observed a curious circumstance, that water is more compressible in cold than in warm weather. Thus, the contraction, under a single incumbent atmosphere, amounted to 4.9 divisions when the thermometer stood at 34°, but was only 4.4 divisions when the heat rose to 64°. This singular fact might afford room for speculation; but it were better, in the mean while, to repeat the experiment again with more delicacy, and on a greater scale.

The compression of some other fluids was likewise measured in the same way. The contraction, under the weight of an atmosphere, and at the ordinary temperature, amounted, in millionth parts of the entire capacity of the ball, to sixty-six with alcohol, to forty-eight with olive oil, to forty with sea-water, and only three when mercury was opposed. We may therefore estimate, in round numbers, the modulus of elasticity belonging to those different substances as under:

| | |
|---|---|
| Alcohol | 580,000 English feet. |
| Distilled water | 700,000 |
| Olive oil | 730,000 |
| Sea-water | 780,000 |
| Mercury | 800,000 |

In liquids of so distinct a nature, we should have expected a greater diversity in their elastic power; nor is it easy to conceive on what conditions or habitudes that quality actually depends. The elasticity of a body, like its other constitutional properties, may result from the peculiar internal structure, or the arrangement of the integral molecules.

Some experiments on the compressibility of water have been since performed with more striking effect, but not equally exempt from all objections. In 1779, Professor Zimmerman of Brunswick printed a short account of some trials made by him and Abich, director of the salt-mines, with a press of a particular construction, consisting of a tight cylinder of very thick brass, with a piston nicely fitted, to be pushed down by means of a long lever, at whose extremity different weights were appended. Rain-water being introduced into the cavity, was subjected to an enormous pressure, equivalent to that of 313 atmospheres, and had its volume then diminished between one thirty-fifth and one thirty-sixth part. This quantity gives, for the effect of a single incumbent atmosphere, a condensation amounting to seventy-five millionth parts, instead of forty-six, as found by Mr Canton. The excess was no doubt owing to the distention of the brass cylinder, which, with all its strength and solidity, would yet partially yield to the action of such prodigious force. This circumstance renders the experiment somewhat unsatisfactory, and the influence of friction must likewise affect the accuracy of the calculation.

The effect of such distention is easily witnessed in the case of glass. If a large bulb of a thermometer be suddenly squeezed between the finger and the thumb, the mercury will start up in the stem perhaps several degrees, and will again sink as quickly after the pressure is removed. To prevent any derangement from communication of heat, the hand may be covered with a thick glove. But the fact can be shown in a less exceptionable way: Let a mercurial thermometer, with a large bulb and a long stem, be first held upright, and then immediately inverted; between these two positions the column of mercury will descend through a visible space. This apparent change of volume has been hastily supposed by some experimenters to mark the compressibility of mercury, which could not be sensible but under the action of a column of incomparably greater height.

It would be most desirable to institute a new set of observations on the condensation of different substances, by means of Bramah's hydraulic press, which is a far more perfect machine, and scarcely subject at all to the disturbance of friction. Having once ascertained the distention of the metallic cavity from pressure, it would be hence easy to correct all the other results. This mode of experimenting promises also the important advantage of enabling us to determine, with ease, the compressibility of solids themselves. It would only be required to give those bodies a cylindrical form nearly adapted to the cavity, and to fill up the interstice with water, or rather with mercury. The contraction which the thin sheet of fluid would undergo, being deducted from the whole contraction, would exhibit the contraction suffered by the solid nucleus.

From all these investigations we may gather, that an impulse, or a sonorous tremor, wonld shoot through a body of fresh water with the velocity of about 4,475 feet each second, being four times swifter than the ordinary flight of sound in the atmosphere. Through the waters of the ocean, the transmission of sound would be still more rapid, by a seventeenth part. It hence follows, that a violent commotion, excited under that vast mass, might reach from pole to pole in the space of three hours and twenty minutes.

The swell of the sea is accordingly always observed to precede the coming storm. The shocks of the famous earthquake at Lisbon, in 1755, were partially felt at very distant points of the ocean, as far even as the West Indies, but after a considerable interval of time.

Respecting the power of ice to conduct sound, we possess not sufficient data for the solution of the problem. The Danish philosophers are indeed said to have lately performed experiments of this kind on a very extensive scale, along the frozen surface of the Baltic. We are not acquainted with the precise results; but it seems probable, from various analogies, that ice has nearly the same faculty of transmission as water itself. If a heavy blow be struck against any part of the frozen surface of a large pool or lake, a person standing at a wide distance from the spot will feel, under foot, a very sensible tremor, some considerable time before the noise conveyed through the atmosphere has reached the ear. It is asserted, that the savage tribes who rove on the icy steppes of Tartary can readily distinguish, from afar, the approach of cavalry, by applying their head close to the frozen surface of the ground.

But the proper and ordinary vehicle of sound is our atmosphere. Aristotle, deriving his information probably from the tenets of the Pythagorean school, seems to have acquired tolerably just notions of the nature of sound and of the theory of harmonics. The language of that philosopher was so much corrupted, however, and disguised by ignorant transcribers, that Galileo, who not only studied music as a science, but practised it as a delightful art, may be fairly allowed to have rediscovered those general doctrines. Mersenne and Kircher afterwards made a variety of most ingenious experiments, which, though rather overlooked at the time, tended greatly to extend the science of harmony. But it was reserved for the genius of Newton to sketch out the theory of sound. In his *Principia* he explained the origin of aerial pulses, and by a fine application of dynamics, conducted with his usual sagacity, he succeeded in calculating their celerity of transmission. The solution which he has given of this intricate problem is far, however, from being unexceptionable in the form and mode of reasoning. Instead of attempting to embrace all the conditions affecting the problem, in a differential equation, for which, indeed, his fluxionary calculus was not far enough advanced, he proceeds less boldly, and only arrives at the conclusion by an indirect process and a sort of compensation of errors. His investigation of the process of sound through the air is chiefly drawn from the analogy of the motion of waves along the surface of water. This comparison greatly assists our conceptions, but it fails in a variety of essential points. Newton further assumed the rising and subsiding of waves to be a reciprocating motion, similar to that of the oscillations of a fluid contained in a wide and long tube, with its ends turned upwards. On this supposition, it was not difficult to prove, that those alternating movements would correspond to the vibrations of a pendulum of half the length of the tube. Transferring the inference, therefore, to the undulations of a fluid, it followed, that the space between two consecutive waves would be described during the sweep of a pendulum having a length equal to this interval. But the conclusion does not very well accord with the phenomena. That a wave travels with a velocity as the square root of its breadth, may be nearly true; and that its reciprocating motions, whatever be the height, are all performed in the same time, is a necessary consequence of the great principle in dynamics first pointed out by Huygens and Hooke,—that when the effort to restore equilibrium is proportioned to the quantity of displacement, the alternations of figure are constantly isochronous. But the velocity of the undulating progression, as calculated from those principles, will not be found to correspond with actual observation. Newton was apparently sensible of this disagreement, and would consider his proposition as only an approximation to the truth ; assigning as the cause of discrepancy, *that the particles of water do not rise and fall perpendicularly, but rather describe arcs of a circle.* The great defect of the hypothesis, however, consisted in supposing all the parts of a wave to rise up and sink together in the same spot. The fact is, that the fore part of a wave is always in the act of ascending, while the hinder part of it is as constantly subsiding ; which combined but contrary movements, without actually transferring any portion of the water, give an appearance of progressive advance to the swell.

In extending this theory to the propagation of sound, Newton was, on the whole, more successful. It resulted from his investigation, that the aerial pulses fly uniformly, spreading themselves equally on every side, and with a celerity equal to what would be acquired by a body in falling through half the height of the modulus of the air's elasticity. This modulus, or the altitude of a column of air, of uniform density, and whose pressure would be equivalent to the ordinary elasticity of that fluid, was computed in the first edition of the *Principia*, which came out in 1687, on the supposition that water is 850 times denser than air, mercury $13\frac{1}{2}$ times denser than water, and that the mean height of the barometer is thirty English inches. The modulus of elasticity, or the height of an equiponderant column of air, was therefore estimated at 29,042 feet, which gave 668 feet each second for the celerity of the transmission of sound through the atmosphere. In the next edition, which did not appear till twenty-six years thereafter, the computation of the modulus was somewhat altered, but certainly not rendered more correct. Assuming the same standard of barometric height as before, and supposing mercury to be $13\frac{2}{3}$ times heavier than water, and water 870 times heavier than air, the modulus would be 29,725 feet, to which the corresponding velocity of sound is 979 feet in the second.

In these successive estimates, there is perhaps betrayed some desire to magnify the result, yet without nearly approaching to the amount of actual observation. Dr Derham had recently determined, from repeated trials made with care, that the ordinary flight of sound is at the rate of 1142 feet each second; and Newton endeavoured, by some very strained hypothesis, to accommodate his calculation to this correct measure. 1. He supposes the particles of air to be perfectly solid spherules, whose diameter is the ninth part of their mutual distance. Sound, being instantaneously communicated through these, would thus have its velocity increased by one-ninth, or 109 feet, or brought up to 1088 feet in the second. 2. He next assumes, that the particles of vapour concealed in the air, and augmenting the common elasticity without partaking of the impression of sound, amount to a tenth part of the whole. This would increase the celerity of the sonorous pulse in the subduplicate ratio of 10 to 11, or as 20 to 21 nearly, and consequently advance the last measure from 1088 to 1142 feet.

But these random and fanciful conjectures hardly require any serious consideration. What may be the size of the ultimate particles of air, or whether they have any sensible magnitude at all, we are utterly without the means of determining. There appears no limit, indeed, to the degree of condensation of which the air is capable, but what proceeds from the imperfection of the engines employed for that purpose. Nay, supposing so large a proportion of absolute matter to exist in the composition of our atmosphere, it really would not affect the result, since the transit of sound, as we have shown, is necessarily progressive, even through the most solid substance. To this principle there could be no exception, unless the particles of air were held to be mere atoms, incapable of further subdivision,—in short, without actual magnitude, and therefore bearing no relation whatever to the space in which they float. The second hypothesis advanced is still more insufficient to rectify the general conclusion. That moisture, in its latent or gaseous form, is united with the air, will be granted; but it by no means constitutes such a notable share of the fluid as Newton has assumed, scarcely exceeding, at the ordinary temperature, perhaps the five-hundredth part of the whole weight. But this diffuse vapour could not in the least derange the original calculation; for, being always combined with the air, the measure of elasticity assigned by experiment was really that of the compound fluid which forms our atmosphere.

We are now enabled, by the help of more perfect data, to rectify the modulus of atmospheric elasticity, or the height of a homogeneous and equiponderant column of the fluid. From the observations made with barometrical measurements, it appears that such a column, exerting a pressure equivalent to the elasticity of the air, has, at the limit of freezing water, an altitude of 26,060 feet, and consequently, that the modulus would, at an ordinary temperature of 62° by Fahrenheit, amount to 27,800 feet. This corrected estimate gives only 943 feet each second for the celerity of sound. And since the elasticity of the medium is exactly proportioned to its density, the result is the same, whatever be the rarefaction or condensation of the air, so long as its temperature continues unaltered. The flight of sound is hence as rapid near the surface as in the higher regions of the atmosphere. It is the conjunction of heat alone that will increase the celerity of transmission, by augmenting the elasticity of the medium without adding to its weight. The acceleration thus produced must amount to rather more than one foot in the second for each degree by Fahrenheit's scale. Such a difference ought to be perceptible under the torrid zone.

But the rate of the transmission of sound must vary in different gases, after the inverse subduplicate ratio of their densities. Thus, through carbonic gas, the communication of the tremor would be about one-third slower than ordinary; but through hydrogen gas, which is twelve times more elastic than common air, the flight would very nearly exceed three and a half times the usual rapidity. An admixture of this gas with the atmosphere would, therefore, greatly accelerate the transmission of sound. The joint combination of heat and moisture, by heightening the elasticity of the air, must likewise produce a similar effect.

These inferences are confirmed by observation, as far as it extends. The velocity of sound was determined with considerable accuracy, and on a great scale, by Cassini and Maraldi, while employed in conducting the trigonometrical survey of France. During the winter of the years 1738 and 1739, these astronomers repeatedly discharged, at night, when the air was calm and the temperature uniform, a small piece of ordnance, from their station on Mont Martre, above Paris, and measured the time that elapsed between the flash and the report, as observed from their signal tower at Mont L'Hery, at the distance of about eighteen miles. The mean of numerous trials gave 1130 feet for the velocity of the transmission of sound.

About the same time, Condamine, who was sent with the other academicians to ascertain the length of a degree in Peru, took an opportunity of likewise measuring the celerity of sound, at two very different points. He found this was 1175 feet on the sultry plain of Cayenne, and only 1120 feet on the frozen heights of Quito. It was obvious, therefore, that the rarefaction of the air in those lofty regions had in no degree affected the result. Compared with what had been observed in France, the velocity of the aerial pulses was somewhat diminished at Quito by the prevailing cold, but was, on the other hand, considerably augmented by the excessive heat and moisture which oppress Cayenne.

But the difference, amounting indeed to one-fifth of the whole, between the velocity of sound as deduced from theory, and as determined by actual experiment, still appeared very perplexing. This want of congruity was the more felt, since the Newtonian system of gravitation, after maintaining a long struggle with the adherents of the Cartesian philosophy, had at last obtained the undisputed possession of the Continent. Its triumph was insured by the admirable dissertations on the subject of tides, transmitted to the Academy of Sciences at Paris in the year 1740, when our celebrated countryman Maclaurin had the honour of sharing the prize with Euler and Daniel Bernoulli. The law of attraction received, indeed, a temporary shock a few years afterwards, from the result of the investigation which Clairaut first gave of the lunar inequalities; but, on resuming his analysis of the problem, and computing the values of the smaller terms of the formula, that great geometer obtained, in 1752, a final product, exactly conformable to the best astronomical observations; and the solidity of the Newtonian system was henceforth placed on the firmest foundation.

It was therefore peculiarly desirable to examine likewise the justness of the hydrodynamical conclusions of Newton. The propositions concerning the propagation of sound were perhaps justly considered as the most obscure part of the whole *Principia*. Some of the first-rate mathematicians abroad, particularly D'Alembert and John Bernoulli, declared their utter inability to comprehend such intricate and disjointed demonstrations. At last the problem of sonorous poules was attacked directly, and in its full extent, by the late Count Lagrange, whose death, although at a ripe age, will be lamented as a most severe loss to mathematical science. That illustrious geometer shone forth at once like a meteor, and before he had completed his twenty-third year, he gave a rigorous and profound analysis of the propagation of sound through the atmosphere, in the first volume of the Turin Memoirs, which appeared in 1759. "He pointed out some mistakes that even Newton had committed in the reasoning; but mistakes, which, by a happy compensation of errors, did not affect essentially the results. Advancing from these discussions, he assigned the dynamical conditions of undulation, which, after the proper limitations, were reduced to an equation involving *partial differences* of the second order. But this refined branch of analysis, invented by D'Alembert and Euler, is still so imperfect, that, in order to integrate the final expression, it had become requisite to omit the higher powers of the differentials. Yet after all this display of accurate research and skilful adaptation of symbols, followed by a lax and incomplete calculus, the same conclusion was obtained as that which Newton had derived chiefly from the force of analogy and sagacity of observation; and philosophers were thus obliged to submit, and to content themselves with recording the variance between theory and experiment in regard to the celerity of sound, or with referring that discrepancy to some extraneous influence."— *Edin. Review,* vol. xv. p. 431.)

M. Poisson, one of those interesting men whose native genius has surmounted all the obstacles of fortune, very lately attempted a more complete analysis of the propagation of sound, in the Papers of the Polytechnic School. The final equation is more fully expressed, and its integration is pushed some few steps farther; but still the result is precisely the same as before. The skill and precaution displayed in framing the conditions of the problem, are afterwards mostly abandoned in the various simplifications adopted to arrive at the conclusion.

A very ingenious and apparently satisfactory method of reconciling theory with observation, in the estimate of the transmission of sound, was not long since suggested by the celebrated Count Laplace. If the heat contained in air had, at every state of the density, been united constantly after the same proportion, the elasticity resulting from the infusion of this subtle and highly distensible element would invariably accord with what observation assigns to the compound aerial fluid. But the capacity of air, or its aptitude to retain heat, varies with its internal condition; being increased by rarefaction, and proportionally diminished by condensation. When air is compressed, therefore, it liberates a portion of its heat; and when it undergoes dilatation, it becomes disposed to abstract more heat from the adjoining bodies. Till the equilibrium of heat is again restored, the air will be sensibly warmer after each act of compression, and colder when suffered to dilate. If the shock given to a portion of air be very sudden and violent, the quantity of heat evolved from it is profuse and powerful. On this principle, M. Mollet, member of the academy of Lyons, led by some facts noticed by artists who manufactured wind-guns, first constructed, in 1804, the curious instrument for producing fire by the rapid condensation of air confined in a tube. But such evolution of heat must besides augment the elasticity of the air, as the contrary abstraction of it will, in a like degree, diminish that force. At every sudden alteration of density, therefore, a new power is infused, which had not entered into the ordinary and undisturbed estimate of the air's elasticity. Consequently, from this consideration alone, the aerial pulses must shoot with some greater celerity than calculation assigns, because the particles of air, which are suddenly condensed, have their elasticity further augmented by the portion of heat evolved, while the corresponding particles, which are simultaneously dilated, have their disposition to contract likewise increased, by the momentary prevalence of cold.

The principle advanced by Laplace must therefore have a real operation, tending to reconcile the calculated velocity of sound with that which is deduced from experiment. The only question is, how far its influence could actually extend. But, according to the formula given in Leslie's *Elements of Geometry*, p. 495, a condensation equal to the 90th part of the volume of air would occasion the extrication of one degree of heat by Fahrenheit's scale. Now, since each degree of heat enlarges the bulk or augments the elasticity of the air by the 450th part, it follows, that the heat, extricated by sudden impulse, will communicate to the air a momentary additional spring, amounting to one-fifth of the whole elastic force. Wherefore the celerity of sound would, by that influence, be increased in the subduplicate ratio of five to six, or nearly as 21 to 23; which gives an addition of only 90 feet each second to the whole quantity, bringing it up to 1033 feet. The correction is thus insufficient, not amounting to half of the discrepancy which it was its object to reconcile.

It may be suspected, therefore, that some inaccuracy or omission infects the investigation itself. Till the integral calculus has arrived at much greater perfection, it will often be requisite for the analyst, in the solution of dynamical problems, to descend from his elevation, and seek to simplify the differential expressions by a sober and judicial application of the principles of physics. "Imagine a string of particles, or physical points, A, B, C, D, E, F, &c. in a state of rest, or mutual balance. If A were pushed nearer to B, and then suddenly abandoned, it would recoil with a motion exactly similar to the oscillation of a pendulum. The time of this relapse might easily be determined, from a comparison of the force of gravity with that of elasticity, or from the number of particles contained in a column of equipoise. The minute interval between the adjacent particles, being now divided by the duration of each fit of contraction, will give the velocity with which the vibratory influence shoots along the chain of communication. This simple investigation leads still to the same result as before. But it proceeds on assumptions which are evidently incorrect; for it supposes the pulses to follow each other in accurate succession, every contraction terminating as the next begins. Since the particles, however, do not exist in a state of insulation, while B repels A, it must likewise press against C; and C in its turn, must gradually affect D. Before the contraction of A and B is completed, that of B and C is therefore partially performed; and this anticipated influence may even extend to the remoter particles. Nor is the system of mutual action at all materially disturbed by such anticipations. Each pulsation is performed in the same way as if it were quite detached; only the succeeding one is partly accomplished before the regular period of its commencement. The velocity of aerial undulations is in this way much accelerated." (*Edin. Review*, vol. xv. p. 433.) Each successive movement among the particles may be viewed as produced by a force not regularly decreasing, but partaking of the uniformity which obtains in projection. Hence the velocity of sound is intermediate between that derived from theory and that with which air would rush into a vacuum. But the arithmetical mean between 943 and 1334 feet is $1138\frac{1}{2}$, and the geometrical is $1121\frac{1}{2}$ feet; neither of which differs much from 1130 feet, the quantity determined by actual experiment.

After the last correction, however, proposed by M. Laplace,

for adjusting theory with observation relative to the celerity of the transmission of sound, the difference will not perhaps be regarded such as longer to present any serious obstacle; especially when the coincidence appears closer than what generally attends the theoretical deductions concerning the motions of fluids. The remaining difficulties affecting the subject refer chiefly to the way in which the aerial pulses are propagated, and the modifications which they are afterwards capable of receiving.

1. No sensation is ever excited, unless the impression made upon our organs be repeated or continued during a certain short space of time. On this principle depends the whole success of the juggler, who contrives to change the situation of the various objects before us with a rapidity exceeding the ordinary exercise of sight or touch. A brand whirled swiftly round the head gives all the appearance of a circle of fire; and if one presses very hard an ivory ball between his fingers, he will seem still to feel it for several instants after it has been withdrawn. To excite the sensation of sound, it is requisite that the aerial pulses should have a certain force and duration. According to some observations, the ear is not affected at all, unless the tremulous impulse communicated to the tympanum lasts during the tenth part of a second. Every pulsation of a more transient kind is lost absolutely and completely to our organ of hearing.

On the other hand, the impression of sound is not prolonged beyond the time of its actual production. If it were otherwise, indeed, all sounds would degenerate into indistinct noises; and articulate discourse, which distinguishes man from the lower animals, and constitutes the charm of social life, would have been utterly impossible. This fact, so obvious, and yet so important, shows indisputably that the propagation of sonorous pulses through the atmosphere is not, in all its circumstances, analogous to the succession of waves on the surface of water. These undulations continue long afterwards to rise and spread from the centre of their production. The pulsations of the air, no doubt, likewise survive their excitement; but such of them as succeed the first impulsion must not have the force and character of those which are directly shot through the fluid. What is the precise discrimination between these different pulses, we are not enabled from mere theory to determine. But such a distinction must undoubtedly exist, otherwise indeed all discourse would continue to fill the ear with a monotonous hum, or an indistinct muttering. It would be difficult to institute conclusive experiments on this subject, yet collateral researches might be devised which could not fail to guide our inquiry.

2. But another defect in the analogy between waves and sonorous pulses is, that the latter, without affecting to spread equally, are capable of acquiring a superior force or tendency in some given direction. Certain unconfined sounds, indeed, are diffused uniformly on every side. Thus, the noise of the explosion of a powder-mill is heard, and often dreadfully felt, at a great distance all round the scene of disaster. But the report of a cannon, though audible in every direction, appears invariably loudest in the quarter to which the engine is pointed. On this principle, a seaman, when he seeks to be heard more audibly, or at a greater distance, is accustomed, if no other help occurs, to apply his spread hands on each side of his mouth, and thus check or diminish the waste of sound by its lateral dispersion. For the same reason, the bent and projecting circular piece annexed to the farther end of a speaking-trumpet is of most decided use, in assisting to give direction to the flight of the aerial pulses.

3. The theory of undulatory movements furnishes some

elucidation, but no adequate explication, of the augmented effect of sound in the direction of a lateral barrier. The extension of such an obstacle might appear to check merely the spread and consequent attenuation of the sonorous pulses; but the great accumulation of impulse always occurs, on either side, at the extremity of the advancing wave. By what system of interior forces this effect is produced, it would be difficult satisfactorily to explain. Yet we perceive something analogous in the swell which runs along the margin of a pool, and in the billow which, flowing from the open sea, heaves against the sides of a projecting mole.

It is hence that sound is made to sweep with such intensity over the smooth surface of a long wall or of an extended gallery. An elliptical figure, though of manifest advantage, is not really essential to a whispering gallery; for the point of sonorous concentration is found beyond the true catoptrical focus, and much nearer to the wall. A fact of the same kind is well ascertained—that sounds are always heard the most audibly, and at the greatest distance, in a level open country, or still better on the smooth surface of a vast lake, or of the ocean itself. The roaring of the cannon in certain naval engagements has been noticed at points so very remote from the scene of action, as might seem, if not perfectly authenticated, to be altogether incredible. On the other hand, again, sound is enfeebled and dissipated sooner in alpine regions. Thus, the traveller, roving at some height above a valley, descries with uncommon clearness, perhaps a huntsman on the brow of the opposite mountain, and while he watches every flash, yet can he scarcely hear the report of the fowling-piece.

On a similar principle, we would explain the operation of the ear-trumpet, which affords such relief to one of the most cheerless maladies that can afflict humanity. The wide mouth of that instrument, it is well known, is turned to catch the stream of sound; the extent of pulsation is gradually contracted as the tide advances; and the same quantity of impulse being probably maintained, the vibratory energy is intensely accumulated at the narrow extremity, where it strikes the cavity of the ear. A trumpet of this form might, in many cases, be found very advantageous, not only for remedying the defects of the organ of hearing, but for assisting the observer to collect feeble and distant sounds. Even an umbrella held close behind the head, with its concavity fronting the sonorous pulses, will, it has been alleged, sensibly heighten their impression.

4. To explain legitimately the reflection of sound, would require some modifications in the theory of atmospheric undulation. Each obstructing point is certainly not the centre of a new system of pulses; for, in many cases, this would occasion unutterable confusion. Nor can the excitement of sound be supposed to dart in straight lines, or to perform the same accurate reflection as the rays of light. In fact, neither smoothness nor exact regularity of surface is required for the production of an echo. A range of buildings, a row of tall trees, a ridge of rocks, or a chain of heights, will, in certain positions, reflect sound with clear and audible effect. It follows, therefore, that the reflection must be formed, not at the immediate surface of those obstacles, which could occasion only an irregular dispersion, but at some boundary at a small distance, and running parallel to the mean direction of the whole barrier. We may conceive the tide of sound accumulating where it stops, and investing the opposite surface like an atmosphere, till a repulsion is exerted, which again rolls it back. What seems to constitute the perfection of an echo is, that the sum of the distances of every point of the reflecting surface from the person who speaks, and from him who listens, should be the same. When this disposition obtains, all the reflected

sounds must reach the ear in due succession, without being intermingled or confused.

We may observe, that echoes are often confounded with the mere *resonance* occasioned by vibrations excited among the obstacles themselves. In a large empty room, with its naked floor, and walls, and benches, the voice quickly throws the whole into a tremulous commotion, and seems drowned in the ringing prolonged sound which is produced; nor does this unpleasant effect cease, until the spectators have occupied the benches, filled the hall, and obstructed by their weight the vibration of the floor. What is called the *deadening* of sound, consists in merely checking or preventing the disturbance of extraneous tremor. For this purpose, the floor is covered with carpets, and the walls lined with wainscot or hangings. Such barriers, we have seen, would not, by their yielding quality, blunt or obstruct the formation of echoes. Their only effect is, to muffle the elastic surfaces which they cover.

The performance of the speaking-trumpet has generally been referred to the concentrated reflection of sound. Some authors have carried the hypothesis even so far as to investigate, from mathematical principles, the best figure of that instrument. Much labour and great ingenuity have been utterly wasted in this fruitless attempt. Kircher proposed the tube to be shaped like a truncated parabolic conoid, the mouth-piece occupying the focus; and he concludes that all the *rays of sound* would, by reflection from such a surface, be sent forward exactly in parallel lines. Other philosophers have imagined, from a fanciful analogy to the property of ivory balls, that the figure described by the revolution of the logarithmic curve about its absciss would be the most proper for the speaking-trumpet. M. Lambert, of the Berlin academy, whose genius and originality were both of the first order, has given a solution still different. But it would be idle to recite the various attempts which have ended in no practical result.

The true physical explication of the speaking-trumpet was first given, as far as we know, in the course of an incidental remark by Professor Leslie, in his *Experimental Inquiry into the Nature and Propagation of Heat.* "In the case of articulate sounds," says he, "the confining of the air does not affect the pitch of voice, but it augments the degree of intonation. The lateral flow being checked, that fugacious medium receives a more condensed and vigorous impulsion. As the breath then escapes more slowly from the mouth, it waits and bears a fuller stroke from the organs of speech. But the speaking-trumpet is only an extension of the same principle. Its performance does certainly not depend upon any supposed repercussion of sound; repeated echoes might divide, but could not augment the quantity of impulse. In reality, however, neither the shape of the instrument, nor the kind of material of which it is made, seems to be of much consequence. Nor can we admit that the speaking-trumpet possesses any peculiar power of collecting sound in one direction; for it is distinctly audible on all sides, and is perhaps not much louder in front, comparatively, than the simple unassisted voice. The tube, by its length and narrowness, detains the efflux of air, and has the same effect as if it diminished the volubility of that fluid, or increased its density. The organs of articulation strike with concentrated force; and the pulses, so vigorously thus excited, are, from the reflected form of the aperture, finally enabled to escape, and to spread themselves along the atmosphere. To speak through a trumpet costs a very sensible effort, and soon fatigues and exhausts a person. This observation singularly confirms the justness of the theory which I have now brought forward."

Nearly about the same time, this theory was confirmed

by some ingenious experiments made by M. Hassenfratz, at Paris. His method of estimating the power of a speaking-trumpet consisted in fixing a small watch in the mouth-piece, and observing at what distance the beats ceased to be distinctly audible. He found that the effects were precisely the same with a trumpet of tinned iron, whether used in its naked form, or after it was tightly bound with linen to prevent any vibration of the metal. Nor could there be the smallest reflection of sound from the internal surface of the tube, for the beating of the watch was heard exactly at the same distance after the whole of the inside had been lined with woollen cloth. These simple experiments prove decisively, that the performance of the speaking-trumpet depends principally on the intenser pulsation which is excited in the column of confined air. In the same way, sound is prodigiously augmented in a long narrow passage. If a musket be fired within the gallery of a mine, the explosion heard in a remote corner will have the loudness and character of thunder.

The progressive motion of sound furnishes the explication of various remarkable facts and striking phenomena. Thus, to a person standing at some distance, and directly in front of a long file of musketry, the general discharge will appear as a single collected sound, the numerous reports all reaching his ear nearly at the same instant. But one stationed at the end of the line will hear only a prolonged rolling noise, not unlike a running fire; because the distinct sounds, from the different distances which they have to travel, will arrive in a continued succession. Hence, likewise, the tremendous rumbling noise of distant thunder, which is not produced, as many have supposed, by the repetition of echoes. In certain situations, indeed, and particularly in hilly tracts, echoes may no doubt contribute to augment the general effect; but their ordinary influence seems to be really insignificant, since it should cause the same modification of sound in the explosion of a cannon, which is essentially different, however, from the muttering and crash of thunder. This lengthened and varied noise must yet be the production of a moment. The rapidity of lightning surpasses conception, and the prolongation of the sound which follows it is owing to the various distances of the chain of points which emit the sonorous impressions. The electrical influence darts with immeasurable swiftness from cloud to cloud, till perhaps it strikes at last into the ground. But from every point of this tortuous path distinct pulses of sound are transmitted, which consequently reach the ear at very different intervals. Sometimes they arrive intermingled, and give the sensation of a violent crash; at other times they seem suspended, and form a sort of pause. It would not be very difficult in any case to imagine the zig-zag track which the lightning must pursue in order to produce a given protracted rumbling noise. The duration of each peal of thunder will evidently be shortened if it chance to shoot athwart, but must continue the longest when it runs in the line of the spectator. As the distance of thunder is estimated by allowing somewhat more than a mile for every five seconds that elapse between the flash and the beginning of the report, so the space traversed by the lightning, if its general direction were known, might be computed by the same rule, from the endurance of the sound.

We will not enter at present on that branch of acoustics which treats of the doctrine of harmony; but a few scattered remarks may trace the general outline of the subject. A musical note, far from being only a repetition of the same simple sound, should be considered as the conjunction of subordinate sounds reiterated at proportional intervals. The sweetness of this compound effect or tone appears to depend on the frequent recurrence of interior unison. The secondary sounds which naturally and invariably accompany the fundamental note are repeated only two, three, or four times faster; nor does the science of music admit of any proportions but what arise from the limited combinations of those very simple numbers. Harmony, again, is created by an artificial union of different notes, analogous to the natural composition of tone.

All tones are produced by the regular vibrations either of solid substances or of confined air itself. Strings of gut or of metal are most generally used; but small plates or pillars of wood, of glass, or even of stone, will answer the same purpose, forming the singular instrument called *staccata* or *harmonica*. In these cases, the quality of the vibrations depends on the joint influence of a variety of circumstances; not only on the length of the fibres, but on their thickness, their elasticity, their density, and the degree of tension to which they are subjected. The motion of a musical stretched chord was first investigated by the very ingenious Dr Brook Taylor, though his solution has been since proved to be incomplete. At the same time, in fact, that the whole chord oscillates, its simpler portions, the half, the third, and the fourth of its length, actually perform a set of intermediate vibrations.

Wind-instruments produce their effect by the vibrations of a column of air confined at one end, and either open or shut at the other. These vibrations are determined merely by the length of the sounding column. Yet interior and subordinate vibrations are found to co-exist with the fundamental one. The whole column spontaneously divides itself into portions equal to the half, the third, or the fourth of its longitudinal extent. We shall more easily conceive these longitudinal vibrations, by observing the contractions and expansions of a long and very elastic string, to the end of which a ball is attached. A spiral spring shows still better the repeated stretching and recoil. If struck suddenly at the one end, it will exhibit not only a total vibration, but likewise partial ones, winding vermicularly along the chain of elastic rings.

But when the air is struck with uncommon force, the subordinate vibrations become predominant, and yield the clearest and loudest tones. This we perceive in the dying sounds of a bell, which rise by one or two octaves, and expire in the shrillest note. On such a very narrow foundation—on the variable force with which it is blown—rests the whole performance of the bugle-horn, whose compass is extremely small, consisting only of the simplest notes. In other wind-instruments, the several notes are caused by the different lengths of the tube, or by the various positions of the holes made in its side.

The longitudinal vibrations of a column of air, contained within a tube open at both ends, are powerfully excited, and very loud and clear tones produced, by the inflammation of a streamlet of hydrogen gas. This curious experiment was made first in Germany, and appears indeed to have been scarcely known, or at least noticed in other countries. Yet it is most easily performed, and will be considered as amusing, if not instructive. A phial, having a long narrow glass pipe fitted to its neck, being partly filled with dilute sulphuric acid, a few bits of zinc are dropt into the liquid. As the decomposition of the water embodied with the acid now proceeds, the hydrogen gas thus generated flows regularly from the aperture, and is capable of catching fire, and of burning for some considerable time, with a small yet steady round flame. This very simple arrangement, frequently styled the philosophic lamp, is in reality of the same nature with the combination, on a large scale, of the gas lights. A glass tube being passed over the exit-pipe, the burning speck at its point instantly shoots into an elongated flame, and creates a continued sharp and brilliant musical sound. This effect is not owing to any vibrations of the tube itself, for it is nowise altered by tying a handkerchief tightly about the glass, or even by substituting a cylinder of paper. The tremor excited in the column of air is therefore the sole cause of the incessant tone, which only varies by a change in the place of the flame, or a partial obstruction applied at the end of the tube. But still it is not easy to conceive how the mere burning of a jet of hydrogen gas within the cavity should produce such powerful vibrations. The exciting force must necessarily act by starts, and not uniformly. The length of the flame might seem to prove, that the hydrogen gas is not consumed or converted into aqueous vapour as fast as it issues from the aperture. A jet of it catches instantaneous fire, but is immediately followed by another, the succession of inflamed portions being so rapid as entirely to escape the keenness of sight. The column of air contained within the tube would thus be agitated by a series of incessant strokes or sudden expansions.

The singular fact now described had occurred incidentally to the writer of this article in the course of his earliest experiments; and he has often thought since, that, on the same principle, an organ might be constructed, which would have a very curious and pleasing effect. A vertical motion of the glass tubes, and the partial shutting or opening of their upper ends, would occasion a considerable variety of notes. By passing the hydrogen gas over different metals, the flame would be made to assume various colours. The apparatus might work by a spontaneous mechanism; and while the eye was gratified by the display of rich and vivid tints, the ear would be charmed with strains of new and melodious symphony.

# AERONAUTICS.

---

In every stage of society men have eagerly sought, by the combination of superior skill and ingenuity, to attain those distinct advantages which nature has conferred on the different tribes of animals, by endowing them with a peculiar structure and a peculiar force of organs. The rudest savage learns from his very infancy to imitate the swimming of a fish, and plays on the surface of the water with an agility and a perseverance which seems to decline with the advancement of civilization. But an art so confined in its exercise, and requiring such a degree of bodily exertion, could not be considered of much avail. It was soon perceived, that the fatigue of impulsion through the water could be greatly diminished, by the support and floating of some light substance. The trunk of a tree would bear its rude proprietor along the stream; or, hollowed out into a canoe, and furnished with paddles, it might enable him even to traverse a river. From this simple fabric, the step was not great to the construction of a boat or barge, impelled by the force of oars. But it was a mighty stride to fix masts and apply sails to the vessel, and thus substitute the power of wind for that of human labour. The adventurous sailor, instead of plying on the narrow seas, or creeping timidly along the shore, could now launch with confidence into the wide ocean. Navigation, in its most cultivated form, may be fairly regarded as the consummation of art, and the sublimest triumph of human genius, industry, courage, and perseverance.

Having achieved by his skill the conquest of the waters that encompass the habitable globe, it was natural for man to desire likewise the mastery of the air in which we breathe. In all ages, accordingly, has ingenuity been tortured in vain efforts at flying. The story of Icarus testifies how fatal such daring attempts had generally proved to their projectors. Trials made with automatons, though less liable to risk and danger, were yet equally fallacious. Archytas, a most eminent Greek geometer and astronomer, who perished by shipwreck on the coast of Calabria, was believed by his admiring contemporaries to have constructed an artificial dove, which, by the action of a system of internal springs, wafted itself through the air. If such a piece of mechanism was ever made, we may be sure that its flight was really produced, as in the scenes of the opera, by means of invisible strings or wires.

So thoroughly were the ancients convinced of the impossibility of men being able to fly, that they ascribed the absolute rule of the sky to divinities of the first order. The supreme Jupiter alone reposed on his empyreal throne, far above the heights of Olympus; and to him was it given, from the region of the clouds, to point the winged lightning, and to hurl the flaming thunderbolt. On special missions he dispatched Mercury, as his messenger, through the wide range of atmosphere. The oriental nations, from whom we have borrowed the greater part of our vulgar mythology, likewise committed such journeys to certain genii, or ministering spirits. But the glowing visions of the East received a darker tinge from the character and climate of our Gothic ancestors. The arch fiend himself was, at no very distant period, firmly believed to have the especial control of the air, and to career in the whirlwind and impel the howling tempest. Those wretched creatures whom the unfeeling credulity of our ancestors, particularly during the prevalence of religious fanaticism, stigmatized and murdered under the denomination of witches, were supposed to work all their enchantments, to change their shapes at will, and to transport themselves through the air with the swiftness of thought, by a power immediately derived from their infernal master. At a period somewhat earlier, every person in possession of superior talents and acquirements was believed to deal in magic, and to perform his feats of skill chiefly through the secret aid granted him by the prince of darkness. In spite of the incurable perverseness of his conduct, it must be confessed that the devil has always had the credit of retaining some little inclination to assist the efforts of genius.

During the darkness of the middle ages, every one at all distinguished by his knowledge in physics was generally reputed to have attained the power of flying in the air. Our famous countryman Friar Bacon, among other dreams engendered in his fervid brain, has not scrupled to claim the invention of that envied and transcendent art. To these pretensions the credulity and indulgent admiration of some authors have lent more credit than they really deserved. Any person who will take the trouble to examine the passages of Bacon's obscure though ponderous works, must soon be convinced, that the propositions advanced by him are very seldom founded on reality, but ought rather to be considered as the sportive illusions of a lively and teeming fancy. Albertus Magnus, who lived about the same period, and was esteemed in Germany as a perfect prodigy, pretended also to the art of flying. More than a century afterwards, John Müller of Königsberg, and thence styled *Regiomon-*

*tanus*, one of the chief restorers of genuine mathematical learning in Europe, was reported by some writers of note to have, like Archytas, fashioned an artificial dove, which displayed its wings, and flew before the emperor Charles V. at his public entrance into Nüremburg. But, unfortunately for the veracity of the story, Regiomontanus died in early life, full sixty years before that visit took place.

While the belief in necromancy prevailed, such tales assumed colours of the most lurid hue. Fiery dragons, created by infernal machination, were imagined to rush impetuous through the sky, vomiting flames, and widely scattering the seeds of pestilence. Grave writers, in those benighted ages, even ventured to describe the method of imitating the composition of such terrific monsters. A mass of large hollow reeds were to be disposed and bound together, then sheathed completely in skin, and smeared over with pitch and other inflammable matters: this light and bulky engine, partially set on fire, and launched in the thickest darkness into the air, might be sufficient, when borne along by the force of the wind, to strike the ignorant populace with affright and horror. But such spectacles would come to lose their terrors by repeated failure, and the insensible progress of knowledge, So late as the year 1750, a small Catholic town in Swabia was almost entirely burnt to ashes by an unsuccessful experiment of that sort, instigated, and probably directed, no doubt for the edification of their flock, by the lowest order of priests. It was attempted to represent the effigy of Martin Luther, whom the monks firmly believed to be the very imp of Satan, under the form of a winged serpent, furnished with all the requisite appendages of a forked tail and hideous claws. Unluckily for the skill of the machinist, this phantom directly fell against the chimney of a house, to which it set fire; and the flames spreading furiously in every direction, the people had soon cause to lament bitterly their intemperate zeal.

The scheme of flying in the air, which men of the first genius had once entertained, appears to have gradually descended to a lower class of projectors. Those who afterwards occupied themselves with such hopeless attempts, had commonly a smattering of mechanics, with some little share of ingenuity, but wrought up by excessive conceit.

In the beginning of the sixteenth century, an Italian adventurer visited Scotland, during the reign of James IV.; and being a man of some address, and at the same time a pretender to alchemy, he contrived to insinuate himself into the favour of that gay and needy prince, by holding out hopes of augmenting his scanty treasury by the acquisition of the philosopher's stone. He was collated by royal favour to the abbacy of Tungland in Galloway; but not having succeeded in creating artificial riches, he resolved, in the height of his enthusiasm, at once to gratify and astonish the courtiers, by the display of a feat still more extraordinary. Having constructed a set of ample wings, composed of various plumage, he undertook, from the walls of Stirling Castle, to fly through the air to France. This experiment he had actually the folly or hardihood to try; but he soon came to the ground, and broke his thigh-bone by the violence of the fall. For his unlucky failure, however, the abbot had the dexterity to draw a very plausible excuse from the wretched sophistry, termed science in that age. "My wings," said the artful Italian, "were composed of various feathers; among them were the feathers of dunghill fowls, and they, by a certain sympathy, were attracted to the dunghill; whereas, had my wings been composed of the feathers of eagles alone, the same sympathy would have attracted them to the region of the air." This anecdote has furnished to Dunbar, the Scottish poet, the subject of one of his rude Satires.

A century afterwards, Fleyder, rector of the grammar-school at Tübingen, entertained, in 1617, the worshipful magistrates of that city with a lecture on the art of flying, which he published at the lapse of eleven years, yet prudently contented himself with barely explaining his theory. A poor monk, however, ambitious to reduce this theory into practice, having provided himself with spacious wings, took his flight from the top of a high tower; but encountering a cross wind, his machinery misgave, and falling precipitously to the ground, he broke both his legs, and perished miserably. An accident of a similar kind is related to have happened not long since near Vienna.

The impossibility of rising, or even remaining suspended in the air, by the action of any machinery impelled by human force, was first demonstrated by Borelli, a most eminent Italian mathematician and philosopher, who lived in the fertile age of discovery, and was thoroughly acquainted with the true principles of mechanics and pneumatics. In his celebrated and excellent work *De Motu Animalium*, published in 1670, he showed, by accurate calculation, the prodigious force which the pectoral muscles of birds must exert and maintain. The same principles, applied to the structure of the human frame, proved how very disproportionate was the strength of the corresponding muscles in man. It is not, therefore, the mere difficulty of contriving and combining machinery which should perform the peculiar motions of wings, that has rendered all attempts of the kind futile, but the utter want of adequate force in the human body to give such impulsion to those extended vanes, as would be necessary for supporting so great a weight in the thin medium of the atmosphere.

Having found by experience the impossibility, from any

application of inherent strength, of ascending into the atmosphere, it was natural for men of ardent minds, who still pursued that dazzling project, to look for some extraneous aid among the varied powers of the elements. The notions entertained by the ancients respecting the composition of the world, might have suggested important hints for realizing the scheme of aerial navigation. The four elements—earth, water, air, and fire or ether, arranged according to their several qualities and tendencies, were supposed to constitute this universal frame. Earth, being heavy and inert, occupies the centre of the system, and above it flowed the waters; air, from its lightness, rose upwards, and invested the globe with an atmosphere; while the diffuse ethereal substance soared, by its extreme buoyancy, to the celestial regions, and filled with splendour their pure expanse. Every portion of these distinct elements, if transported from its place, was conceived as having a natural and constant appetency to return to its original situation. Earth and water sink downwards by their gravity, while air and fire, endued with an opposite principle, as invariably rise to the higher spaces. A portion of fire, joined to water or to air, communicates, in a corresponding degree, its levity or disposition to ascend. Thus, warm air always rises; water, subdued by excessive heat, flies upwards in the form of vapour; and the volatile parts of inflamed bodies are borne to the sky in smoke.

The first person that seems to have formed a just idea of the principle on which a balloon could be constructed, was Albert of Saxony, a monk of the respectable order of St Augustin, who lived in the fourteenth century, and wrote a learned commentary on the physical works of Aristotle. Since fire is more attenuated than air, and floats above the region of our atmosphere, this ingenious person conceived that a portion of such ethereal substance, inclosed in a light hollow globe, would raise it to a certain height, and keep it suspended in the sky. But the same philosopher rightly subjoined, that a greater mixture of air introduced into the balloon, by rendering this heavier than before, would cause it to descend proportionally, in the same way precisely as water admitted through the seams of a ship makes the vessel to sink in the ocean. It is evident that nothing was wanted for completing Montgolfier's discovery, but to carry those fine views into execution.

The ideas of Albert of Saxony were long afterwards zealously embraced by Francis Mendoza, a Portuguese Jesuit, who died at Lyons in the course of a tour through France in 1626, at the age of 46. He maintained that the combustible nature of fire was no real obstacle to its application in balloons, since its extreme levity, and the exclusion of the air, would hinder it from supporting inflammation. Caspar Schott, a Jesuit likewise, pursued more soberly the same speculation in Germany. He stated that no air of these lower regions is ever light enough to produce an ascent, and that the lucid ethereal matter which swims above our atmosphere is alone fitted for aerial navigation. Were any superhuman power, therefore, to bring down a store of that buoyant substance, to be inclosed in a hollow ball of wood or thin lead, the vessel, being furnished with a rudder and sails, might then, he conceived, boldly navigate the sky.

Similar notions have been renewed at different times. They were likewise often blended with the alchemical tenets so generally received in the course of the fifteenth, sixteenth, and part of the seventeenth centuries. Conceiving with the ancients, that the dew which falls during the night is of celestial origin, and shed by the stars, speculative men still imagined this pure humidity to be drawn up again to the heavens by the sun's rays in the heat of the day. Many persons, imbued with the wretched learning of that age, had the simplicity to believe that an egg-shell filled with the morning dew, and placed at the foot of a ladder leaning against the roof of a house, would, as the day advanced, spontaneously rise along the bars, and mount to the chimney tops. This whimsical fancy is confidently related as an observed fact by Father Lauretus Laurus. "Take," says he, very gravely, "a goose egg, and having filled it with dew gathered fresh in the morning, expose it to the sun during the hottest part of the day, and it will ascend, and rest suspended for a few moments." To perform the experiment on a greater scale, however, he proposed to employ the largest swan's egg, or a bag artificially prepared from the thinnest and lightest skin, into which, instead of dew, he would introduce the three alchemical elements, nitre, sulphur, and mercury; and he imagined that these active bodies, expanded and sublimed by the mere heat of the sun, must spring powerfully upwards. In this way he thought the dove of Archytas might be constructed. But the visionary priest had yet another scheme to advance for effecting the ascent of the automaton: he proposed to cram the cavity of the dove with highly condensed air; and was so grossly ignorant of the principles of motion as to suppose that this imprisoned fluid would impel the machine in the same manner as wind does a sail. Should such a force be found not sufficiently efficacious, he finally recommended the application of fire; not, however, on account of its buoyant property, but because of the propulsive power which it exerts. To prevent the fire from consuming the wooden machine, he recommends lining the inside with cloth of asbestos, or other incombustible materials; and to feed and support steadily this fire, he suggested a compound of butter, salts, and orpiment, lodged in metallic tubes, which he imagined would at the same time heighten the whole effect, by emitting a variety of musical tones like an organ.

Influenced by the same views, other authors, and particularly the famous Cardan, have proposed, for aërial ascents, to apply fire acting as in a rocket. Still later, but in the same country, Honoratus Fabry, penitentiary of the pope, and teacher in the gymnasium at Rome, who died about the end of the seventeenth century, described a huge apparatus, consisting of very long tin pipes, in which air was compressed by the vehement action of fire below. In a boat suspended from the machine, a man was to sit and direct the whole, by the opening or shutting of valves.

The projects and vagaries of learned men about the misty period of the restoration of science, were finely ridiculed by Cyrano de Bergerac, a very witty and eloquent French writer, in a philosophical romance, entitled, *The Comical History of the States and Kingdoms in the Sun and Moon.* This eccentric genius, born at Perigord in 1620, was noted for his impetuous temper and boiling courage. He spent his youth in dissipation and feats of arms; but afterwards, in riper life, he quitted the military profession, and betook himself to the study of poetry and philosophy, which he prosecuted with great ardour and success till he died, at the early age of 35. In his romance, from which perhaps Swift borrowed the idea of Gulliver's voyage to Laputa, Bergerac introduced a good deal of the Cartesian philosophy, then just coming into vogue; but lashed severely the pedantry and ignorance of various pretenders to science. To equip himself for performing the journey to the moon, the French traveller fastens round his body a multitude of very thin flasks filled with the morning's dew. The heat of the sun, by its attractive power exerted on the dew, raised him up to the middle region of the atmosphere; where some of his flasks happening unluckily to break, the adventurer sunk again to the ground, and alighted in Canada. There he constructed a new machine, acting by a train of wheels, with which he mounted to some height; but falling down, he had the misfortune to break his leg. He crept aside, in search of ox-marrow to compose a salve, with which he instantly healed his bruises; and returning again, he found his engine in the possession of some soldiers, who had fixed to it a number of sky-rockets. Replacing himself now in the car, he applied fire to the rockets, and darted upwards with inconceivable swiftness; the earth retired gradually from view, while the orb of the moon appeared proportionally to expand, till, approaching the sphere of her activity, he was borne softly along, and descended on the lunar surface into a most delicious and luxuriant grove. Here, of course, he met with angelic personages, endowed with every perfection of body and mind, and far exalted above the mean vices and the rancorous passions which poison and inflame the inhabitants of this blood-stained globe. In the conversations which Bergerac held with those supernal beings, he was informed that a native of our planet, utterly disgusted at the crimes which pollute its "sin-worn mould," had once on a time provided himself with a pair of very large and thin metallic vessels, which he filled with smoke, and sealed in the light; and having attached himself below them, the buoyant power of the confined smoke carried him to the highest region of our atmosphere, where the attraction of the moon at length prevailing, drew him to her surface, while the great extent of the machinery, by opposing resistance, served to break the force of his fall. The moment, however, those slender capacious vessels were liberated from his weight, they rose again by the action of the smoke, till they reached a medium of the same density, and finally took their station in the bright fields of ether, where they form the constellation now called the *Balance*.

In further discourse with his sublime instructor, our romantic voyager was shown how to obtain the power of ascension from the loadstone. He was directed to take two magnets, each about a foot square, to roast them in the fire, to separate their impurities by solution, and thus concentrate their attractive virtue in a mere calx, which could be formed into a ball. Aided by such counsels, he now resolved to visit the sun. With much labour and perseverance he constructed a chest of very thin steel, six feet high and three feet wide; an icosahedron of crystal, the highest of all the regular solids, being fitted into the top, and the bottom having a small valve which opened outwards. Into this chest he shut himself, while the sun's rays, concentrated and multiplied by reflection from the numerous facets of the crystal, heated the air intensely, and drove a great part of it out below; and he ascended rapidly towards the glorious luminary, breathing ecstaticly in divine light, which gleamed with the richest tints of enamelled gold and purple. But it would be foreign to our purpose to follow the rest of the narrative, which, though disguised and mingled with fantastic visions, evidently contains the true principles of aeronautics.

The most noted and elaborate scheme for navigating the atmosphere was proposed by the Jesuit Francis Lana, in a book written in the Italian language, and printed at Brescia in 1670, with the aspiring title of *Prodromo dell' Arte Maestra*. His project was to procure four copper balls of very large dimensions, yet so extremely thin, that, after the air had been extracted, they should become, in a considerable degree, specifically lighter than the surrounding medium. He entered into some calculations to prove that the buoyant power thus obtained would be fully adequate to produce the desired effect. Yet he seems to have had only a slender knowledge of geometry, and but little acquaintance with the progress of physical science. For instance, he founds his computations entirely on the pneumatical discoveries of Galileo and Toricelli, without making any reference to those important facts which the invention of the air-pump by Otto Guerické had successively detected in the course of near 30 years. He assumes that air is 640 times lighter than water; that a cubic foot of water weighs 80 pounds, and consequently that the weight of the same bulk of air is an ounce and a half. If we rectify the estimate of Lana, and reduce it to English measures, each of his copper balls had about 25 feet in diameter, with the thickness of only the 225th part of an inch, the metal weighing 365 pounds avoirdupois, while the weight of the air which it contained must amount to 670 pounds, leaving, after a vacuum had been formed, an excess of 305 pounds, for the power of ascension. Those four balls would therefore rise together into the atmosphere with a combined force of 1220 pounds, which was thought sufficient by the projector to transport a boat completely furnished with masts, sails, oars, and rudders, and carrying several passengers. To extract the air from their cavities, the method proposed was, to procure a Toricellian vacuum, by connecting each globe, fitted with a stop-cock, to a tube of at least thirty-five feet long; the whole being filled with pure water, and raised gently into a vertical position, the mass of liquid, exceeding the pressure of the atmosphere, would flow out, and subside to some point below the cock, which could then be shut.

Lana enumerates the different objections which might be urged against his scheme, and endeavours to answer them. He thinks that the spherical and perfectly arched form of the shell of copper would, notwithstanding its extreme thinness, enable it, after the exhaustion was effected, to sustain the enormous pressure of the external air, which, acting equally on every point of the surface, would rather tend to consolidate than to crush or tear the metal. As the atmosphere becomes always lighter in the upper regions, the machinery could only rise to a certain limit; and if this were found too high for easy breathing, the ascent could be regulated by opening occasionally the cocks to admit some air into the cavity of the balls, and thus increase their specific gravity. There seemed to him no very great difficulty in directing and impelling the aerial bark, by means of rudders, oars, and sails; but the objection was more serious on account of the hazards of tremendous shipwreck, from the violence of winds and tempests. Yet what most alarmed the insinuating Jesuit, and which he earnestly prays God to avert, was the danger that would result, from the successful practice of the art of aeronautics, to the existence of civil government, and of all human institutions. No walls or fortifications could then protect cities, which might be completely subdued or destroyed, without having the power to make any sort of resistance, by a mere handful of daring assailants, who should rain down fire and conflagration from the region of the clouds.

So sanguine was Lana, as to conceive that the very moderate sum of a hundred ducats would be sufficient to defray the expense of all this huge and delicate apparatus. But his poverty, fortunately no doubt for his credit as a man of learning, prevented him proceeding farther than mere speculation; and none of the foreign princes, who about that period often squandered, like gamesters, much of their wealth in the dark and chimerical search after the philosopher's stone, seemed any way disposed to engage in the magnificent scheme of aerial navigation.

The project of Lana appears to have in some degree excited the attention of the learned, though it was at the same time very generally condemned. Hook, Borelli, Leibnitz, and Sturmius, examined it, and severely exposed its defects. Indeed, any person at all acquainted with actual experiment must see that it was absolutely impracticable. Passing over other circumstances, the attenuated shell of copper, from its size and excessive thinness, could not have strength enough to support even its own weight, far less the slightest pressure of the atmosphere. The plate, however, that Lana has given of his whole combined apparatus appears very striking; and after Montgolfier's discovery, it could not fail to attract a greater share of notice than it was otherwise entitled to claim.

So late as the year 1755, and not very long before the final invention of balloons, a very fanciful scheme, yet on the grandest scale, for navigating the atmosphere, was published with most circumstantial detail, in a small pamphlet, by Joseph Galien, a Dominican friar, and professor of philosophy and theology in the papal university of Avignon. This visionary proposed to collect the fine diffuse air of the higher regions, where hail is formed, above the summit of the loftiest mountains, and to inclose it in a bag of a cubical shape, and of the most enormous dimensions, extending more than a mile every way, and composed of the thickest and strongest sail-cloth. With such a vast machine, far outrivalling in boldness and magnitude the ark of Noah, it would be possible, he thought, to transport a whole army, and all their ammunitions of war. But we need not stop one moment to consider a project so perfectly chimerical, which involves, besides, the erroneous supposition that the air of the upper regions is, independently of its diminished compression, essentially thinner and more elastic than the air below.

It cannot fail to strike the reader, that the persons who have occupied themselves the most with attempts at aerial navigation, were all of them Catholic priests;—whether this pursuit is to be explained from their habits of seclusion and their ignorance of the affairs of real life, or from their familiar acquaintance with the relations of miracles and other legendary tales, which might lead them to see nothing very extraordinary in the art of flying through the air. The various schemes of that kind, produced at different times, contain a few just principles, generally mixed up, however, with a large portion of absurdity. But very wide is the distance from such speculations to the real exhibition of the experiment itself.

Some writers have stated that Lord Bacon first published the true principles of aeronautics. This round assertion we cannot help noticing, because it has really no foundation, except in the propensity, fostered by indolence, which would gladly refer all the discoveries ever made to a few great names. They mistake, indeed, the character of Bacon, who seek to represent him as an inventor. His claim to immortality rests chiefly on the profound and comprehensive views which he took of the bearings of the different parts of human knowledge; for it would be difficult to point out a single fact or observation with which he enriched the store of physical science. On the contrary, being very deficient in mathematical learning, he disregarded or rejected some of the noblest discoveries made in his own time.

We can find only two passages in Lord Bacon's works which can be considered as referring to aeronautics, and they both occur in that collection of loose facts and inconclusive reasonings which he has entitled *Natural History*. The *first* is styled *Experiment Solitary, touching Flying in the Air,* and runs thus: "Certainly many birds of good wing (as kites and the like) would bear up a good weight as they fly; and spreading feathers thin and close, and in great breadth, will likewise bear up a great weight, being even laid, without tilting upon the sides. *The further extension of this experiment might be thought upon.*" This hint is not in fairness obnoxious to stricture, since the ingenious Bishop Wilkins, twenty years afterwards, still believed that men could acquire the art of flying. Nor was there any reason to despair, till Borelli at length demonstrated its absolute impossibility.—The second passage is more diffuse, but less intelligible: it is styled *Experiment Solitary, touching the Flying of unequal Bodies in the Air.* "Let there be a body of unequal weight (as of wool and lead, or bone and lead); if you throw it from you with the light end forward, it will turn, and the weightier end will recover to be forwards, unless the body be over long. The cause is, for that the more dense body hath a more violent pressure of the parts from the first impulsion; which is the cause (though heretofore not found out, as hath been often said) of all violent motions; and when the hinder part moveth swifter (for that it less endureth pressure of parts) than the forward part can make way for it, it must needs be that the body turn over; for (turned) it can more easily draw forward the lighter part." The fact here alluded to is the resistance that bodies experience in moving through the air, which depending on the quantity of surface merely, must exert a proportionally greater effect on rare substances. The passage itself, however, after making every allowance for the period in which it was written, must be deemed confused, obscure, and unphilosophical.

That a body must remain suspended in a fluid denser than itself, was first established by Archimedes, whose propositions in hydrostatics were farther extended in modern times by Stevinus and other early mathematicians. But the principles on which a balloon could be made to rise in the atmosphere were scarcely understood till very long afterwards, when chemistry, near the latter part of the last century, had succeeded in ascertaining the properties of the different kinds of aeriform substances. The Greeks of the lower empire knew that air is greatly dilated by warmth; and Sanctorio, the ingenious medical professor at Padua, by applying this expansion, about the year 1590, to the construction of the thermometer, had happily placed it in a strong light. His countryman Borelli remarked, almost a century afterwards, that a heated iron or a burning taper brought near one of the scales of a well-poised balance, by exciting a vertical current, will cause it to mount up with force,—a fact which affords the only true explication of the numerous experiments of Buffon with the weighing of red-hot balls, whose regular and constant results appeared to that eloquent philosopher to exhibit a conclusive demonstration of the actual ponderability of heat. Yet warm air, alone and unassisted, has still no very great power of ascension. The buoyancy communicated to that fluid by the distensible vapour of water and other more volatile liquids is in some cases considerable, especially when combined at the same time with heat. But those aeriform substances which are more elastic than common air, display the most steady and powerful tendency to rise in the atmosphere. Such, in a remarkable degree, is the hydrogen gas, owing probably to the expansive force communicated by the very large share of heat which is embodied with it. The late most ingenious and accurate Mr Cavendish, in 1766, found, by a nice observation, this fluid to be at least seven times lighter than atmospheric air. It therefore occurred to Dr Black of Edinburgh, that a very thin bag filled with hydrogen gas would rise to the ceiling of a room. He provided, accordingly, the allantois of a calf, with the view of showing, at a public lecture, such a curious experiment before his numerous auditors; but, owing to some unforeseen accident or imperfection, it chanced to fail, and that celebrated professor, whose infirm state of health and cold or indolent temper more than once allowed the finest discoveries, when almost within reach, to escape his penetration, did not attempt to repeat the exhibition, or seek to pursue the project any farther. Several years afterwards, a similar idea occurred to Mr Cavallo, who found, however, that bladders, though carefully scraped, are too heavy, and that China paper is permeable to the gas. It is rather singular that he did not think of gold-beater's skin, which had for like purposes been recommended two centuries before by the grammarian Joseph Scaliger and some other writers. But in 1782 this ingenious person succeeded with the pretty experiment of elevating soap-bubbles, by inflating them with hydrogen gas.

To construct an aeronautic machine, it is only required, therefore, to provide a thin bag, of sufficient capacity, and to fill it with hydrogen gas, or with air which is kept in a rarefied state. The form and strength of the material are not so essential as in Lana's project, since it here suffers an equal pressure on both its outer and its inner side. Nor is it an absolute condition that the substance of the bag should be quite impervious to the gas or confined air; though such a defect, by allowing the partial escape of the buoyant fluid, must inevitably diminish the vigour and abridge the duration of the power of the balloon's ascent. This power is evidently the excess of the weight of an equal bulk of atmospheric air above the aggregate weight of the included gas, joined to that of the bag, and of all its appendages: in other words, the final power of the ascent is the difference between the weight of the included gas and of that of an equal volume of external air, further diminished by the weight of the whole apparatus. But supposing the form of the balloon to remain the same, this counteracting load, as it depends on the quantity of surface contained in the bag, must be proportioned to the *square* of the diameter; whereas the difference between the internal and external volume of fluid, which constitutes the whole of the buoyant force, increases in a faster ratio, being proportioned to the capacity of the bag, or the *cube* of its diameter. It hence follows, that however small the excess may be of the specific gravity of the external air above that of the collected fluid, there must always exist some corresponding dimension which would enable a balloon to mount in the atmosphere.

The theory of aeronautics, considered in its detail, includes three distinct things: *first,* the power of a balloon to rise through the air; *second,* the velocity of its ascent; and, *third,* the stability of its suspension at any given height in the atmosphere. These points we shall examine separately.

I. *The buoyant force of balloons.* Since balloons in their shape generally approach to the spherical form, it will be more convenient to ground our calculations on that figure. A globe of common air at the level of the sea, and of the mean density and temperature, is found to weigh about the 25th part of a pound avoirdupois. Consequently, if a perfect vacuum could be procured, a balloon of ten feet diameter must rise with a force of 40 pounds; one of twenty feet diameter, with that of 320 pounds; and a balloon of thirty feet in diameter would mount in the atmosphere with the power of 1080 pounds:—thus augmenting always in the ratio of the cube of the diameters. But air expands by heat about the 450th part of its bulk for each degree on Fahrenheit's scale. Supposing, therefore, that the air included within the bag were heated 50 degrees, which is as much perhaps as could be well supported, it would follow that one-ninth part of this fluid would be driven out by the warmth, and consequently, that the tendency of the balloon to rise upwards would be equal only to the ninth part of the entire power of ascension. Were it possible to maintain a heat of 75 degrees within the balloon, the buoyant force would yet not exceed the sixth part of the absolute ascensional power.

The dilatation which the presence of humidity communicates to air will, during fine weather in this climate, amount generally to one eighteenth part, though it may sometimes reach to more than the double of this quantity. But, in the tropical regions, such dilatation will commonly exceed the twentieth part of the volume of fluid. Hence moist air thrown into a bag, likewise wetted, and sufficiently large, would cause it to rise in the atmosphere. To succeed, however, in this way, the balloon constructed of coarse linen would require enormous dimensions; not less than three hundred feet in diameter.

But it is the union of heat and moisture that gives to air the greatest expansion. The white smoke with which the balloons are filled on Montgolfier's plan, was found, by computation, to be at least one third specifically lighter than the external air. This purer sort of smoke is scarcely any thing but air itself charged with vapour, being produced by the burning of chopped straw or vine twigs in a brasier, under the orifice of the bag. It would have required no fewer than 150 degrees of heat alone to cause the same extent of rarefaction.

We have therefore sufficient data for calculating the buoyant force of the common fire, or rather smoke balloons. This force, being estimated about 12½ pounds avoirdupois when the diameter of the bag is ten feet, would amount to 1562½ pounds if the diameter were fifty feet, and to 12,500 pounds if it were a hundred feet. The weight of the linen case may be reckoned at two-fifths of a pound for a sphere of one foot in diameter. Consequently, a balloon of ten feet diameter, would, without its appendages, weigh 40 pounds; one of fifty feet diameter, 1000 pounds; and one of a hundred feet diameter, 4000 pounds. Such a balloon of ten feet diameter would need 27½ pounds to make it rise, but one of fifty feet diameter would ascend with a force of 562½ pounds, and one of a hundred feet diameter would exert an ascending power of not less than 8500 pounds. There is besides to be deducted the weight of the cordage, the car, the ballast, and the passengers. It would require, on these estimates, a diameter of 33½ feet, to procure merely an equilibrium between the weight of the canvass and the buoyant force of the rarefied air.

The hydrogen gas obtained from the action of dilute sulphuric acid upon iron filings is only six times lighter than atmospheric air; but the gas evolved during the solution of zinc in that acid is not less than twelve times lighter than the standard fluid. The ordinary way of examining the specific gravity of the different gases requires a very delicate operation of weighing with the most exquisite balance; a serious difficulty, which long retarded our knowledge of their comparative densities. In one of the notes to his *Treatise on Heat,* Professor Leslie has pointed out a very simple method, founded on the principles of pneumatics, for discovering the relative specific gravities of the aeriform fluids. This consists in observing the time that a given portion of the gas, under a determinate pressure, takes to escape through a very small aperture. The density of the gaseous fluid must be inversely as the square of the interval elapsed. Thus, the hydrogen gas procured from zinc, but without any depuration, was found, under a pressure of the same column of water, to flow thrice as fast as atmospheric air. This experiment is very striking, and requires no more apparatus than a cylindrical glass jar, open below, and surmounted by a cap terminating in a fine tubular orifice.

On a very moderate supposition, therefore, and after making every allowance for imperfect operation, we may consider the hydrogen gas which fills a balloon as six times lighter than the like bulk of common air. Consequently, such a balloon must exert five-sixths of the whole buoyant force corresponding to its capacity, or will have a tendency to mount in the atmosphere, that is equal to the thirtieth part of a pound avoirdupois for a globe of one foot diameter. A spherical balloon of fifteen feet diameter would hence have a buoyancy of $112\frac{1}{2}$ pounds; one of thirty feet, 900 pounds; and one of sixty feet no less than 7200 pounds. But thin silk, varnished with caoutchouc or elastic gum, to render it impervious to air, is found to weigh only the twentieth part of a pound when formed into a globe of one foot diameter. A silk balloon of fifteen feet diameter would hence weigh $11\frac{1}{4}$ pounds; one of thirty feet diameter, 45 pounds; and one of sixty feet diameter, 180 pounds. Wherefore, the power of ascension exerted by such balloons, would, in pounds avoirdupois, be respectively $101\frac{1}{4}$, 855, and 7020. It follows, also, that a balloon of a foot and a half in diameter would barely float in the atmosphere, the weight of its varnished silk being then exactly balanced by the buoyant effort of the body of hydrogen gas.

But the calculations now given would in strictness require a small modification. The weight of the bag and of all the appendages must evidently compress the included gas, and thereby render it in some degree denser. To compute this minute effect, we have only to consider, that the pressure of a column of atmosphere, at the mean temperature, and near the level of the sea, is 1632 pounds, on a circle of a foot diameter. Thus, in the large balloon of sixty feet diameter, if we suppose the whole load to have been 6000 pounds, the compression of the bag would only amount to five-thirds of a pound for each circle of a foot diameter in the horizontal section, or correspond to the 979th part of the entire pressure of the atmosphere. But the weight of the confined gas being 1200 pounds, its buoyancy must have suffered a diminution of somewhat more than a pound, or $\frac{7}{31}$, from the encumbrance opposed to it. This correction is therefore a mere theoretical nicety, which may be totally disregarded in practice.

II. The next circumstance to be considered in aeronautics is, *the celerity with which balloons make their ascent.* It is obvious that the efficient power of ascension, or the excess of the whole buoyant force above the absolute weight of the apparatus, would, by acting constantly, produce always an accelerated motion. But this acceleration is very soon checked, and a uniform progress maintained, by the increasing resistance which the huge mass must experience in its passage through the air. The velocity which a balloon would gain from unobstructed acceleration must, from the theory of dynamics, be to that which a falling body acquires in the same time, as the efficient buoyancy is to the aggregate weight of the apparatus and of the contained fluid. Thus, if the balloon were to rise with a force equal to the eighth part of its compound weight, the celerity resulting from a constant acceleration would be expressed by multiplying four feet into the number of seconds elapsed since it was launched into the air. Its accelerating advance, however, being opposed, the balloon may to all appearance attain, though still affected with partial oscillations, the final velocity in perhaps little more than double the time required without such obstruction.

This final velocity, or the velocity at which the ascent becomes uniform, the resistance from the air being then equal to the efficient buoyancy of the balloon, is easily calculated. The resistance a circle encounters in moving through any fluid in the direction perpendicular to its plane, is measured by the weight of a column of that fluid, having the circle for its base, and an altitude equal to the height from which a heavy body in falling would acquire the given celerity. But near the level of the sea, and at the mean temperature, a column of atmospheric air 17 feet high, and incumbent on a circle of one foot diameter, weighs a pound avoirdupois; which is therefore the resistance that such a circle would suffer if carried forwards with the celerity of 33 feet each second. According to the same theory, however, which we owe to the sagacity of Newton, the resistance of a sphere is just the half of that of its generating circle, and consequently a velocity of $46\frac{2}{3}$ feet in a second through the air would, in ordinary cases, create a resistance of one pound to a ball of one foot diameter. In other circumstances, the quantity of resistance must be proportional to the squares of the velocities and of the diameters. Whence, if the buoyant force were always the same, the velocity of the ascent of a balloon would be inversely as its diameter.

Suppose a balloon to have thirty feet in diameter, and an ascensional power of 100 pounds. This effort is evidently the same as the ninth part of a pound for a globe of a foot diameter, and would therefore be countervailed by the resistance corresponding to a velocity of $46\frac{2}{3}$ divided by 3, the square root of 9, or $15\frac{5}{9}$ feet each second. The balloon would therefore reach the altitude of a mile in about six minutes. Its accelerating force being only the sixteenth part of its total weight, it might have acquired the uniform motion of ascent in twenty seconds, or before it had attained the height of 200 feet. This example differs very little from reality, and the method of computation will easily be transferred to other cases.

But the resistance of the air assigned by theory is, from the circumstances omitted in the simplification of the problem, generally somewhat less than the results of observation. In low velocities this difference amounts seldom to the fifth part of the whole effect; but in the high velocities it increases considerably, exceeding even the third part in certain extreme cases. From the numerous and accurate experiments of Dr Charles Hutton, we may, however, deduce a simple formula for expressing the terminal velocity of balloons, or the celerity of their uniform ascent. Let $a$ denote the diameter of the balloon in English feet, and $f$ its ascensional power, measured in pounds avoirdupois; then $\frac{40}{a}\sqrt{f}$ will very nearly represent, in feet, the velocity each second of its regular ascent, or that velocity which would cause a resistance from the air precisely equal to the buoyant force. Or, to express the rule in words: As the diameter of the balloon in feet is to the constant number 40, so is the square root of the ascensional power in pounds to the terminal or uniform velocity of ascent each second. To illustrate the application of the formula by an easy example; suppose the balloon to have a diameter of 60 feet, with an accelerating power of 144 pounds; the corresponding rate of uniform ascent becomes $\frac{40}{60}\sqrt{144}$, or $\frac{2}{3} \times 12$, that is, 8 feet each second, or about a mile in eleven minutes.

III. The last point which demands attention in aeronautics, is *the stability of the suspension of a balloon at any given height in the atmosphere.* The circumstances which might regulate or determine that stability, requiring some little exercise of thought, have been commonly neglected, and very seldom examined with due care. It will be proper to consider, *first*, the fire or smoke balloons; and, *secondly*, the balloons filled with hydrogen gas.

1. The warm humified air of the balloon constructed after Montgolfier's plan, suffering less external compression as it approaches the upper strata of the atmosphere, must at the

same time necessarily expand, and partly escape by the orifice above the brasier. The weight of the included fluid, and that of the part expelled, constituting its buoyant force, will hence be reduced, in proportion to the diminished density of the medium in which it floats. The balloon will continue to ascend till its enfeebled buoyancy is no longer able to support the incumbent load. At the height of a mile above the surface, the power of ascension would be diminished rather more than one fifth part; but at an altitude of three miles and a half it would be reduced to one half. At the ordinary temperature, this buoyancy would suffer a reduction of the hundredth part for each ascent of 278 feet. Resuming the data formerly stated, and supposing the balloon to have a spherical shape, its actual power of ascension, estimated in pounds avoirdupois, will be denoted by $\frac{a^3}{80}$, where $a$ signifies the diameter in feet, or the cube of the diameter divided by the constant number 80. If $m : n$ express the ratio of atmospheric density at the surface and at any given height, then will $\frac{n}{m}.\frac{a_3}{80}$ denote the diminished buoyant force at that altitude.

We shall select, for example, a balloon of 100 feet diameter, which is one of the largest dimensions ever actually constructed. Near the level of the sea, and at the ordinary temperature, its power of ascension would be 12,500 pounds; but at the height of 8000 feet, or somewhat more than a mile and a half, when the density is diminished one-fourth, or $\frac{n}{m}=\frac{3}{4}$, that power becomes reduced to $\frac{3}{4}\times 12{,}500$, or 9375 pounds, being a deficiency of 3125 pounds. On the supposition that the balloon was at first so much loaded as to rest just suspended at the ground, a ballast of 3125 pounds must have been thrown out, to make it rise to the altitude of a mile and a half. Hence also the rejection of 125 pounds would have been sufficient to give the balloon an elevation of 278 feet. For the same reason, 10 pounds of ballast heaved out would raise it 22 feet at the surface, 29 feet at the height of a mile and a half, and 44 feet at that of three miles and a half.

2. The stability of the suspension of balloons filled with hydrogen gas, must depend on principles which are very different and less marked. In these aeronautic machines, after the gas has been once introduced, it is closely shut up; and therefore, having constantly the same absolute weight, it should likewise, in all situations, exert the same buoyant force. Hence, if the balloon were capable of indefinite extension, it would still continue its ascent through unbounded space. The determinate capacity of the bag alone can oppose limits to its rise in the atmosphere. The upper strata, being rarer than those below, will have less power to keep any given bulk suspended; and the actual buoyancy being diminished from that cause, the balloon will find its station at a corresponding height in the diffuse medium. But this diminution of the buoyant force, and the consequent increase in the density of the hydrogen gas, must necessarily be confined within very moderate limits, otherwise the thin silk case would be torn to shreds by the expansive efforts of the imprisoned fluid. A safety-valve is accordingly placed at the top of the balloon, calculated to give vent to the gas before the distention has become such as to endanger the bursting of the case.

A balloon should not at first be filled completely with hydrogen gas, but allowed to begin its ascent in rather a flaccid state. As it mounts into the rarer atmosphere, it will gradually swell, till it has attained its full distention, when the safety valve may come to act. But such dissipation of the gas ought, by a previous arrangement, to be as much as possible avoided. If the balloon were intended to rise to the height of four miles, it would not be requisite to fill more than half its capacity with the elastic fluid. To push the charge any farther in this case, would only occasion a superfluous waste of materials. By throwing out part of his ballast, the aeronaut may raise himself higher; and by opening the valve to permit some of the imprisoned gas to escape, he may descend again: but both those expedients are attended by a wasteful expenditure of power.

It is evident that a balloon can have no stability of equipoise, so long as it remains in a loose or flaccid state. The slightest action would then be sufficient to make it rise or fall, since, under such circumstances, any change of its station could not in the smallest degree affect the measure of its buoyant force. The general elevation to which the balloon will ascend must be determined by its quantity of ballast, conjoined with the regulation of the safety-valve; but the strain of the silk case itself would be sufficient to confine the ascent within certain limits, and to procure the stability of the floating mass. Thus, if a balloon fully distended had yet a slight disposition to rise, the imprisoned gas, suffering more and more compression as it gradually ascends, would become proportionally denser, and therefore lose a corresponding part of its previous buoyancy. An equilibrium would hence soon obtain, which must arrest the floating machine at a determinate height in the atmosphere.

Suppose a balloon to be capable, without any danger of bursting, of sustaining an expansion equal to the hundredth part of the elasticity of the included fluid; the whole buoyancy would, by such an alteration, be diminished one five-hundredth part, or this floating machine would subside 55 feet near the surface, and sink proportionally more in the upper regions. To produce the effect, it would only be requisite to throw common air into the bag, without suffering any portion of the hydrogen gas to escape. On this principle Meunier, an ingenious French chemist, very soon after the first discovery of balloons, proposed to regulate with nicety their ascent and position of equilibrium in the atmosphere. The mode which he suggested was, to place within the principal balloon a much smaller one, to be filled occasionally with common air by help of bellows, or emptied again by opening an exterior valve. The aeronaut would thus have it in his power, without expending the charge of hydrogen gas, either to sink gently through a short space, or to rise again at will, by inflating the inner balloon, or allowing it to collapse. The adjustment of the height of a balloon could hence be managed with great precision.

The command possessed by the aeronaut of raising or depressing his machine at pleasure, might afford him the means of influencing the direction of its course. From the various motions of the several ranges of clouds, we may infer that different currents exist at the same time in the atmosphere. The aeronaut has, therefore, in his ascent, only to seek the current best suited to his purpose, and, taking his station in that stratum, to commit his flimsy vessel to the guidance of the stream.

Any other attempts to direct or control the flight of a balloon are altogether fallacious. Since it is carried along with the swiftness of the wind, no rudders or sails can have any action whatever. The aeronaut might fancy himself to float in a perfect calm, unless he chances to encounter irregular currents. The application of oars may turn a balloon, but can have no sensible effect in diverting or impelling its course. How vastly disproportionate is the force of the human arm to the overwhelming pressure of the wind against so huge a machine! To adapt machinery under these circumstances would be preposterous, and to look for help from such a quarter is visionary in the extreme. It must be admitted, however, that after a balloon has once gained its station of equilibrium, or passively floats in the air, the vigorous action of broad vanes, downwards or upwards, might serve to raise or depress the machine through a small space. Thus, a vertical force, exerted equal to nine pounds, would lift a balloon of thirty feet in diameter 278 feet higher. The application of ballast is hence infinitely preferable to any such bulky and unmanageable apparatus.

At the period where we left our narrative, the principles on which a balloon could be constructed were, therefore, pretty generally known to men of science. But to reduce these principles to complete effect, was still an enterprize of the most dazzling kind. This experiment seemed unfit for a cabinet or a laboratory, and it could succeed only on a large scale, exposed to the gaze of the multitude. Without the toil of investigation, or indeed any exercise of thought, all the world might witness the result, and admire the magnificent spectacle which it would present. This triumph over matter was at length achieved by the skill and perseverance of Stephen and Joseph Montgolfier, sons of the proprietor of an extensive and very celebrated paper manufactory established at Annonay, on the banks of a rivulet which flows into the Rhone, near forty miles below Lyons. These remarkable persons, though bred in a remote provincial town, possessed in a high degree ingenuity and the spirit of observation. Without having the benefit of a liberal education, their active curiosity had led them to acquire a more extensive and accurate stock of knowledge than is usually found in the same condition of life. Stephen was more attached to mathematics, but Joseph directed his studies chiefly to chemistry and natural philosophy. They were associated in business with their father, who passed his quiet days, like a patriarch, amidst a large family, and a numerous body of workmen, and reached the very advanced age of 93. Of the younger brother, who survived the other, and lived to make the very valuable, yet much neglected, discovery of the hydraulic ram, we may venture to speak from personal acquaintance. He was a man of great modesty and simplicity of character, yet firm and undaunted, of a calm and sedate aspect, tall and athletic in his person, and of a swarthy complexion, not unlike the celebrated Mr Watt, whom he resembled in some other particulars of his fortune. He was too speculative, perhaps, to succeed in the details of business; for, after trying various schemes of improvement, he quitted his paper manufactory, and repaired to the capital, where he obtained a situation of trust under the late imperial government, at the Chamber of Models, as inspector of patents and internal improvements. In 1809, he had a stroke of palsy, which induced him to resort to the waters of Bourbonne; but receiving no benefit from them, he gladly preferred those of Balaruc, near his old friends, where he died on the 26th of June in the following year, at the age of 60.

The two brothers, who were accustomed to form their plans in concert, had long contemplated the floating and ascent of clouds in the atmosphere. It seemed to them, that a sort of factitious cloud, formed of very thin vapour, inclosed in a light bag of immense size, would mount to the higher regions. In pursuit of this idea, they selected a fluid specifically lighter than atmospheric air; and, accordingly, introduced hydrogen gas into large bags of paper and of thin silk, which rose up, as had been expected, to the ceiling, but fell down in a few seconds, owing to the rapid escape of the gas through the cracks and pores of the case. This great facility with which hydrogen gas makes its way through any substance of a loose and incompact texture, is partly due to its extreme fluidity, but is chiefly occasioned by its strong and obstinate attraction for common air. The mode of preventing, or at least checking that escape, by the application of a proper varnish, was yet unknown. The prospect of overcoming the difficulty was so discouraging, that our experimenters had recourse to another scheme, more analogous to their original ideas, and it rewarded their continued efforts with the most complete success. In the month of November 1782, Joseph Montgolfier, happening, in the course of his frequent excursions, to be then at Avignon, procured a small silk bag, of the form of a parallelopipedon, open below, like a lady's hoop, and having a capacity of about forty-five cubic feet; under its orifice he burnt some paper, and saw, with inexpressible transport, the bag quickly swell, and mount rapidly to the height of seventy-five feet, where it remained till by cooling it lost its buoyancy. Returning to Annonay, he communicated the happy result to his brother, and it was resolved by them to prosecute the experiment on a large scale. Having provided a large quantity of coarse linen, they formed it into the shape of a globe, about thirty feet in diameter, which they lined with paper. On lighting a fire within its cavity, to warm and expand the air, they had the delightful satisfaction of seeing the bag ascend with a force equivalent to 500 pounds.

It was very natural that the brothers should now desire an occasion for exhibiting this grand experiment in their native town. They invited the members of the provincial meeting of the states of the Vivarais, then assembled at Annonay, to witness the first public aerial ascent. On the 5th June 1783, amidst a very large concourse of spectators, the spherical bag or balloon, consisting of different pieces of linen, merely buttoned together, was suspended from cross poles; two men kindled a fire under it, and kept feeding the flames with small bundles of chopped straw; the loose bag gradually swelled, assuming a graceful form, and in the space of five minutes it was completely distended, and made such an effort to escape, that eight men were required to hold it down. On a signal being given, the stays were slipped, and the balloon instantly rose with an accelerating motion, till it reached some height, when its velocity continued uniform, and carried it to an elevation of more than a mile. All was admiration and transport. Amidst the shouts of unbounded acclamation, the progress of the artificial cloud retiring from sight arrested every eye. It was hurried along by the wind; but, its buoyant force being soon spent, it remained suspended only ten minutes, and fell gently in a vineyard, at the distance of about a mile and a half from the place of its ascension. So memorable a feat lighted up the glow of national vanity, and the two Montgolfiers were hailed and exalted by the spontaneous impulse of their fellow-citizens.

Of this splendid experiment a very hasty and imperfect account was transmitted to Paris, and quickly circulated over Europe. In those halcyon days, during the transient calm of political turmoils, and the happy absence of all military events, the prospect of navigating the atmosphere excited a very general ferment, and engrossed the conversation of all ranks. Yet the tale appeared so extraordinary as to leave some doubts of its veracity. In many places, and especially in this country, the more ignorant class of men, and those who affected superior wisdom, both agreed in considering the relation of Montgolfier's discovery as nothing but an imposition practised on the public credulity. To dispel the suspicions which infected the subject, it was necessary to repeat the experiment in every large capital.

When the intelligence of the first ascent of a balloon reached St Petersburg, it found the venerable Euler in a state of great debility, worn out with years and unremitting intellectual toil. Having lost, in the middle of his career, the sight of an eye, he had been for several years visited with total blindness. But in this afflicting situation his mind was still entire, and found delightful exercise in his former habits of calculation. It was in training a domestic to act as his amanuensis, that this great genius now condescended to dictate, in the German language, to his humble pupil, a work of the highest merit, *The Elements of Algebra.* During his last illness, Euler made an expiring effort, and applied his favourite analysis to determine the ascending motion of a balloon. He dictated the preliminary steps of the problem to one of his grand-children: but the hand of death was already stretched over the patriarch;—no farther could he proceed with his investigation;—and composing himself for nobler scenes, he calmly expected the moment of dissolution.

The virtuosi at Paris were eager to repeat the experiment of the ascension of a balloon. M. Faujas de St Fond, an active and zealous naturalist, set on foot a subscription for defraying the expense, which was soon filled up. The construction of the machine was intrusted to the skill of two brothers of the name of Robert, under the superintendence of M. Charles, an ingenious lecturer in natural philosophy. It had at first been proposed merely to copy the process of Montgolfier, but Charles preferred the application of hydrogen gas; a resolution which afterwards occasioned much difficulty and delay. The balloon consisted of thin silk or tiffany, varnished with a solution of elastic gum, disposed into a globular shape, of about thirteen feet in diameter. The hydrogen gas was procured from the action of dilute sulphuric acid upon iron-filings, and was introduced through leaden pipes. But this gas, being rapidly formed, without having been made to pass through a body of cold water, entered the cavity of the balloon excessively hot, and charged with acid fumes, which afterwards condensed against the inside of the bag, injuring its texture, and loading it with superfluous humidity. Not fewer than 500 pounds of sulphuric acid were used, and twice this weight of iron-filings; yet several days were spent in abortive attempts to fill the balloon completely. At last it rose, and was kept suspended at the height of 100 feet above the ground. In this state it was conveyed with acclamations to the Place des Victoires, where it rested, and underwent some repair. About midnight it was thence transported in silent procession, preceded by torch-lights, and guarded by a detachment of horse and foot soldiers, to the Champ de Mars, at the distance of near two miles. The few passengers found at that still hour on the streets gazed with astonishment on the floating mass; and the very coachmen, filled with a sort of awe, respectfully saluted it as they passed.

Next day, being the 27th of August 1783, an immense concourse of people covered the Champ de Mars, and innumerable spectators had planted themselves along the banks of the Seine and the amphitheatre of Passy. By three o'clock every avenue was filled with carriages, and all the beauty and fashion of Paris flocked towards the Ecole Militaire. The preparations being finished, a cannon was discharged as the signal of ascent. The balloon, liberated from its stays, shot upwards with such rapidity as in two minutes to reach, according to calculation, the height of 3000 feet, where it seemed lost in a dark cloud. It re-appeared at a greater elevation, but was soon obscured again amidst other clouds; and after performing a flight of about fifteen miles in the space of three quarters of an hour, it sunk to the ground in a field near Ecouen, where the peasants secured it, having noticed a rent in the upper part of the bag, to which its fall might be imputed. The success of the experiment was complete. The incredulous were sadly mortified; but every minor reflection was drowned in the tumult of excessive joy and exultation. It began to rain immediately after the balloon was launched, yet this unlucky circumstance had no effect to abate the curiosity of the spectators. Regardless of the torrents that fell, they were wholly absorbed in following with eager gaze the progress of the machine through the air. Even elegant ladies, dressed in their finest attire, stood exposed, looking intently the whole time, and were drenched to the skin. This small balloon weighed only thirty pounds, and had at first a buoyant force of forty pounds avoirdupois. If we employ the formula before given, the terminal velocity would be $\frac{40}{13}\sqrt{40} = 19\frac{6}{13}$ feet in a second, or 1168 feet each minute; which appears to correspond very well with fact.

About this time Joseph Montgolfier visited Paris, and was invited by the Royal Academy of Sciences to repeat his experiment of Annonay on a larger scale. He constructed, with coarse linen and a paper lining, a balloon of a pear shape, and about 43 feet wide and 75 feet high. The smoke of fifty pounds of dry straw, in small bundles, joined to that of twelve pounds of wool, was found sufficient to fill it, in the space of ten minutes. The bag duly swelled, and made an effort to rise equivalent to the weight of 500 pounds; but being reserved for exhibition the next day, it was totally destroyed, by its exposure during the night to incessant and violent rain. It became necessary, therefore, to prepare another balloon; and such was the expedition of the artist, that in five days he got the whole completed. Early on the morning of the 19th of September, it was placed upon an octagon scaffold, in front of the palace of Versailles. It had a very showy appearance, being painted with ornaments in oil colours. By ten o'clock the road from Paris was crowded with carriages of all descriptions. Every person of any note or fashion hurried from the metropolis to view the experiment; ladies of distinguished rank filled the windows; and the spacious courts and walks, and even the tops of the houses, were covered with impatient spectators. The royal family and their attendants came forth, and examined the details of the apparatus. About one o'clock the discharge of a mortar gave notice that the filling of the balloon was to commence. In eleven minutes another discharge announced that it was completely inflated; and on the third discharge of the mortar, the cords were cut, and the balloon instantly liberated. After balancing at first in a moment of anxious expectation to the spectators, it rose majestically, in an oblique direction, under the impulse of the wind, till it reached the height of 1500 feet, where it appeared for a while suspended; but in the space of eight minutes it dropped to the ground, at the distance of two miles from the point of its ascent. A sheep, a cock, and a duck, which had been put into the basket, the first animals ever carried up into the air, were found perfectly safe and unhurt by the journey, and the sheep even feeding at perfect ease.

This successful experiment encouraged Montgolfier to prepare, on a more solid construction, another balloon, of a spheroidal form, 45 feet wide and 75 feet high. While it was filling with smoke, Pilatre de Rozier, a young naturalist of great promise, and full of ardour and courage, leaped into the car, and was borne up to the height of 300 feet, where he continued some minutes suspended, the balloon being held down by long cords till it gently descended. The dangers of navigating the balloon being thus brought to a more correct estimate, it was resolved speedily to attempt the daring but sublime experiment. The badness of the weather, however, at this late season of the year, made the project be deferred several days. At last, on the 21st of November, every thing was ready for the ascent in the spacious gardens of the chateau of Muette, belonging to the court of the Dauphin. The sky had a lowering aspect, being loaded with heavy clouds, driven about by irregular winds. But the adventurers were not to be easily discouraged. After a first trial, which had nearly proved fatal to them, the balloon was again filled; and Rozier, with the Marquis d'Arlandes, a major of infantry, who had volunteered to accompany him, took their seats in the car, having a store of ballast, and a provision of straw to supply the fire. About two o'clock the machine was launched, and it mounted with a steady and majestic pace. Wonder, mingled with anxiety, was depicted in every countenance; but when, from their lofty station in the sky, the navigators calmly waved their hats, and saluted the spectators below, a general shout of acclamation burst forth on all sides. As they rose much higher, however, they were no longer discernible by the naked eye.

> ——— in the surging smoke
> Uplifted spurn the ground; thence many a league,
> As in a cloudy chair ascending, ride
> Audacious.———

This balloon soared to an elevation of more than 3000 feet, and traversed, by a circuitous and irregular course, the whole extent of Paris, whose gay inhabitants were all absorbed in admiration and amazement. A curious circumstance occurred during the passage of the floating mass: to the gazers planted on the towers of the metropolitan church of Notre Dame, it chanced to intercept the body of the sun, and thus gave them, for a few seconds, the spectacle of a total eclipse. It has been alleged, that when the balloon had reached so high that the objects on earth were no longer distinguishable, the Marquis d'Arlandes began to think that his curiosity and ambition were sufficiently gratified. He was therefore anxious to descend, and murmured against his companion, who still kept feeding the fire. At last, on hearing some cracks from the top of the balloon, and observing holes burning in the sides, the major became outrageously alarmed at his imminent danger, and applying wet sponges to stop the progress of combustion, he compelled the *savant* to desist from his officious operations. As they now descended too fast, however, M. d'Arlandes was not less anxious and diligent in throwing fresh straw upon the fire, in order to gain such an elevation as would clear the different obstacles. The navigators dexterously avoided the lofty buildings of Paris, by supplying fuel as occasion required; and, after a journey of 20 or 25 minutes, they safely alighted beyond the Boulevards, having described a track of six miles.

Such was the prosperous issue of the first aerial navigation ever achieved by mortals. It was a conquest of science which all the world could understand; and it flattered extremely the vanity of that ingenious people, who hailed its splendid progress, and enjoyed the honour of their triumph. The Montgolfiers had the annual prize of six hundred livres adjudged to them by the Academy of Sciences; the elder brother was invited to court, decorated with the badge of St Michael, and received a patent of nobility; and on Joseph a pension was bestowed, with the further sum of forty thousand livres, to enable him to prosecute his experiments with balloons.

The facility and success, however, of the smoke or fire balloons, appeared to throw into the shade the attempts made by the application of hydrogen gas. M. Charles, the promoter of this plan, was keenly reproached by M. Faujas de St Fond, for departing from the method practised by the original inventor; and he was, moreover, with his associates the Roberts, held up to public derision in the smaller theatres of Paris. To silence the cavils and insinuations of his antagonists, he resolved, therefore, on making some new efforts.

A subscription was opened to defray the expense of a globe twenty-eight feet in diameter, and formed of tiffany, with elastic varnish. After repeated accidents and delays, this balloon was planted, on the 1st of December 1783, at the entrance of the great alley of the Tuilleries; and the diffuse fluid was this time introduced into it from a sort of gasometer. The dilute sulphuric acid and the iron-filings being put into wooden casks, disposed round a large cistern, the gas was conveyed in long leaden pipes, and made to pass through the water under a glass bell plunged in it. The whole apparatus cost about £400 sterling, one half of which was expended on the production of the gas alone. An immense concourse of spectators had collected from all parts. The discharge of a cannon at intervals announced the progress in filling the balloon. To amuse the populace, and quiet their impatience, M. Montgolfier was desired to let off a small fire-balloon, as a mark of his precedence. At last, the globe being sufficiently inflated, and a quantity of ballast, consisting of small sand bags, lodged in the car, leaving only 22½ pounds for the measure of the buoyant force, MM. Charles and Robert placed themselves in the appended boat or car, and the machine was immediately disengaged from its stays. It mounted with a slow and solemn motion. According to the formula given, the terminal velocity of ascension must have been only about 400 feet each minute, or at the rate of somewhat less than five miles in the hour. "The car, ascending amidst profound silence and admiration," to borrow the warm and exaggerated language of the reporter, "allowed, in its soft and measured progress, the bystanders to follow with their eyes and their hearts two interesting men, who like demigods soared to the abode of the immortals, to receive the reward of intelligence, and carry the imperishable name of Montgolfier. After the globe had reached the height of 2000 feet, it was no longer possible to distinguish the aerial navigators; but the coloured pennants which they waved in the air testified their safety and their tranquil feelings. All fears were now dissipated; enthusiasm succeeded to astonishment; and every demonstration was given of joy and applause." The balloon, describing a tortuous course, and rising or sinking according to the fancy of its conductors, was, after a flight of an hour and three quarters, made to alight on the meadow of Nesle, about twenty-five miles from Paris. For the space of an hour, the buoyancy of the machine had been sensibly augmented by the sun's rays striking against the surface of the bag, and heating up the contained gas to the temperature of 55 degrees by Fahrenheit's scale.

After this prosperous descent, the globe, though become rather flaccid and loose by its expenditure, yet still retained a great buoyant force when relieved from the weight of the travellers. The sun had just set, and the night was beginning to close; but M. Charles formed the resolution of making alone another aerial excursion. His courage was rewarded by the spectacle of one of the most novel and enchanting appearances in nature. He shot upwards with such celerity as to reach the height of near two miles in ten minutes. The sun rose again to him in full orb; and, from his lofty station in the heavens, he contemplated the fading luminary, and watched its parting beams, till it once more sunk below the remote horizon. The vapours rising from the ground collected into clouds, and covered the earth from his sight. The moon began to shine, and her pale rays scattered gleams of various hues over the fantastic and changing forms of those accumulated masses. This scene had all the impressive solemnity of the true sublime. No wonder that the first mortal eye that ever contemplated such awful grandeur, could not refrain from shedding tears of joy and admiration. The region in which M. Charles hovered was now excessively cold; and as he opened the valve occasionally during his ascent, to prevent the violent distention of the balloon, the hydrogen gas, not having time to acquire the temperature of the exterior air, rushed out like misty vapour, with a whistling noise. But prudence forbade the voyager to remain long at such an elevation, while darkness was gathering below. He therefore descended slowly to the earth, and, after the lapse of 35 minutes, alighted near the wood of Tour du Lay, having, in that short interval, travelled about nine miles.

This balloon, with its passengers and ballast, weighed at first 680 pounds; but, notwithstanding the care taken in filling it, the hydrogen gas must have been mixed with a large proportion of common air, since it was only $5\frac{1}{4}$ times lighter than this fluid. The barometer, which stood at 29.24 English inches at the surface of the ground, subsided to 20.05 at the greatest elevation to which M. Charles had reached. This gives by calculation an altitude of 9770 feet. The thermometer, which was at 41° by Fahrenheit's scale at the first ascent, fell to 21° at the highest flight; giving a difference of one degree for every $488\frac{1}{2}$ feet of ascent.

The next voyage through the air was performed in the largest balloon ever yet constructed. The elder Montgolfier had been persuaded to open a subscription at Lyons for the sum of £180 sterling, to construct an aeronautic machine capable of upholding a great weight, and of carrying a horse or other quadruped. It had an elongated shape, 109 feet wide, and 134 feet high, and was formed of two folds of linen, having three layers of paper laid between them, and quilted over with ribands. It showed at first enormous buoyant power. A truss of straw, moistened with spirit of wine, was found, when set on fire, to yield humid smoke sufficient to inflate the balloon, and the burning of five pounds weight of alder faggots kept it in full action. Though loaded with a ballast of eighteen tons, it yet lifted up six persons from the ground. Unfortunately, it was very much damaged one night, in consequence of being exposed to rain, frost, and snow. However, on the 19th of January 1784, the balloon was charged in seventeen minutes, by the combustion of 550 pounds of alder. Joseph Montgolfier, accompanied by the ardent Pilatre de Rozier, and four other persons of note, with the proper ballast, took their seats in a wicker gallery, and were launched into the atmosphere. They manœuvred over the city of Lyons, and near the course of the Rhine, for the space of forty minutes; but a large rent having been observed in the upper part of the balloon, they were compelled to descend abruptly, though without any further accident.

The difficulties and dangers of aerial navigation being at length surmounted, the ascents of balloons were now multiplied in all quarters. It will therefore be sufficient henceforth to notice very succinctly some of the more distinguished attempts of that kind.

The Chevalier Paul Andreani of Milan had a spherical balloon, of 70 feet in diameter, formed after Montgolfier's plan, at his own charge, in which, accompanied by two companions, he ascended from that capital on the 25th of February 1784. The machine rose to the height of 1300 feet; but after having described, in twenty minutes, a very circuitous track, it settled upon a large tree, from which, however, the voyagers, by applying fresh fuel, extricated themselves, and alighted on clear ground, without receiving any hurt.

On the 2d of March, Blanchard, who had been for some years before occupied with the chimerical project of flying in the air, and who fancied that the same principles and contrivances might be applied to direct the motion of balloons, mounted alone, and with great intrepidity, at Paris, in a silk balloon 40 feet in diameter, constructed by subscription, and filled with hydrogen gas. He darted rapidly to the height of above a mile, and after being driven about by cross winds for an hour and three quarters, he descended in the plain of Billancourt.

On the 28th of June in the same year, an ascent was made at Lyons before the King of Sweden, who then travelled under the name of Count Haga, with a fire-balloon, having somewhat of a pear shape, and 75 feet in height. Two passengers, M. Fleurant and a young lady, Madame Thiblé, the first female that ever adventured on such a daring voyage, entered the car, and ascended with great velocity. In four minutes the noise of the multitude was no longer audible, and in two more the eye could not distinguish them. It was inferred, from a trigonometrical calculation, that they had reached the altitude of 13,500 feet. Their flag, with its staff of 14 pounds weight, being thrown down, took seven minutes to fall to the ground. The thermometer had dropt to 43° on Fahrenheit's scale; and to the sensation of cold which they felt was joined that of a ringing in the ears. Different currents were found to occupy distinct strata of the atmosphere; and in passing from one stratum to another, the balloon was affected by a sensible undulation. The travellers continued to feed their fire with the loppings of vines, till this provision being nearly spent, they safely alighted in a corn field, having traversed about six miles in three quarters of an hour.

About a fortnight afterwards, the same prince was gratified by a more splendid ascent, commanded for his entertainment by the French monarch. A large fire balloon, carrying the naturalists Rozier and Proust, was launched from the outer court of Versailles. It soared to the height of 12,520 feet, and might appear to float in a vast congregation of extended and towering white clouds. The thermometer stood at 21° of Fahrenheit, and the flakes of snow fell copiously on the voyagers, while it only rained below. Descending again from that chaotic abyss, they were charmed with the lively aspect of a rich and populous district. They alighted at the entrance of the forest of Chantilly, about thirty-six miles from Versailles, after a flight of an hour and five minutes.

We omit the relation of a prosperous ascent performed at Rhodes on the 6th of August, by the Abbé Carnus and his companion, with a fire balloon, of a globular shape, and 57 feet in diameter.—The longest aerial journey yet made was accomplished at Paris on the 19th of September. The Duke de Chartres, afterwards Orleans and the noted Egalité, employed Robert to construct for him a silk balloon, which should be filled with hydrogen gas. It had 56 feet in height, and 36 feet diameter, being composed of a cylinder terminated by two hemispheres; a construction which was rightly supposed to give much additional solidity to the machine. A small bag, on Meusnier's plan, had been introduced within it, and the boat was, besides, furnished with a helm and four oars. This balloon, bearing the duke himself, the two artists, and another companion, and having 500 pounds of ballast, was allowed to rise very slowly, with a buoyancy of only 27 pounds. At the height of 1400 feet, the voyagers perceived, not without uneasiness, thick dark clouds gathering along the horizon, and threatening the approach of a thunder-storm. They heard the distant claps, and experienced something like the agitation of a whirlwind, although they had not felt the slightest concussion in the air from the discharge of cannon. The thermometer suddenly dropped from 71° Fahrenheit to 61°; and the influence of this cold caused the balloon to descend within 200 feet of the tops of the trees near Beauvais. To extricate themselves, they now threw out more than forty pounds of ballast, and rose to an elevation of 6000 feet, where it was found that the confined gas had so obstinately retained its heat, as to be no less than 42° warmer than the external air. The duke became alarmed, and betrayed such impatience to return again to the earth, that he is said to have pierced the lower part of the silk bag in holes with his sword. After narrowly escaping the dangers from wind and thunder, the balloon at last descended in safety near Bethune, having performed a course of 135 miles in the space of five hours.

On the 25th of April in the same year, the celebrated chemist Guyton-Morveau, with the Abbé Bertrand, ascended from Dijon in a balloon, nearly of a globular shape, 29 feet in diameter, composed of the finest varnished tiffany, and filled with hydrogen gas. They did not start till five o'clock in the evening, the barometer being at 29.3 inches, and the thermometer at 57° on Fahrenheit's scale; and, after surmounting some accidents, they rose to an altitude of 10,465 feet, or very nearly two English miles, where the barometer had sunk to 19.8 inches, and the thermometer to 25°. They felt no inconvenience, however, except from the pinching of their ears with cold. They saw an ocean of clouds below them, and in this situation they witnessed, as the day declined, the beautiful phenomenon of a parhelion, or mock sun. The real luminary was only ten degrees above the horizon, when, all at once, another sun appeared to plant itself within six degrees of the former. It consisted of numerous prismatic rings, delicately tinted, on a ground of dazzling whiteness. At half-past six o'clock, after a voyage of an hour and a half, they safely alighted near Magny, about fifteen miles distant from Dijon.

With the same balloon, M. Guyton-Morveau made a second ascent on the 12th of June, accompanied by the President De Virly. It was launched at seven o'clock in the morning, the barometer being then at 29.5 inches, the thermometer at 66°, and Saussure's hygrometer at $83\frac{1}{5}$°. It swelled very fast, however, owing to the effect of the sun's increasing heat; and the upper valve being at intervals opened, to give vent to excess of the gas, this escaped with a noise like the rushing of water. As the voyagers did not mount to any very great elevation, they enjoyed an agreeable temperature, and could easily, by observing the situation of the different villages scattered below them, trace out their tortuous route on the surface of the map. By nine o'clock they had reached the height of 6030 feet, the barometer now standing at 24.7 inches, the thermometer at 70°, and the hygrometer at $65\frac{1}{2}$. Three quarters of an hour afterwards they descended at the village of Etevaux, only twelve miles from Dijon, having described at least double this distance in the air. The heat had increased so much since the morning, that, notwithstanding the loss of elastic fluid, the balloon seemed yet nearly inflated on touching the ground.

An aerial voyage, most remarkable for its duration and its adventures, was performed on the 18th of June 1786, from Paris, by M. Testu, with a balloon 29 feet in diameter, constructed by himself, of glazed tiffany, furnished with auxiliary wings, and filled, as usual, with hydrogen gas. It had been much injured by wind and rain during the night before its ascension; but having undergone a slight repair, it was finally launched, with its conductor, at four o'clock in the afternoon. The barometer then stood at 29.68 inches, and the thermometer as high as 84°, though the day was cloudy, and threatened rain. The balloon had at first been filled only five-sixths, but it gradually swelled as it became drier and warmer, and acquired its utmost distention at the height of 2800 feet. But, to avoid the waste of gas or the rupture of the silk, the navigator endeavoured to descend by the re-action of his wings. This force being insufficient, however, he threw out some ballast; and, at half-past five o'clock, he softly alighted on a corn-field in the plain of Montmorency. Without leaving the car, he began to collect a few stones for ballast; when he was surrounded by the proprietor of the field and a troop of peasants, who insisted on being indemnified for the damage occasioned by his idle and curious visitors. Anxious now to disengage himself, he persuaded them that, his wings being broken, he was wholly at their mercy. They seized the stay of the balloon, which floated at some height, and dragged their prisoner through the air in a sort of triumph towards the village. But M. Testu, finding that the loss of his wings, his cloak, and some other articles, had considerably lightened the machine, suddenly cut the cord, and took an abrupt leave of the clamorous and mortified peasants. He rose to the region of the clouds, where he observed small frozen particles floating in the atmosphere. He heard thunder rolling beneath his feet, and, as the coolness of the evening advanced, the buoyant force diminished, and, at three quarters after six o'clock, he approached the ground near the abbey of Royaumont. There he threw out some ballast, and in the space of twelve minutes rose to a height of 2400 feet, where the thermometer was only 66 degrees. He now heard the blast of a horn, and descried huntsmen below in full chace. Curious to witness the sport, he pulled the valve, and descended, at eight o'clock, between Etouen and Varville, when, rejecting his oars, he set himself to gather some ballast. While he was thus occupied, the hunters galloped up to him. He mounted a third time, and passed through a dense body of clouds, in which thunder followed lightning in quick succession.

With fresh alacrity and force renew'd,
Springs upward, like a pyramid of fire,
Into the wild expanse, and through the shock
Of fighting elements, on all sides round
Environ'd wind his way.

The thermometer fell to 21°, but afterwards regained its former point of 66°, when the balloon had reached the altitude of 3000 feet. In this region the voyager sailed till half-past nine o'clock, at which time he observed from his "watch-tower in the sky" the final setting of the sun. He was now quickly involved in darkness, and enveloped in the thickest mass of thunder-clouds. The lightnings flashed on all sides, and the loud claps were incessant. The thermometer, seen by the help of a phosphoric light which he struck, pointed at 21°, and snow and sleet fell copiously around him. In this most tremendous situation the intrepid adventurer remained the space of three hours, the time during which the storm lasted. The balloon was affected by a sort of undulating motion upwards and downwards, owing, he thought, to the electrical action of the clouds. The lightning appeared excessively vivid, but the thunder was sharp and loud, preceded by a sort of crackling noise. A calm at last succeeding, he had the pleasure to see the stars, and embraced this opportunity to take some necessary refreshment. At half-past two o'clock the day broke in; but his ballast being nearly gone, and the balloon again dry and much elevated, he resolved to descend to the earth, and ascertain to what point he had been carried. At a quarter before four o'clock, having already seen the sun rise, he safely alighted near the village of Campremi, about 63 miles from Paris.

At this period, ascents with balloons had been multiplied, not only through France, but all over Europe. They were very seldom, however, directed to any other object than amusement, and had soon degenerated into mere exhibitions for gain. The first balloon seen in England was constructed by an ingenious Italian, the Count Zambeccari. It consisted of oiled silk, in a globular shape, about ten feet in diameter, and weighed only eleven pounds. It was entirely gilt, which not only gave it a beautiful appearance, but rendered it less permeable to the gas. On the 25th of November 1783 it was filled about three-fourths, and launched at one o'clock from the artillery ground, and in the presence of a vast concourse of spectators. At half-past three in the afternoon it was taken up at Petworth, in Sussex, about the distance of 48 miles.—It was not till the following year, on the 21st of September, that a countryman of his, named Lunardi, first mounted in a balloon at London. He afterwards repeated the experiment in different parts of England, and during the following year in Scotland. This active person took an expeditious but careless method to fill his balloon with gas. He had two large casks, sunk into the ground for their better security, in which he deposited 2000 pounds of the borings of cannon, divided by layers of straw, to present a larger surface. An equal weight of sulphuric acid, or common oil of vitriol, diluted with six times as much water, was poured upon the iron, and the hydrogen gas now formed, without being cooled or washed, was immediately introduced into the balloon.—To Lunardi succeeded Blanchard, who possessed just as little science, but had greater pretensions, and some share of dexterity and skill. This adventurer is said to have performed not fewer than thirty-six voyages through the air, and to have acquired a large sum of money by those bold and attractive exhibitions. His most remarkable journey was across the British Channel, in company with Dr Jeffries, an American gentleman. On the 7th of January 1785, in a clear frosty day, the balloon was launched from the cliff of Dover, and after a perilous course of two hours and three quarters, it alighted in safety on the edge of the forest of Guiennes, not far beyond Calais. By the magistrates of this town were the two aerial travellers treated with the utmost kindness and hospitality; and their wondrous arrival was welcomed as a happy omen, alas! how fallacious, of the lasting harmony to subsist between rival nations, now cemented by the conclusion of the famous Commercial Treaty.

The original smoke balloon of Montgolfier appears to have gradually fallen into disrepute, and the more elegant and expensive, but far more powerful construction, which employs varnished silk to contain hydrogen gas, came to be generally preferred. With due precaution and management, the sailing through the atmosphere is perhaps scarcely more dangerous now than the navigating of the ocean. Of some hundred ascents made at different times with balloons, not above two or three cases are recorded to have had a fatal termination. The first was rendered memorable by the shocking death of the accomplished and interesting Rozier, who perished a martyr to his ardent zeal for the promotion of science. Being anxious to return the visit which Blanchard and Jeffries had paid to the French coast, by crossing the channel again and descending in England, he transported his balloon, which was of a globular shape, and forty feet in diameter, to Boulogne; and after various delays, occasioned chiefly by adverse winds, he mounted on the 15th June 1785, with his companion M. Romain. From some vague idea of being better able to regulate the ascent of the balloon, he had most incautiously suspended below it a small smoke one of ten feet diameter; a combination to which may be imputed the disastrous issue. Scarcely a quarter of an hour had elapsed, when the whole apparatus, at the height of above three thousand feet, was observed to be on fire; and its scattered fragments, with the unfortunate voyagers, were precipitated to the ground. They fell near the sea-shore, about four miles from Boulogne, and were instantly killed by the tremendous shock, their bodies being found most dreadfully mangled. The next fatal accident with balloons happened in Italy, several years later, when a Venetian nobleman and his lady, after having performed successfully various ascents, fell from a vast height, and perished on the spot.

Balloonists have no doubt been often exposed, in their aerial excursions, to the most imminent hazard of their lives. The chief danger consists in the difficulty of preventing sometimes a rapid and premature descent. To guard, in some degree, against the risks arising from the occurrence of such accidents, the *Parachute* was afterwards introduced; being intended to enable the voyager, in cases of alarm, to desert his balloon in mid air, and drop, without sustaining injury, to the ground. The French language, though not very copious, has yet supplied this convenient term, signifying *a guard for falling*, as it has likewise furnished the words of analogous composition, *parapluie, paravent,* and *parasol,* to denote *an umbrella, a door-screen,* and *a shade for the sun.* The *parachute* very much resembles the ordinary umbrella, but has a far greater extent. The umbrella itself, requiring such strength to bear it up against a moderate wind, might naturally have suggested the application of the same principle to break the force of a fall. Nothing was required but to present a surface having dimensions sufficient to experience from the air a resistance equal to the weight of descent, in moving through the fluid with a velocity not exceeding that of the shock which a person can sustain without any danger. Accordingly, in the East, where the umbrella, or rather the parasol, has been from the remotest ages in familiar use, this implement appears to be occasionally employed by vaulters, for enabling them to jump safely from great heights. Father Loubere, in his curious account of Siam, relates that a person, famous in that remote country for his dexterity, was accustomed to divert exceedingly the king and the royal court by the prodigious leaps which he took, having two umbrellas with long slender handles fastened to his girdle. He generally alighted on the ground, but was sometimes carried by the wind against trees or houses, and not unfrequently into the river. Not many years since, the umbrella was, at least on one occasion, employed in Europe with similar views, but directed to a very different purpose. In the early part of the campaign of 1793, the French general Bournonville, having been sent by the National Convention, with four more commissioners, to treat with the Prince of Saxe-Cobourg, was, contrary to the faith or courtesy heretofore preserved in the fiercest wars that have raged among civilized nations, detained a prisoner with his companions, and sent to the fortress of Olmutz, where he suffered a rigorous confinement. In this cruel situation he made a desperate attempt to regain his liberty. Having provided himself with an umbrella, he jumped from a window at the height of forty feet; but being a very large heavy man, this screen proved insufficient to check his precipitate descent. He struck against the opposite wall, fell into the ditch, and broke his leg, and was carried in this condition back again to his dungeon.

Blanchard was the first who constructed parachutes, and annexed them to balloons, for the object of effecting his escape in case of accident. During the excursion which he undertook from Lisle, about the end of August 1785, when this adventurous aeronaut traversed, without halting, a distance not less than 300 miles, he let down a dog from a vast height in the basket of a *parachute,* and the poor animal, falling gently through the air, reached the ground unhurt. Since that period the practice and management of the *parachute* have been carried much farther by other aerial travellers, and particularly by M. Garnerin, who has dared repeatedly to descend from the region of the clouds with that very slender machine. This ingenious and spirited Frenchman visited London during the short peace of 1802, and made two fine ascents with his balloon, in the second of which he threw himself from an amazing elevation with a parachute. This consisted of thirty-two gores of white canvass formed into a hemispherical case of twenty-three feet diameter, at the top of which was a truck or round piece of wood ten inches broad, and having a hole in its centre, admitting short pieces of tape to fasten it to the several gores of the canvass. About four feet and a half below the top, a wooden hoop of eight feet in diameter was attached by a string from each seam; so that when the balloon rose, the parachute hung like a curtain from this hoop. Below it was suspended a cylindrical basket covered with canvass, about four feet high, and two and a quarter wide. In this basket the aeronaut, dressed in a close jacket and a pair of trousers, placed himself, and rose majestically from an inclosure near North Audley Street at six o'clock in the evening of the 2d of September. After hovering seven or eight minutes in the upper region of the atmosphere, he meditated a descent in his parachute. Well might he be supposed to linger there in dread suspense, and to

> ——————— look a while
> Pondering his voyage; for no narrow frith
> He had to cross———
> He views the breadth, and, without longer pause,
> Downright into the world's first region throws
> His flight precipitant, and winds with ease,
> Through the pure marble air, his oblique way.

He cut the cord by which his parachute was attached to the net of the balloon; it instantly expanded, and for some seconds it descended with an accelerating velocity, till it became tossed extremely, and took such wide oscillations, that the basket or car was at times thrown almost into a horizontal position. Borne along likewise by the influence of the wind, the parachute passed over Mary-le-bonne and Somers Town, and almost grazed the houses of St Pancras. At last it fortunately struck the ground in a neighbouring field; but so violent was the shock as to throw poor Garnerin on his face, by which accident he received some cuts, and bled considerably. He seemed to be much agitated, and trembled exceedingly at the moment he was released from the car. One of the stays of the parachute had chanced to give way; which untoward circumstance deranged the apparatus, disturbed its proper balance, and threatened the adventurer, during the whole of his descent, with immediate destruction. The feeling of such extreme peril was too much for human nature to bear.

From the principles before explained, we may easily determine the descent of a parachute. When, with its attached load, it is abandoned in the air, it must, from the continued action of gravity, proceed at first with an accelerated motion, till its increasing velocity comes to oppose a resistance equal to the force of attraction, or to the combined weight of the whole apparatus. After this counterpoise has taken place, there existing no longer any cause of acceleration, the parachute should descend uniformly with its acquired rapidity. This perfect equilibrium will not, however, be attained at once. The accumulation of swiftness produced by the unceasing operation of gravity, is not immediately restrained by the corresponding increased resistance of the atmosphere. The motion of a parachute must hence, for some short time, be subject to a sort of interior oscillation, alternately accelerating and retarding. It first shoots beyond the terminal velocity, and then, suffering greater resistance, it relaxes, and contracts within the just limits. This unequal and undulating progress which a parachute exhibits subsequent to the commencement of its fall, is calculated to excite disproportionate alarms of insecurity and danger.

The terminal velocity of a parachute, or the uniform velocity to which its motion tends, would, according to theory, be equal, if its surface were flat, to the velocity that a heavy body must acquire in falling through the altitude of a column of air incumbent on that surface, and having, under the usual circumstances, the same weight as the whole apparatus. But we have already seen, that a cylinder of air, one foot in diameter and height, weighs only, in ordinary cases, the seventeenth part of a pound avoirdupois. Wherefore, if the square of the diameter of a parachute be divided by 17, the quotient will give the number of pounds equivalent to the weight of an atmospheric column of one foot; and the weight of the apparatus being again divided by this quotient, the result will express the entire altitude of an equiponderant column. Of this altitude, the square root multiplied by 8 will denote the final velocity, or that with which the parachute must strike the ground. Suppose, for example, that the diameter of the parachute were 25 feet: then $25 \times 25 = 625$; and this, divided by 17, gives $36\frac{13}{17}$. Consequently, if the parachute with its load weighed only $36\frac{13}{17}$ pounds, the shock received at the surface of the earth would be precisely the same as that which is felt in dropping from an elevation of one foot. Had the weight of the apparatus, therefore, been four times greater, or $147\frac{1}{17}$ pounds, the shock sustained would be the same as that from a fall of four feet; which is near the limit, perhaps, of what a person can bear without suffering injury from the violence of concussion.

The velocity of descent, on this latter supposition, would be 8 $\sqrt{4}$, or sixteen feet each second.

But the actual resistance of the air is rather greater than what theory would give; and it is besides augmented by the concavity of the opposing surface, which occasions an accumulation of the fluid. Let $a$ denote the diameter of a parachute, and $f$ the total weight of the apparatus abandoned to its gravity in the atmosphere; if we take Dr Hutton's valuable experiments for the basis of the calculation, the terminal velocity of descent may be expressed in round numbers, by $\frac{26}{a}\sqrt{f}$, in feet each second, and consequently the length of fall which would occasion the same shock, is $\left(\frac{13}{4a}\right)^2 . f$, or very nearly $\frac{10\frac{1}{2}}{a^2} . f$. Thus, if the parachute had thirty feet in diameter, and weighed, together with its appended load, 225 pounds; then $\frac{26}{30}\sqrt{225} = \frac{26}{30} \times 15 = \frac{13}{15} \times 15 = 13$ feet, or the velocity with which it would strike the ground; and $\left(\frac{13}{120}\right)^2 225 = 2\frac{5}{8}$ feet, being the height from which a person dropping freely would receive the same shock.

Since the resistance which air opposes to the passage of a body is diminished by rarefaction, it is evident, that the parachute disengaged from a balloon, in the more elevated regions of the atmosphere, will at first acquire a greater velocity than it can afterwards maintain as it approaches the ground. Resuming the notation employed before, the ratio of the density of the air at the surface, and at any given height, being expressed by $n : m$; then the velocity of counterpoise at that elevation would be $\frac{26}{a}\sqrt{\frac{m}{n} . f}$, or it would be equal to what is accumulated in falling freely a height of $\frac{10\frac{1}{2}}{a^2} \cdot \frac{m}{n} \cdot f$ feet. It is the final velocity, however, that must be chiefly considered in parachutes, being what determines the shock sustained in alighting. The violence of the rushing through the air will seldom be attended with any serious inconvenience. If we suppose the mean velocity with which a parachute descends to be twelve feet each second, this will correspond to the rate of a mile in $7\frac{1}{2}$ minutes; not more than that of a very gentle trot. We are not told from what height Garnerin dropped; but if he took four minutes in his descent, it was probably about half a mile.

The practice of aeronautics has not realized those expectations of benefit to mankind which sanguine projectors were at first disposed to entertain. It was soon found, that a balloon, launched into the atmosphere, is abandoned, without guidance or command, to the mercy of the winds. To undertake to direct or impel the floating machine by any exertion of human strength, was evidently a chimerical attempt. All the influence which the aeronaut really possesses consists in a very limited power of raising or depressing it, according to circumstances. He cannot hope to shape his course, unless by skilfully adapting his elevation to catch the prevailing currents.

Almost the only purpose to which balloons have hitherto been applied with success, had for its object that of military *reconnaissance*. In the early part of the French revolutionary war, when ingenuity and science were so eagerly called into active service, a balloon, prepared under the direction of the *Aerostatic Institute* in the Polytechnic School, and entrusted to the command of two or three experienced officers, was distributed to each of the republican armies. The decisive victory which General Jourdan gained, in June 1794, over the Austrian forces in the plains of Fleurus, has been ascribed principally to the accurate information of the enemy's movements before and during the battle, communicated by telegraphical signals from a balloon which was sent up to a moderate height in the air. The aeronauts, at the head of whom was the celebrated Guyton-Norveau, mounted twice in the course of that day, and continued, about four hours each time, hovering in the rear of the army at an altitude of 1300 feet. In the second ascent, the enterprize being discovered by the enemy, a battery was opened against them; but they soon gained an elevation above the reach of the cannon. Another balloon, constructed by the same skilful artist, M. Conté, was attached to the army sent on the memorable expedition to Egypt. What service it rendered the daring invader in the wide plains and sandy deserts of Africa, we are not informed; but, after the capitulation of Cairo, it was brought back, with the remains of the army, to France, and employed in the sequel, as we shall find, more innocently in philosophical research. These balloons, being calculated for duration, were of a more solid and perfect construction than usual. Originally they were filled with hydrogen gas, obtained from the decomposition of water on a large scale. For this purpose, six iron cylinders had been fixed by masonry in a simple kind of furnace, each of their ends projecting, and covered with an iron lid. Two sets of metal tubes were also inserted, the one for conveying hot water, and the other for carrying off the gas which was formed. The cylinders being charged with iron-turnings, and brought to a red heat, the humidity was instantly converted into steam, and decomposed, the oxygen uniting with the iron, while the hydrogen gas was discharged, and made to deposit any carbonic gas that might adhere to it, by passing through a reservoir filled with caustic lye before it entered the balloon.

By this method there was procured, at a very moderate expense, and in the space of about four hours, a quantity of hydrogen gas sufficient to inflate a balloon of thirty feet in diameter.

The ascents with balloons should appear to furnish the readiest means of ascertaining important facts in meteorology and atmospheric electricity, departments of science which are still unfortunately in their infancy. Some aeronauts have asserted that the magnetic needle ceased to traverse at very great elevations in the atmosphere ; a statement which received some countenance from the observations made by Saussure on the lofty summit of the Col du Geant, where that celebrated naturalist thought he had found the magnetic virtue to be diminished one fifth part. It has been pretended by others, that the air of the higher regions is not of the same composition as at the surface of the earth, and is, independently altogether of its rarity, less fitted for the purpose of respiration. To determine these, and other relative points, was, therefore, an object interesting to the progress of physical science. A few years since, two young and ardent French philosophers, MM. Biot and Gay-Lussac, proposed to undertake an aerial ascent, in order to examine the magnetic force at great elevations, and to explore the constitution of the higher atmosphere and its electrical properties. For such a philosophical enterprize they were eminently qualified, having been educated together at the Polytechnic School, and both of them deeply versed in mathematics ; the former indulging in a wide range of study, and the latter concentrating his efforts more on chemistry, and its application to the arts. Their offer to government was seconded by Berthollet and Laplace, and the celebrated chemist Chaptal, then minister of the interior, gave it his patronage and warm support. The balloon which had once visited Egypt was delivered to the custody of Biot and Gay-Lussac; and the same artist who constructed it was, at the public expense, ordered to refit and prepare it under their direction. Besides the usual provision of barometers, thermometers, hygrometers, and electrometers, they had two compasses, and a dipping-needle, with another fine needle, carefully magnetized, and suspended by a very delicate silk thread, for ascertaining by its vibrations the force of magnetic attraction. To examine the electricity of the different strata of the atmosphere, they carried several metallic wires, from sixty to three hundred feet in length, and a small electrophorus freely charged. For galvanic experiments they had procured a few discs of zinc and copper, with some frogs; to which they added insects and birds. It was also intended to bring down a portion of air from the higher regions, to be subjected to a chemical analysis; and for this purpose a flask, carefully exhausted, and fitted with a stop-cock, had been prepared.

The balloon was placed in the garden of the *Conservatoire des Arts*, or Repository of Models, formerly the Convent of St Martin; and no pains were spared by Conté in providing whatever might contribute to the greater safety and convenience of the voyagers. Every thing being now ready for their ascent, these adventurous philosophers, in presence of a few friends, embarked in the car at ten o'clock in the morning of the 23d of August 1804. The barometer was then at 30.13 inches, the thermometer at 61°.7 on Fahrenheit's scale, and Saussure's hygrometer pointed at 80°.8, or very near the limit of absolute humidity. They rose with a slow and imposing motion. Their feelings were at first absorbed in the novelty and magnificence of the spectacle which opened before them; and their ears were saluted with the buzz of distant gratulations, sent up from the admiring spectators. In a few minutes they entered the region of the clouds, which seemed like a thin fog, and gave them a slight sensation of humidity. The balloon had become quite inflated, and they were obliged to let part of the gas escape, by opening the upper valve; at the same time, they threw out some ballast, to gain a greater elevation. They now shot through the range of clouds, and reached an altitude of about 6500 English feet. These clouds viewed from above, had the ordinary whitish appearance; they all occupied the same height, only their upper surface seemed marked with gentle swells and undulations, exactly resembling the aspect of a wide plain covered with snow.

MM. Biot and Gay-Lussac now began their experimental operations. The magnetic needle was attracted, as usual, by iron; but they found it impossible at this time to determine with accuracy its rate of oscillation, owing to a slow rotatory motion with which the balloon was affected. In the meanwhile, therefore, they made other observations. A Voltaic pile, consisting of twenty pairs of plates, exhibited all its ordinary effects,—gave the pungent taste, excited the nervous commotion, and occasioned the decomposition of water. By rejecting some more ballast, they had attained the altitude of 8940 feet, but afterwards settled to that of 8600 feet. At this great elevation, the animals which they carried with them appeared to suffer from the rarity of the air. They let off a violet bee, which flew away very swiftly, making a humming noise. The thermometer had fallen to 56.4° by Fahrenheit, yet they felt no cold, and were, on the contrary, scorched by the sun's rays, and obliged to lay aside their gloves. Both of them had their pulses much accelerated: that of Biot, which generally beat seventy-nine times in a minute, was raised to one hundred and eleven; while the pulse of his friend Gay-Lussac, a man of a less robust frame, was heightened from sixty to eighty beats in the minute. Notwithstanding their quickened pulsation, however, they experienced no sort of uneasiness, nor any difficulty in breathing.

What perplexed them the most was the difficulty of observing the oscillations of a delicately suspended magnetic needle. But they soon remarked, on looking attentively down upon the surface of the conglomerated clouds, that the balloon slowly revolved, first in one direction, and then returned the contrary way. Between these opposite motions there intervened short pauses of rest, which it was necessary for them to seize. Watching, therefore, the moments of quiescence, they set the needle to vibrate, but were unable to count more than five, or very rarely ten oscillations. A number of trials, made between the altitudes of 9500 and 13,000 feet, gave 7″ for the mean length of an oscillation, while at the surface of the earth it required $7\frac{1}{20}$″ to perform each oscillation. A difference so very minute as the hundred and fortieth part could be imputed only to the imperfection of the experiment; and it was hence fairly concluded, that the force of magnetic attraction had in no degree diminished at the greatest elevation they could reach. The direction of this force, too, seemed, from concurring circumstances, to have continued the same; though they could not depend on observations made in their vacillating car with so delicate an instrument as the dipping needle.

At the altitude of 11,000 feet they liberated a green linnet, which flew away directly; but, soon feeling itself abandoned in the midst of an unknown ocean, it returned and settled on the stays of the balloon. Then mustering fresh courage, it took a second flight, and dashed downwards to the earth, describing a tortuous yet almost perpendicular track. A pigeon which they let off under similar circumstances afforded a more curious spectacle. Placed on the edge of the car, it rested a while, measuring as it were the breadth of that unexplored sea which it designed to traverse:

now launching into the abyss, it fluttered irregularly, and seemed at first to try its wings in the thin element; till, after a few strokes, it gained more confidence, and, whirling in large circles or spirals, like the birds of prey, it precipitated itself towards the mass of extended clouds, where it was lost from sight.

It was difficult, in those lofty and rather humid regions, to make electrical observations; and the attention of the scientific navigators was besides occupied chiefly by their magnetical experiments. However, they let down from the car an insulated metallic wire of about 250 feet in length, and ascertained, by means of the electrophorus, that the upper end indicated resinous or negative electricity. This experiment was several times repeated; and it seemed to corroborate fully the previous observations of Saussure and Volta relative to the increase of electricity met with in ascending the atmosphere.

The diminution of temperature in the higher regions was found to be less than what is generally experienced at the same altitude on mountains. Thus, at the elevation of 12,800 feet, the thermometer was at 51° by Fahrenheit, while it stood as low as 63½° at the observatory; being only a decrement of one degree for each 1000 feet of ascent. This fact corresponds with the observations made by former aeronauts, and must have been produced, we conceive, by the operation of two distinct causes. First, the rays from the sun, not being enfeebled by passing through the denser portion of the atmosphere, would act with greater energy on the balloon and its car, and consequently affect the thermometer placed in their vicinity. Next, the warm current of air, which during the day rises constantly from the heated surface of the ground, must augment the temperature of any body which is exposed to its influence. During the night, on the contrary, the upper strata of the atmosphere would be found colder, we presume, than the general standard, owing to the copious descent of chill portions of air from the highest regions.

The hygrometer, or rather hygroscope, of Saussure, advanced regularly towards dryness, in proportion to the altitude which they attained. At the elevation of 13,000 feet it had changed from 80°.8 to 30°. But still the conclusion, that the air of the higher strata is drier than that of the lower, we are inclined to consider as fallacious. In fact, the indications of the hygroscope depend on the relative attraction for humidity possessed by the substance employed, and the medium in which it is immersed. But air has its disposition to retain moisture always augmented by rarefaction, and consequently such alteration alone must materially affect the hygroscope. The only accurate instrument for ascertaining the condition of air with respect to dryness is founded on a property of evaporation. But we shall afterwards have occasion to discuss this subject at due length.

The ballast now being almost quite expended, it was resolved to descend. The aeronauts therefore pulled the upper valve, and allowed part of the hydrogen gas to escape. They dropped gradually, and when they came to the height of 4000 feet, they met the stratum of clouds, extending horizontally, but with a surface heaved into gentle swells. When they reached the ground, no people were near them to stop the balloon, which dragged the car to some distance along the fields. From this awkward and even dangerous situation they could not extricate themselves without discharging the whole of their gas, and therefore giving up the plan of sending M. Gay-Lussac alone to explore the highest regions. It has been reported that his companion, M. Biot, though a man of activity and not deficient in personal courage, was so much overpowered by the alarms of their descent, as to lose for the time the entire possession of himself. The place where they alighted, at half-past one o'clock, after three hours and a half spent in the midst of the atmosphere, was near the village of Meriville, in the department of the Loiret, and about fifty miles from Paris.

It was the desire of several philosophers at Paris, that M. Gay-Lussac should mount a second time, and repeat the different observations at the greatest elevation he could attain. Experience had instructed him to reduce his apparatus, and to adapt it better to the actual circumstances. As he could only count the vibrations of the magnetic needle during the very short intervals which occurred between the contrary rotations of the balloon, he preferred one of not more than six inches in length, which therefore oscillated quicker. The dipping needle was magnetized and adjusted by the ingenious M. Coulomb. To protect the thermometer from the direct action of the sun, it was inclosed within two concentric cylinders of pasteboard covered with gilt paper. The hygrometers, constructed on Richer's mode, with four hairs, were sheltered nearly in the same way. The two glass flasks intended to bring down air from the highest regions of the atmosphere had been exhausted till the mercurial gage stood at the 25th part of an inch, and their stopcocks were so perfectly fitted, that after the lapse of eight days they still preserved the vacuum. These articles, with two barometers, were the principal instruments which M. Gay-Lussac took with him. The skill and intelligence of the artist had been exerted in further precautions for the safety of the balloon.

At forty minutes after nine o'clock on the morning of the 15th of September, the scientific voyager ascended, as before, from the garden of the Repository of Models. The barometer then stood at 30.66 English inches, the thermometer at 82° by Fahrenheit, and the hygrometer at $57\frac{1}{2}°$. The sky was unclouded, but misty. Scarcely had the observer reached the height of 3000 feet, than he observed spread below him, over the whole extent of the atmosphere, a thin vapour, which rendered the distant objects very indistinct. Having gained an altitude of 9950 feet, he set his needle to vibrate, and found it to perform twenty oscillations in 83″, though it had taken 84″.33 to make the same number at the surface of the earth. At the height of 12,680 feet he discovered the variation of the compass to be precisely the same as below; but with all the pains he could take, he was unable to determine with sufficient certainty the dip of the needle. M. Gay-Lussac continued to prosecute his other experiments with the same diligence, and with greater success. At the altitude of 14,480 feet he found that a key, held in the magnetic direction, repelled with its lower end, and attracted with its upper end, the north pole of the needle of a small compass. This observation was repeated, and with equal success, at the vast height of 20,150 feet; a clear proof that the magnetism of the earth exerts its influence at remote distances. He made not fewer than fifteen trials at different altitudes, with the oscillations of his finely suspended needle. It was generally allowed to vibrate twenty or thirty times. The mean result gives 4″.22 for each oscillation, while it was 4″.216 at the surface of the earth; an apparent difference so extremely small, as to be fairly neglected.

During the whole of his gradual ascent, he noticed, at short intervals, the state of the barometer, the thermometer, and the hygrometer. Of these observations, amounting in all to twenty-one, he has given a tabular view. We regret, however, that he has neglected to mark the times at which they were made, since the results appear to have been very considerably modified by the progress of the day. It would likewise have been desirable to have compared them with a register noted every half hour at the Observatory. From the surface of the earth to the height of 12,125 feet, the temperature of the atmosphere decreased regularly from 82° to 47°.3 by Fahrenheit's scale. But afterwards it increased again, and reached to 53°.6, at the altitude of 14,000 feet; evidently owing to the influence of the warm currents of air which, as the day advanced, rose continually from the heated ground. From that point the temperature diminished, with only slight deviations from a perfect regularity. At the height of 18,636 feet the thermometer subsided to 32°.9, on the verge of congelation; but it sunk to 14°.9 at the enormous altitude of 22,912 feet above Paris, or 23,040 feet above the level of the sea, the utmost limit of the balloon's ascent.

From these observations no conclusive inference, we think, can be drawn respecting the mean gradation of cold which is maintained in the higher regions of the atmosphere; for, as we have already remarked, the several strata are during the day kept considerably above their permanent temperature, by the hot currents raised from the surface through the action of the sun's rays. If we adopt the formula given by Professor Leslie at the end of his Elements of Geometry, which was the result of some accurate and combined researches, the diminution of temperature corresponding to the first part of the ascent, or 12,125 feet, ought to be forty degrees of Fahrenheit. It was actually 34°.7, and would no doubt have approached to 40°, if the progressive heating of the surface, during the interval of time, were taken into the account. In the next portion of the voyage, from the altitude of 14,000 to that of 18,636 feet, or the breadth of 4636 feet, the decrement of temperature according to the formula should only have been $16\frac{1}{4}$°, instead of 20°.7, which was really marked; a proof that the diurnal heat from below had not yet produced its full effect at such a great height. In the last portion of the balloon's ascent, from

18,636 feet to 22,912, a range of 3276 feet, the decrease of heat ought to be 15½°, and it was actually 18°; owing most probably to the same cause, or the feebler influence which warm currents of air from the surface exert at those vast elevations. Taking the entire range of the ascent, or 22,912 feet, the diminution of temperature, according to the same formula, would be for the gradation of temperature in ascending the atmosphere 85°.4. The decrease actually observed would be 67°.1, which might be raised to 80°, if we admit the very probable supposition, that the surface of the earth had become heated from 82° to 94°.9 during the interval between ten o'clock in the morning and near three in the afternoon, when the balloon floated at its greatest elevation.

After making the fair allowances, therefore, on account of the operation of deranging causes, the results obtained by M. Gay-Lussac, for the gradation of temperature in the atmosphere, appear, on the whole, to agree very nearly with those derived from the formula which theory, guided by delicate experiments, had before assigned. This gradation is evidently not uniform, as some philosophers have assumed; but proceeds with augmented rapidity in the more elevated regions. The same conclusion results from a careful inspection of the facts which have been stated by other observers.

The hygrometers, during the ascent of the balloon, held a progress not quite so regular, but tending obviously towards dryness. At the height of 9950 feet they had changed from 57°.5 to 62°; from which point they continued afterwards to decline, till they came to mark 27°.5, at the altitude of 15,190 feet. From this inferior limit the hygrometers advanced again, yet with some fluctuations, to 35°.1, which they indicated at the height of 18,460 feet. Above this altitude the variation was slight, though rather inclining to humidity. There can exist no doubt, however, that, allowing for the influence of the prevailing cold, the higher strata of the atmosphere must be generally drier than the lower, or capable of retaining, at the same temperature, a larger share of moisture.

At the altitude of 21,460 feet M. Gay-Lussac opened one of his exhausted flasks; and, at that of 21,790 feet, the other. The air rushed into them through the narrow aperture, with a whistling noise. He still rose a little higher, but, at eleven minutes past three o'clock, he had attained the utmost limit of his ascent, and was then 22,912 feet above Paris, or 23,040 feet, being more than four miles and a quarter, above the level of the sea. The air was now more than twice as thin as ordinary, the barometer having sunk to 12.95 inches. From that stupendous altitude, sixteen hundred feet above the summit of the Andes, more elevated than the loftiest pinnacle of our globe, and far above the heights to which any mortal had ever soared, the aerial navigator might have indulged the feelings of triumphant enthusiasm. But the philosopher, in perfect security, was more intent on calmly pursuing his observations. During his former ascent, he saw the fleecy clouds spread out below him, while the canopy of heaven seemed of the deepest azure, more intense than Prussian blue. This time, however, he perceived no clouds gathered near the surface, but remarked a range of them stretching, at a very considerable height, over his head: the atmosphere, too, wanted transparency, and had a dull, misty appearance. The different aspect of the sky was probably owing to the direction of the wind, which blew from the north-north-west in his first voyage, but in his second from the south-east.

While occupied with experiments at this enormous elevation, he began, though warmly clad, to suffer from excessive cold, and his hands, by continual exposure, grew benumbed. He felt likewise a difficulty in breathing, and his pulse and respiration were much quickened. His throat became so parched from inhaling the dry attenuated air, that he could hardly swallow a morsel of bread; but he experienced no other direct inconvenience from his situation. He had indeed been affected, through the whole of the day, with a slight headache, brought on by the preceding fatigues and want of sleep; but though it continued without abatement, it was not increased by his ascent.

The balloon was now completely distended, and not more than 33 pounds of ballast remained; it began to droop, and M. Gay-Lussac, therefore, only sought to regulate its descent. It subsided very gently, at the rate of about a mile in eight minutes; and after the lapse of thirty-four minutes, or at three quarters after three o'clock, the anchor touched the ground, and instantly secured the car. The voyager alighted with great ease near the hamlet of St Gourgon, about sixteen miles north-west from Rouen. The inhabitants flocked around him, offering him assistance, and eager to gratify their curiosity.

As soon as he reached Paris, he hastened to the laboratory of the Polytechnic School, with his flasks, containing air of the higher regions, and proceeded to analyze it in the presence of Thenard and Gresset. Opened under water, the liquid rushed into them, and apparently half filled their capacity. The transported air was found, by a very delicate analysis, to contain exactly the same proportions as that collected near the surface of the earth, every 1000 parts holding 215 of oxygen. From concurring observations, therefore, we may conclude that the atmosphere is essentially the same in all situations.

The ascents performed by MM. Biot and Gay-Lussac are memorable, for being the first ever undertaken solely for objects of science. It is impossible not to admire the intrepid coolness with which they conducted those experiments, operating, while they floated in the highest regions of the atmosphere, with the same composure and precision as if they had been quietly seated in their cabinets at Paris. Their observations on the force of terrestrial magnetism, show most satisfactorily its deep source and wide extension. The identity of the constitution of the atmosphere to a vast altitude was likewise ascertained. The facts noted by Gay-Lussac, relative to the state of the thermometer at different heights, appear generally to confirm the law which theory assigns for the gradation of temperature in the atmosphere. But many interesting points were left untouched by this philosopher. We are sorry that he had not carried with him the *cyanometer*, which enabled Saussure to determine the colour of the sky on the summits of the Swiss mountains. Still more we regret that he was not provided with an hygrometer and a photometer, of Leslie's construction. These delicate instruments could not have failed, in his hands, to furnish important data for discovering the relative dryness and transparency of the different strata of air. It would have been extremely interesting, at such a tremendous height, to have measured with accuracy the feeble light reflected from the azure canopy of heaven, and the intense force of the sun's direct rays, and hence to have determined what portion of them is absorbed in their passage through the lower and denser atmosphere.

Since that time numerous ascents have been performed in different countries, generally by adventurers guided by no philosophical views, nor leading to any valuable results. It would therefore be superfluous to recount such repeated attempts.

Balloons have at different times been thought capable of useful application. It has been even proposed to employ their power of ascension as a mechanical force. This might be rendered sufficient, it was believed, to raise water from mines, or to transport obelisks, and place them on great elevations. We can easily imagine situations where a balloon could be used with advantage; such as to raise, without any scaffolding, a cross or a vane to the top of a high spire. But the power would then be purchased at a very disproportionate expense. It would require $4\frac{1}{2}$ pounds of iron, or 6 of zinc, with equal quantities of sulphuric acid, to yield hydrogen gas sufficient to raise up the weight of one pound.

The proposal of employing balloons in the defence and attack of fortified places appears truly chimerical. They have rendered important service, however, in reconnoitring the face of a country, and communicating military signals; and it is rather surprising that a system, which promised such obvious benefits, has not been carried much farther.

But to a skilful and judicious application of balloons, we may yet look for a most essential improvement of the infant science of meteorology. Confined to the surface of this globe, we have no direct intimation of what passes in the lofty regions of the atmosphere. All the changes of weather, which appear so capricious and perplexing, proceed, no doubt, from the combination of a very few simple causes. Were the philosopher to penetrate beyond the seat of the clouds, examine the circumstances of their formation, and mark the prevailing currents, he would probably remove in part the veil that conceals those mighty operations. It would be quite practicable, we conceive, to reach an elevation of seven miles, where the air would be four times more attenuated than ordinary. A silk balloon, of forty feet diameter, if properly constructed, might be sufficient for that enormous ascent, since its weight would only be 80 pounds, while its buoyant force, though not more than a quarter filled with hydrogen gas, would amount to $533\frac{1}{3}$, leaving $453\frac{1}{3}$ pounds for the passenger and the ballast. The balloon could be safely charged, indeed, to the third part of its capacity, on account of the contraction which the gas would afterwards suffer from the intense cold of the upper regions; and this gives it an additional buoyancy of $177\frac{8}{9}$ pounds. The voyager would not, we presume, suffer any serious inconvenience from breathing the very thin air. The animal frame adapts itself with wonderful facility to external circumstances. Perhaps the quickened pulse and short respiration, which some travellers have experienced on the summits of lofty mountains, should be attributed chiefly to the suddenness of their transition, and the severity of the cold. The people of Quito live comfortably 9560 feet above the level of the sea; and the shepherds of the hamlet of Antisana, the highest inhabited spot in the known world, who breathe, at an elevation of 13,500 feet, air that has only three-fifths of the usual density, are nowise deficient in health or vigour. But the intenseness of the cold is probably what the resolute observer would have most to dread, at the height of seven miles. This decrease of temperature, perhaps equal to 148 degrees, might extend below the point at which mercury freezes. Yet several circumstances tend to mitigate such extreme cold, and proper clothing might enable an experimenter for a short time to resist its effects.

Much could be done, however, without risk or material expense. Balloons from fifteen to thirty feet in diameter, and carrying register thermometers and barometers, might be capable of ascending alone to altitudes between eight and twelve miles. Dispatched from the centres of the great continents, they would not only determine the extreme gradation of cold, but indicate by their flight the direction of the regular and periodic winds which doubtless obtain in the highest regions of the atmosphere. But we will not enlarge. In some happier times, such experiments may be performed with the zealous concurrence of different governments;—when nations shall at last become satisfied with cultivating the arts of peace, instead of wasting their energies in sanguinary, destructive, and fruitless wars.

# BAROMETER.

---

Our dictionaries contain such an account of the discovery and construction of this most valuable instrument, as can be drawn from the popular treatises of natural philosophy in the English language. But, unfortunately, the compilers of elementary works have seldom taken the trouble to remount to the original sources of information, and have frequently, by substituting their own fancies, or servilely copying the mistakes of others, contrived to disfigure egregiously the relation of facts, and the history of the progress of invention. We purpose, therefore, as far as our limits will admit, to discuss the subject with more careful research; but, passing rather lightly over the description of the different kinds of barometers, and other practical details, to dwell more especially on the successive steps which led to the fine discovery of atmospheric pressure, and its application to physical science.

The opinions entertained by the ancients concerning physical subjects, appear at best only splendid visions. They speculated boldly in cosmological theories, but were easily satisfied with those conclusions which merely soothe the fancy. Many of the philosophical notions adopted in remote ages have, however, left a durable impression in the structure of language, and still continue to exercise a visible influence in moulding the current sentiments of mankind. The early sages of Greece distinguished matter into the four primary elements of *earth*, *water*, *air*, and *fire*, which, by their various combinations, were supposed to produce the animated spectacle of the universe. With these elements were associated corresponding qualities, in a binary conjunction; *hot* and *cold*, *dry* and *moist*. *Earth* and *water* were considered as ponderous and inert; but *air* and *fire*, endued with elastic virtue, were imagined to possess lightness and activity. *Fire*, though extracted from all bodies by the operations of nature or of art, was yet conceived to be derived, by invisible emanation, from that diffuse lambent fluid which, under the name of *Æther*, occupied the highest heavens, and furnished the substance and nutriment of the celestial bodies. While the earthy matter would, therefore, naturally settle towards the centre, and the aqueous fluids roll along the surface of the solid globe, the air and fire soared aloft; the former occupying the whole of the region below the moon, and the latter streaming through the boundless extent of space. This *sublunary* scene is exposed to incessant change, calamity, and decay; but above it was supposed to reign a perpetual calm, the seat of bliss, and of divine and imperishable essence.

Aristotle and some other philosophers, viewing *æther* as altogether distinct from *culinary fire*, were disposed to consider it as a *fifth* element, of a pure, divine, and incorruptible nature; an opinion which afterwards gave occasion to the famous *Quinta Essentia*, or *Quintessence* of the schoolmen. The alchemists, who sprung up nearly about the same benighted period, in adopting these notions, modified them to suit their own peculiar views. To the elements commonly received, they joined the active auxiliaries of *mercury* and *sulphur*. For *quintessence* they substituted *spirit* and *elixir*; the former, drawn off by the application of fire, being conceived to represent the animating principle of each body; while the latter, extracted by the combined action of heat and moisture, was supposed to exhibit its concentrated and most select qualities.

Some of the ancient cosmologists imagined a vacuum beyond the shining expanse of æther, destined to receive the exhalations from this nether world. Others denied the existence of a separate void, but admitted small vacuities interspersed through bodies. Aristotle, however, maintained the necessity of a plenum, asserting that our idea of space or extension is inseparable from that of body. To this principle he ascribed the suspension of water in a tube, when the finger is applied to shut the upper end. Yet the very contempt in which that philosopher, from a consciousness of his own superiority, was accustomed to hold the received opinions, might have led him to take juster views. He rejected the notion that air has levity inherent in its nature; nor would he admit the more plausible idea, that a fluid so easily moved must possess the quality of perfect indifference, and be neither light nor heavy. Aristotle not only maintained that air is ponderous, but did not scruple to appeal to direct experiment in support of this assertion. A bladder, he says, will be found to gain some weight, on being blown or filled with air. But this was evidently a mere random assertion, betraying his ignorance of the constitution of fluids. A bag filled with air, and suspended in a like medium, it is obvious, from the laws of hydrostatics, must weigh exactly the same as before. If it be alleged that, in blowing up the bladder, a portion of air would be introduced immediately from the lungs, and containing therefore a small admixture of carbonic acid gas, which is specifically heavier than the common atmospheric fluid; the additional weight, amounting scarcely perhaps to a grain, would be too minute to be detected by any of the jewellers' balances constructed in ancient Greece.

The mutual opposition of the leading philosophical sects of antiquity had, in general, most fatally discouraged the application of mathematical reasoning to the system of the material world. The Academicians, or the disciples of Plato, who cultivated geometry with ardour and brilliant success, were disposed to regard that science as a pure intellectual contemplation; and resigning themselves to the illusion of their lofty dreams, they turned with disdain from the investigation of individual facts, and all the vulgar realities of life. The mind of Aristotle was of a more sober and practical cast; acute, profound, and discriminating, it ranged with incredible industry over an immense field of inquiry. That judicious philosopher recommended a careful and constant appeal to external observation, as the only sure ground on which to erect the structure of physics; but unfortunately his scholars neglected too much the study of mathematics, the most powerful instrument for conducting physical research. The precepts of Aristotle, though excellent in some respects, were hence in the sequel unproductive of any genuine fruit. On the contrary, the weight of his opinions, during a long course of ages, repressed the efforts of human genius.

It must be gratefully acknowledged, that the alchemists, styled also *philosophers by fire*, were the first among the moderns who dared to explore new paths of science. Their projects were, indeed, highly chimerical, but they had the merit, at least, of setting the example of investigating the properties of matter by actual experiment. They likewise formed associations among individuals for the more effectual prosecution of such researches. Hence the origin of that obscure sect, known by the fanciful title of *Rosicrucians*, who sprung up in Germany, and insensibly spread their influence over the Continent. Those principles were afterwards transplanted into the matured soil of Italy, where philosophy, succeeding to the cultivation of letters, wore a more attractive garb. Baptista Porta, a Neapolitan nobleman, who flourished about the latter part of the sixteenth century, was particularly distinguished by his zeal in promoting such pursuits. Having spent many years in travelling over Europe to gain information respecting natural objects, he invited a few individuals of a congenial taste to assemble at stated times in his house, and assist him in making new experiments. These meetings, however, gave umbrage to the watchful jealousy of the clergy, and they were soon suppressed by a mandate from the court of Rome. But the example was imitated in other parts of Italy, where the papal authority enjoyed less respect; and academies for the promotion of natural science were successfully instituted under the patronage of different princes, especially those of the illustrious house of Medici.

In this ferment of inquiry, Galileo arose, a man fitted alike by the gifts of nature, and the lights of education, to be the founder of experimental science. His elegant genius was invigorated by the study of the Greek geometry; and he conceived the happy and prolific idea of employing that refined instrument to explore facts, and combine the results. Archimedes, indeed, among the ancients, had anticipated this road of discovery, having most successfully applied the powers of geometrical analysis to the investigation of some parts of mechanics and hydrostatics. But his was a solitary instance, unheeded by succeeding ages. The ingenuity of Galileo prepared a complete revolution in science. By means of a few simple but striking experiments, performed on the lagoons of Venice, he established the laws of motion, which he now transferred from the surface of our globe, to direct the revolutions of the heavenly bodies. The publication of his *Dialogues*, which unfold the right process of induction, and are not less distinguished by fineness of conception than beauty of diction, form a new era in the annals of philosophy. He was the first who attempted to ascertain the weight of air by actual experiment; and considering the nicety of the operation, and the rudeness of the instruments constructed at that period, he made a very tolerable approach to the truth. It had been known for many ages that air is capable of being highly condensed; and Ctesebius of Alexandria had invented an engine, which, by the force of the sudden expansion of this compressed fluid, hurled missile weapons. This was afterwards improved into the wind or air-gun, which seems to have been not uncommon in Europe as early as the fifteenth century, though soon afterwards generally superseded in practice by the introduction of fire-arms. Galileo, being led by a different path from that pursued at present, set himself to examine the weight which air acquires by condensation. Having fitted a large copper ball with a valve, he injected air into its cavity by means of a syringe, and then suspended it to a balance. The additional increase of weight being thus found, he opened the valve under an inverted glass receiver full of water, and measured, by the displacement of this liquid, the surplus quantity of air which had been injected into the copper vessel. He thence concluded that air is 400 times lighter than water, being about the double of the true estimate.

After he had, by such researches, acquired celebrity in the scientific world, Galileo accepted an invitation, with a very handsome appointment, from Cosmo de Medici; and devoting himself intensely to astronomical observations, aided by the telescope, which, from an obscure hint, he had recently constructed, yet occasionally unbending his mind with elegant recreation, he spent almost the whole of the evening of his life at the villa of Arcetri, near Florence, in a style of comfort and even splendour. But, while occupied with those delightful pursuits, exploring the planetary phases, and discovering new worlds, he was for a moment recalled to his early studies by an incident destined to form an epoch in the history of physical science. Some artisans, in the service of the grand duke, having been empleyed to construct a lifting or sucking pump for a very deep well, found, with equal surprise and vexation, that, in spite of all the pains they had taken in fitting the piston and valves, the water could by no effort be made to rise higher in the barrel than eighteen palms, or thirty-two feet. In this dilemma they applied to Galileo for an explication of the cause of a failure then so unexpected and perplexing. But the philosopher was not yet prepared to encounter such a discordant fact. The Aristotelian tenet of the impossibility of the existence of a void, was, at this period, universally received as an unquestionable truth. It had become a favourite axiom of the schoolmen, deceiving themselves, as Leibnitz did afterwards in proposing his principle of *sufficient reason*, by the glimmer of a metaphorical expression, the *fuga vacui*, or *nature's horror of a void.* To create a vacuum, they gravely maintained, would require the hand of Omnipotence, transcending the utmost power of men or even devils. But Galileo, though borne along by the current of opinion, saw the necessity of at least modifying the general principle. Without questioning nature's abhorrence of a vacuum, he supposed the influence of this horror to be confined within certain limits, not exceeding the pressure of a column of water eighteen palms in height. This was evidently evading, rather than meeting, the difficulty proposed for his solution. Yet, in the last of his Dialogues, he actually mentions an experiment to ascertain this power, or *virtu*, as he calls it, of a vacuum. A piston with a valve, exactly fitted into a smooth hollow cylinder, was rammed quite to the end, and this carefully shut up; then placing the cylinder in an upright but inverted position, successive weights were appended to the rod, till it was drawn from the close end, and pulled down. It may seem strange that the Tuscan philosopher, after advancing so far, should have stopt on the verge of a great discovery. He had already weighed the air, and it was only another small step thence to infer the effect of its incumbent mass. But the atmosphere was still believed to reach to the moon, and the pressure of columns of such enormous altitude seemed to mock all calculation, and overwhelm the imagination.* Yet, on reconsidering the subject, Galileo began to suspect the solidity of the explication which he had given; but it was now too late for him in his advanced age, loaded with bodily infirmities, and dispirited by clerical persecution, to attempt any further innovation in science. Recommending it earnestly to his friend and pupil Torricelli to resume the investigation, this illustrious precursor of Newton expired in 1642, the very year in which the English philosopher was born. His uniform kindness and urbanity rendered him extremely beloved; and his disciples, particularly Torricelli,

---

* This narrative, which marks so well the slow and timid steps whereby men, even of the highest intellectual endowments, usually advance in the search after truth, is drawn from the writings of Galileo himself. The carelessness of some authors in mis-stating facts, and imputing unworthy motives to those patriarchs of science who could not open their eyes all at once to the bright effulgence of day, deserves severe reprehension. We may remark, in passing, that M. Biot, who ranks now among the first mathematicians and philosophers in France, has not scrupled, in an elaborate compilation on physics, to allege that Galileo merely joked with the artizans who asked him the reason of the failure of their pump; that he had an idea of the true explication, but chose to keep his secret, and suffer it to die with him. Such conduct would certainly have been a reproach to Galileo's acknowledged candour.

Viviani, and Ricci, venerating his memory, caught the same taste, and followed the same pursuits.

Torricelli now conceived the happy idea of exhibiting the action of a pump on a contracted scale, by means of a column of mercury, which is nearly fourteen times heavier than water. This experiment he first communicated to his friend Viviani, who performed it with success in 1643; and he afterwards repeated and varied it himself. The method which he took brought very neatly under one view all the circumstances affecting the question. Having selected a tube about a quarter of an inch wide, and four feet long, he sealed one of the ends hermetically, or closed it under the flame of a lamp; he then filled the cavity of the tube with mercury, and applying his finger to the open end, he inverted it in a basin likewise containing mercury, though covered with a portion of water. The mercury instantly sunk to nearly thirty inches above the lower surface; but on raising the tube, till its orifice communicated with the layer of water, the mercury ran all out, and the water now sprung up to the top, and occupied the whole of the cavity. It was thus proved, that the water and mercury are each supported by the same equipoise, which Torricelli, after some hesitation, at last concluded to be the pressure of the external atmosphere. He next converted the mercurial column into a form adapted for observation, by bending the lower end of the tube, and constructed what has since received the name of the *syphon barometer*. (Plate I. Fig. 1.) Thus provided with a commodious instrument, he soon detected the variation of atmospheric pressure, which depends on the change of weather. These important results were published in the year 1645; but Torricelli did not live to enjoy the fame of his great discovery, for this most promising genius was snatched away by a putrid fever in the flower of his age.

The report of Torricelli's first experiments having been carried to France before he had ventured to draw his capital conclusion, set philosophers to speculate on the cause of such an unexpected fact. Descartes, with his usual rapidity and boldness of conception, did not hesitate, in his correspondence with Mersenne, to refer the suspension of the mercury in the tube at once to the pressure of the external atmosphere. But this inference appears not very consistent with his system, which assumed the existence of an absolute plenum, and only supplied the place of a void by the diffusion of subtile abraded particles of matter. He suspected also the accuracy of Galileo's estimate of the weight of the air, which he thought was scarcely appreciable by experiment.

But, in the same country, the subject was now pursued with deliberate caution, and through all its details, by another genius of the highest order; one of the finest and most original that France has ever produced. Pascal had shown premature and extraordinary talents, which were encouraged by his father, a man of learning, who lived in habits of intimacy with the literati of Paris. The young philosopher happened to be residing at Rouen, in 1646, when he was informed of the famous Italian experiment. Having access, fortunately, to a glass-house, he resolved immediately to repeat the observations on a large scale. He had already suspected the justness of the principle, that "nature abhors a vacuum," and thought that the condensation and rarefaction of the air point to a different, or at least a modified conclusion. With a view to clear up this subject, Pascal performed a number of satisfactory experiments, of which we shall cite a few of the more striking, nearly in his own language, tinctured evidently with the prevailing opinions of the age:—1. Having fitted a piston to an open glass tube, and rammed it quite down, he applied his finger close to the lower end, and plunged the whole under water; then drawing back the piston, which was done with ease, the finger felt strongly and rather painfully *attracted,* while an *apparent vacuity* was formed above it, and continued to enlarge; but instantly on removing the finger, the water, *contrary to its nature,* darted up and filled the whole of the cavity. 2. A glass tube, about fifty feet long, sealed hermetically at one end, and filled with water, or rather red wine, as a more visible fluid, was inverted perpendicularly in a basin of the same. The liquid immediately subsided, leaving a vacant space of thirty-five feet; but on gradually reclining the tube, the liquid rose again, and continued to mount, till it struck a sharp blow against the top of the glass. 3. A syphon, having one leg fifty-five feet high, and the other only fifty, being filled with water, and planted in two basins containing the same, such that the shorter branch had a perpendicular position, the water sunk in both to the same level, without being *attracted,* as usual in syphons, to the longer branch; but, on leaning the syphon back, the columns rose till they united at the top, and then the water began to flow towards the lower basin. The same experiment was also performed with mercury, the syphon with one leg ten feet, and the other only nine feet and a half in length, the mercury being found to divide itself into two columns, which continued suspended at an altitude of about thirty inches. 4. Having nicely fitted a piston to a long glass syringe, and pushed it down to the end, he immersed this in a basin of mercury, and held the tube in a vertical position; on gently drawing up the piston, the mercury closely followed it to the height of twenty-nine inches, but then stopt, leaving the piston to form above it an apparent vacuity. In this state, also, the syringe weighed exactly the same, whatever was the magnitude of the vacant space.

From these and other similar experiments, Pascal led his inductive process, with a degree of caution that might seem to border on timidity. He inferred that all bodies have a *reluctance* to a visible separation, or that nature abhors an apparent void; that this reluctance is exactly the same for a small as for a great vacuity; and that the force is limited, and exceeds not the pressure of a column of water thirty-three feet in height. He next ventured one step farther, and concluded, that this apparent vacuity was not filled by air lodged in the pores of the glass, or derived from external filtration; that it contained no subtile matter secreted from the atmosphere, and was not occupied by mercurial vapours or spirituous exhalations; in short, that a real and absolute vacuum had been formed.

Pascal, then only twenty-four years of age, proposed to write a treatise on the subject of those inquiries; but thought proper, in the mean time, to publish a short abstract of it, which appeared in 1647, and involved him in a wretched controversy. Father Nöel, rector of the Jesuits' College at Paris, keenly attacked it, armed with all the miserable sophisms of the schools, and the absurd dogmas of the Romish church. He contended, that the space above the mercurial column was corporeal, because it was visible and admitted light; that a void being a mere non-entity, cannot have different degrees of magnitude; that the separation produced in the experiments was violent and unnatural; and he presupposed that the atmosphere, like blood, containing a mixture of the several elements, the fire and the finer part of the air were detached from it, and violently forced through the pores of the glass, to occupy the deserted space. To enforce these puerile arguments, the reverend Jesuit did not scruple to employ the poisoned weapon which his order has often wielded with deadly effect, namely, hinting an oblique charge of heresy. This rude attack only roused Pascal, and disposed him boldly to throw off the fetters of inveterate opinion. He began to perceive that "abhorrence" cannot, in strict logic, be applied to nature, which is a mere personification, and incapable of passion; and was inclined, by degrees, to adopt the clear disembarrassed explication of Torricelli, referring the suspension of the mercurial column to the pressure of the external atmosphere. In stating this conclusion, he makes some remarks which would deserve the serious attention of philosophers in the present age. "When the weakness of men is unable to find out the true causes of phenomena, they are apt to employ their subtlety in substituting imaginary ones, which they express by specious names that fill the ear, without satisfying the judgment. It is thus that the sympathy and antipathy of natural bodies are asserted to be the efficient and unequivocal causes of several effects, as if inanimate substances were really capable of sympathy and antipathy. The same thing may be said of the antiperistasis, and various other chimerical causes, which afford only a vain relief to the avidity of men to know hidden truths, and which, far from discovering them, only serve to conceal the ignorance of those who invent such explications, and nourish it in their followers." These remarks, equally judicious and profound, are the more striking, since Lord Bacon, while he proposed to reform and new-model the whole structure of human learning, yet complied with the taste of the age in retaining much of the jargon and barbarous distinctions of the schools.

But Pascal did not rest satisfied with mere reasoning, however strictly conducted; and he soon devised an experiment which should palpably mark, under different circumstances, the varying effects of atmospheric pressure. It occurred to him, that, if the mercury in the Torricellian tube were really supported by the counterpoising weight of the atmosphere, it would be affected by the mass of superincumbent fluid, and must therefore partially subside in the higher elevations. He was impatient to have his conjecture tried in a favourable situation; and, in November 1647, he wrote a letter communicating those views to his brother-in-law, Perier, who held an office of considerable trust in the province, and commonly resided at Clermont in Auvergne, in the immediate vicinity of the Puy de Dôme, a lofty conical mountain, which rose, according to estimation, above the altitude of 500 toises. Various avocations, however, prevented that intelligent person from complying with his instructions till the following year. Early in the morning of the 19th September 1648, a few curious friends joined him in the garden of a monastery, situated near the lowest part of the city of Clermont, where he had brought a quantity of mercury, and two glass tubes hermetically sealed at the top. These he filled and inverted, as usual, and found the mercury to stand in both at the same height, namely, 26 inches and $3\frac{3}{4}$ lines, or 28 English inches. Leaving one of the tubes behind, in the custody of the sub-prior, he proceeded with the other to the summit of the mountain, and repeated the experiment, when his party were surprised and delighted to see the mercury sink more than three inches under the former mark, and remain suspended at the height of 23 inches and 2 lines, or 24.7 English inches. In his descent from the mountain, he observed, at two several stations, that the mercury successively rose; and, on his return to the monastery, he found it stood exactly at the same point as at first. Encouraged by the success of this memorable experiment, Perier repeated it on the highest tower of Clermont, and noted a difference of two lines at an elevation of twenty toises. Pascal, on his part, as soon as the intelligence reached him at Paris, where he then chanced to be, made similar observations on the top of a high house, and in the belfry of the church of St Jacques des Boucheries, near the border of the Seine; and so much was he satisfied with the results, that he proposed already the application of the barometer for measuring the relative altitudes of distant places on the surface of the globe.

The investigation of the existence and effects of atmospheric pressure was now completed, and it threw a sudden blaze over the whole contexture of physical science. The fame of the experiments performed in Italy and in France, quickly spread over Europe. Yet such is the force of habit and early prejudice, that, after the first moments of surprise and confusion, few of the learned at this period had the courage to open their eyes to the light which had so unexpectedly burst upon them; but, secretly cherishing their inveterate notions, they sought to comfort themselves by starting a variety of captious objections. Father Mersenne, though a man of some abilities, conceived that suction was occasioned by certain hooked particles dispersed through the atmosphere, which laid hold of any fluid in contact with them, and drew it towards the general mass. Father Linus, plunging still deeper in mysticism and absurdity, gravely proposed the *funicular hypothesis*, which attributes the suspension of the mercurial column to the agency of certain small invisible threads. But others of the clergy attacked Pascal with envenomed bitterness. The Jesuits of the college of Montserrand scrupled not, in their public theses, to pervert his expressions, and even contest the originality of his experiments. The philosopher was justly incensed at their base conduct; and those repeated provocations served, no doubt, to give a keener edge to his wit, when he afterwards directed it with such overwhelming energy against that insidious and formidable order of priesthood. He composed in 1653, though they were not published till after his death, two short treatises, *On the Equilibrium of Liquors*, and *On the Weight of the Mass of Air*, remarkable for their neatness, perspicuity, and lucid order. The laws of the equilibrium of fluids are there beautifully deduced from a single principle, which suggests a variety of original views and admirable remarks. In those tracts he likewise gives a description of the Hydraulic Press, a most useful and powerful machine, which has lately been revived in this country, and considered as a new invention.

A similar discovery, which was made about the same time in Germany, came seasonably to support the triumph of innovation. Otto Gürické, a wealthy magistrate of Magdeburg, who amused his leisure by constructing pieces of mechanism, and instituting curious physical inquiries, finding that the belief in the impossibility of a vacuum, with other scholastic tenets, was on the gradual decline, had the boldness to conceive that the forming of a void was a task perhaps within the reach of human ingenuity. Fired with the idea of accomplishing what for ages had been deemed unattainable, he directed all his efforts to compass that end. In his first trials he failed, as might be expected; but, by perseverance, he was enabled to surmount every obstacle. Having filled a wooden cask with water, he attempted to extract this again, by means of a small sucking pump, introduced at the bottom of the cask, and worked vigorously by three stout men; a hissing noise was heard like that of boiling water, the air entered from above through the interstices of the wood, and the water flowed out. The more effectually to exclude the air, he next took a smaller cask, with a sucker attached to it, and placed it within a larger one, having filled up the space between them with water. On working the pump as before, the water was forced through the pores of the wood into the inner cask, but none was extracted by the action of the piston. Foiled in these attempts with wooden casks, he had recourse to a copper ball, to the under part of which he screwed an inclining sucker; and, with this apparatus, he at last succeeded in extracting the air. He continued the operation, till no farther portion of air was perceived to issue from the vent. On opening the cock again, the air rushed into the cavity of the ball with violence; and the same effect took place, with scarcely any diminution of power, after an interval of a day or two. The construction of the machine was afterwards rendered more perfect, by substituting a large inclined metal sucker, with its joints secured by immersion in water.

Such was the origin of that most valuable addition to philosophical apparatus, the air-pump, which long retained its earliest rude and simple form on the Continent. By help of this new and powerful instrument, Gürické was enabled to perform some interesting and very important experiments. One of these, which demonstrates in a very striking way the pressure of the atmosphere, has since been deservedly styled the Magdeburg Experiment. It was performed with two hollow copper hemispheres, closely fitted together, and the air exhausted from their cavity. This singular experiment Gürické had the honour of exhibiting, in the year 1654, before the princes of the empire and the foreign ministers, assembled at the diet of Ratisbon. The force of two teams, each consisting of a dozen of horses, made to pull in opposite directions, was found insufficient to separate the hemisphere. It was now that the burgomaster of Magdeburg heard, for the first time, of Torricelli's great discovery, and the intelligence must have appeared quite delightful to him, who, by a path so different, had yet arrived at a similar conclusion.

After his return from this splendid assembly, Gürické pursued at home various pneumatical researches. He showed the diminished pressure of the atmosphere at an elevation above the surface, by means of a hollow ball fitted with a stop-cock; having carried this to a height, a portion of the contained air rushed out on turning the cock; but when it was brought down again and opened, the same measure of air apparently flowed into its cavity. He actually weighed the air by ascertaining, with the help of a nice balance, the loss which a large bottle sustained on being exhausted, and found that the air is 970 times lighter than water; a very near approximation, if allowance be made for the residuum of air still left in the bottle. He was the first who proposed the *Statical Balance* for measuring the variations of atmospheric density, consisting of a hollow glass ball about a foot in diameter, hermetically sealed, and freely suspended in the air, to indicate by its different buoyancy the changes which take place in the gravity of the external fluid.

But Gürické took great pleasure in a huge water barometer erected in his house. It consisted of a tube above thirty feet high, rising along the wall, and terminated by a tall and rather wide tube hermetically sealed, containing a toy, of the shape of a man. The whole being filled with water, and set in a basin on the ground, the column of liquid settled to the proper altitude, and left the toy floating on its surface; but all the lower part of the tube being concealed under the wainscoting, the little image, or weather mannikin, as he was called, made its appearance only when raised up into view in fine weather. This whimsical contrivance, which received the name of *anemoscope*, or *semper vivum*, excited among the populace vast admiration; and the worthy magistrate was in consequence shrewdly suspected by his townsmen of being too familiar with the powers of darkness.

Before the taste for experimental science was imported from the Continent into England, the great struggle for the security of private rights had called forth the national energy, and its triumphant success had infused among all classes of men a spirit of boldness and enterprise most favourable to the reception of the new philosophy. The parliamentary commissioners, by removing the more violent and bigoted members of the universities, contributed, on the whole, to encourage a more liberal tone of thinking in those opulent seminaries. Near the close of the civil war, and during the vigorous administration of Cromwell, the philosophy by experiment found some proselytes at last in the cloisters of Oxford, where the mass of antiquated opinions had lain so long embalmed and protected by religious awe. A small association was there formed, for combining together the efforts of individuals in the prosecution of such inquiries; and the fruits of this mutual compact were afterwards visible in the composition of various philosophical works. But the Restoration, by which the nation, in a burst of inconsiderate loyalty, surrendered the privileges which it had purchased with torrents of blood, threw the government of the universities again into the hands of men decidedly hostile to the very shadow of improvement. Experimental science withdrew to a more congenial soil, and sought shelter and support in the wider scope of the capital. The college, founded by the munificence of Sir Thomas Gresham, for the benefit of the citizens of London, though now unfortunately sunk in absolute neglect, had the merit of being the first to extend its protection to the pursuits of inductive philosophy. It produced a succession of professors, eminent in mathematical learning, which is so closely allied with experimental research. A more extensive association was accordingly formed in London, which regularly met at the apartments within the Exchange, and was afterwards, at the suggestion of Oldenburg, the resident from the city of Hamburg, and in imitation of the foreign academies, constituted by charter into the Royal Society. Such was the humble beginning of that illustrious body, and such all the countenance it received from a needy and profligate government. The institution, however, proved at first eminently useful, by its influence in directing public opinion, and the shelter it afforded to experimental philosophy, against the jealousy and declared hostility of the clerical and scholastic seminaries. The union of rank, or wealth, or talent, though still very limited in its range, bestowed a degree of lustre on the infant society, that was quite necessary for its defence against the attacks of ignorance and the mining of bigotry.

One of the most active members of the Royal Society was the honourable Mr Boyle, who having become acquainted with experimental researches in the course of his travels, devoted, after his return home, his time and his fortune to such calm but engaging pursuits. In this occupation he derived the most essential aid from Dr Hooke, whom he had the discernment to engage as his assistant,—the most skilful mechanician, and the best practical philosopher, of the age. The same ingenious person was likewise employed as operator to the society, and undertook to produce at each meeting some new experiments for the instruction and entertainment of the members. One of the favourite subjects was to exhibit the properties of the atmosphere. Dr Hooke, at the instance of Mr Boyle, had given a more convenient form to the air-pump, and had materially improved its construction, especially by the application of oil to the joints and valves. With this improved machine, a more perfect vacuum was procured than Gürické had obtained; and the English philosophers were thus enabled to perform a variety of delicate and interesting experiments, which extended the influence of the original discovery.

In those early meetings, too, of the Royal Society, the suspension of the mercury in the Torricellian tube had still the attraction of novelty. The famous Italian experiment, as it was called, was frequently repeated and varied in the presence of a few of the more assiduous members, who, delighted with the exhibition, still continued to argue and to doubt concerning the cause of the phenomenon. These doubts acquired new force from a singular experiment which the celebrated Huygens some years afterwards communicated, during a visit he made to London. Having filled a glass tube eighty inches long with mercury, and carefully expelled whatever air was lurking about the sides, he gently inverted it, as usual, in a basin; when the mercury notwithstanding remained still hanging from the top of the tube, and did not subside to the proper height till it was struck with a slight blow. This anomalous fact appeared then extremely puzzling. The experiment, indeed, requires great nicety and address on the part of the operator, and evidently depends on a concurrence of circumstances which have not yet been sufficiently explained. There can, at present, exist no doubt that this extraordinary suspension of the mercury is occasioned by its obstinate adhesion to the inside of the tube, which, in the process of purging the air, becomes probably lined with a very thin film of mercurial oxyd. But Huygens, who had embraced the leading principles of the Cartesian philosophy, was inclined to draw a very different conclusion. He thought that the fact proved the existence of another fluid besides the atmosphere, and one possessed of such extreme subtilty and power, as to be capable of permeating the grosser bodies. In ordinary cases, this fine ethereal substance might be supposed to escape through the pores of the glass, and leave the mercurial column to the mere pressure of the atmosphere. Such was the unfortunate introduction of that ideal being—an æther—into experimental science, which it has continued to infest with mysticism, and to dazzle with a false glare. Similar notions are perpetually renewed by a certain class of superficial inquirers, and have exercised a visible and most pernicious influence in retarding the progress of sound philosophy.

It was soon perceived that the syphon barometer of Torricelli has a disadvantageous form. Both branches of the tube being supposed to be of the same width, the mercury must evidently sink as much in the one as it will rise in the other; so that the variations in the height of the column are thus reduced to half the true quantity. A small basin, or semicircular wooden box, to hold the surplus mercury, was therefore attached to the frame of the instrument; and this construction, with very little change, was adopted, during the course of a century, by the makers of the ordinary barometer. But the syphon barometer itself was afterwards materially improved by having its lower branch blown into a wide bulb for holding the charge of mercury. (Plate I. Fig. 2.) This form of the barometer is not quite accurate, owing to the smallness and unequal shape of the round bulb; but being very convenient for carriage, it has grown into general use, at least for the cheaper and more common sort of instruments.

As soon as the barometer came to be regarded as a weather-glass, ingenuity was set at work to devise the means of enlarging its scale of variations. Descartes first proposed a simple method for effecting that object, by combining a mercurial with a water barometer; which arrangement, though subject to imperfection, has led to many of the subsequent improvements. (Plate I. fig. 4.) He directed two short barometric tubes to be cemented, the one into the bottom, and the other to the neck of a phial; or, still better, that the tubes should be joined, by the flame of a lamp, to the opposite ends of a wide and regular cylinder. The lower tube, and a portion of the cylinder, were then to be filled with mercury, and above it was to be introduced pure water, reaching to the top of the upper tube, and there sealed close. When this compound tube was inverted in a basin of mercury, it is evident that the columns both of mercury and of water would sink, till their joint pressure became just equal to the superincumbent weight of the external atmosphere. But the variation of this weight would afterwards be indicated chiefly by the large motion of the water; since the mercurial column, spreading out above into a broad surface, must, in any case, experience a very slight difference of altitude. Thus, suppose the cylinder to have eight times the diameter of the upper tube, or a section sixty-four times greater, mercury being 13.6 times denser than water: for each inch of increase of altitude which the ordinary mercurial column gains, the top of the water would be raised in the tube 11.4 inches, its own rise being 11.18 inches, and that of the wide mercurial cylinder only .18 of an inch, yet equal in pressure to 2.4 inches of water. But Descartes, generally satisfied with mere theory and speculation, did not live to see his construction of the barometer carried into effect; and Chanut, the French resident at Stockholm, to whom he had imparted his views, met with such difficulty in the execution of the project, that, after some fruitless attempts, he abandoned it altogether.

Huygens was more fortunate, and succeeded, by dint of perseverance and skill, in constructing the Cartesian barometer. But he had the mortification to find that, in spite of all the pains he could take, the water, after it was relieved from the pressure of the atmosphere by the sealing up of the tube, constantly discharged a portion of air, which collected at the top, and by its elasticity depressed the compound column below its due altitude. Convinced that this source of imperfection is irremediable, he sought to rectify the construction of the instrument, and produced his *Double Barometer;* a form of combination frequently used, especially when the object is rather to make the variations very sensible than to obtain delicate results. (Plate I. fig. 5.) He joined a barometric tube of the usual length, by the flame of a blow-pipe, to two wide cylinders, the one sealed at the top, and the other annexed likewise hermetically to a tall and narrow tube, open at its extremity; he then bent the thicker tube a little above the lower cylinder, and brought the two branches to be parallel. The instrument being thus formed, he filled the first branch with mercury, and introduced above, in the second branch, some liquid of comparative lightness. Alcohol would, in this respect, answer extremely well, if it were not so liable to waste by evaporation. An alkaline lye, or the deliquiate salt of tartar, which also readily admits of being coloured, was therefore, on the whole, preferred.

The principle of this construction is evidently the same as in that of Descartes; but the vacuum, lying contiguous to the mercury itself, can have no admixture of disengaged air or of aqueous vapour. Since the cylinders are made very much wider than the bore of the annexed tube, the variation of pressure will be produced almost entirely by the change of altitude which the alkaline liquor undergoes, the mercury suffering only a very minute alteration of ascent or descent. The divisions of the ordinary scale will be about tenfold enlarged, if a section of each cylinder should exceed twenty times that of the tube in which the liquor plays.

A barometer of this construction has decided advantages with respect to the extent of its changes, but still it is not exempt from considerable defects. The moisture on the inner surface of the cylindrical reservoir increases the adhesion of the mercury, and retards its movements. But a much greater source of error proceeds from the influence of heat in extending the volume of liquor contained in that reservoir, and rising into the narrow stem. This instrument, therefore, to a certain extent, blends the indications of the barometer with those of the thermometer, which are essentially different, and can seldom accord.

About the same period, Dr Hooke likewise proposed a double barometer, of a similar construction. He afterwards resumed the subject, and with a view to correct the defect of the former arrangement, he produced, in 1685, an instrument of a more complex form, but very ingeniously conceived. (Plate I. fig. 6.) To the upper end of the open stem he joined a third cylinder of the same dimensions as the two former, but tapering away to a fine orifice at the top. The principal tube being filled as usual with mercury, extending to occupy the bottoms of both the connected cylinders, he introduced a liquor immediately over the mercury in the second cylinder, rising partly into the stem; above this, again, he poured another liquor specifically lighter, and differently coloured, filling up the rest of the stem, and mounting into the third cylinder. By this artificial and delicate combination, the mercury is left perfectly stationary, and all the movements corresponding to the atmospheric pressure are performed by the counterpoising liquors, and marked by their line of mutual separation. Since the stem or narrow tube remains constantly full, the variation of its pressure must depend on the different proportions of its length occupied by the two fluids. If the weight of external atmosphere should, for instance, increase, the denser liquor will rise, and consequently cause the lighter liquor to contract its column. The action of this compound barometer being thus produced merely by the difference of the gravity of the two fluids, might, therefore, be augmented indefinitely. Suppose the liquid resting on the mercury to be pure water, and the superincumbent liquid to be olive oil, which is about one-twelfth part lighter, the scale would be enlarged no less than 163 times, or an alteration of one-tenth in the altitude of the common mercurial column would be marked by a motion through 12 × 1.36 inches, or 16.3 inches. But such a vast enlargement of the scale is far greater than would ever be desirable in practice. It were better, therefore, to introduce next the mercury some fluid which is denser than water. If oil of sassafras were combined with oil of oranges, the divisions of the scale would be augmented only sixty-eight times, and consequently the whole range might not exceed ten or twelve feet. Those oils, however, would move rather sluggishly, especially in cold weather, and might, from their incessant shiftings, during a lengthened period, become insensibly mixed. On the other hand, fluids of distinct characters are seldom free from chemical action; they expand differently with heat, and by coating with other traces the inside of the tube, they are the more apt to retard the motion of the column. In general, the advantage of any very great augmentation of the scale is counterbalanced, as the fluids then work by irregular starts; and the instrument loses in delicacy whatever it has gained in extent of action.

Another method of augmenting the variations of the barometer was invented by the same fertile genius, which has the advantage of uniting great simplicity with tolerable accuracy. (Plate I. fig. 7.) Resuming the syphon barometer, he made a small float of iron or glass to rest on the exterior surface of the mercury, and suspended by a slender thread passed round a small wheel or cylindrical axis that carried an index. Though the varieties of the height of the mercurial column are, in a tube of this form, reduced to half the ordinary measure, yet, from the great length of the index compared with the diameter of its axis, the divisions on the circumference of the circle in which it travels are much amplified. The little machinery being concealed within the frame of the instrument, the index only is brought into view, protected by a circular plate of glass.

Thus fitted up, the whole forms rather a handsome piece of furniture. The *Wheel Barometer*, as it is called, has long maintained its reputation among ordinary observers.

A very simple method of enlarging the divisions of the barometer is commonly ascribed to Sir Samuel Moreland, the same person who had invented, or perhaps only revived, the *Speaking Trumpet.* (Plate I. fig. 8.) It consisted in merely bending the upper part of the tube into a very oblique position. By this plan, however, the scale which depends on the perpendicular altitude, cannot be augmented beyond three or four times without incurring evident risk of inaccuracy. The instrument is called the *Inclined* or *Diagonal Barometer*. The form has been sometimes varied by the fancy of artists, who, repeating the inclination of the tube, have occasionally given the upper part a zig-zag appearance.

The most ingenious barometer, filled with mercury only, and yet admitting a scale of any extent, was invented by Cassini and by John Bernoulli, who first gave the description of it in 1710. (Plate I. fig. 9.) A wide cylinder is annexed to the top of the main tube, at the bottom of which there is joined at the right angles another long and narrow tube. The mercury, in ascending or descending within the wide cylinder, must, therefore, run along this horizontal tube. If that cylinder have a diameter only four times greater than the bore of the tube, the scale of variation will be augmented sixteen times. This instrument is, from its shape, called the *Square Barometer*. It is not found in practice to answer so well as the theory might lead us to suppose. The mercury creeps along the horizontal tube with difficulty, and by desultory advances ; and these irregularities increase as it becomes, from its motion and exposure, covered with dust and partial oxidation.

The simplest of all the barometers, with an enlarged

scale, and at the same time one of the most ingenious, is the *Conical* or *Pendant Barometer*, invented and described in 1695 by Amontons, a French philosopher, who being afflicted with total deafness, in consequence of a fever in his infancy, had devoted himself to mechanical contrivances. (Plate I. fig. 3.) This instrument consists merely of a tube, four feet or more in length, with a bore narrower than ordinary, and tapering regularly to the top. The width at the bottom must hardly exceed three-twentieth parts of an inch, while near the top it may be contracted to about one-tenth. A column of thirty-one inches of mercury being introduced, the tube is gently inverted, and held perpendicular; the cohesion of such a narrow column is sufficient to prevent it from dividing and admitting air unless it be shaken; but, overpowering the atmospheric pressure, it descends till it has contracted into the equiponderant altitude, by passing into a wider part of the tube. To obtain equal divisions on the scale, it is necessary that the tube should have a uniform taper. The most accurate construction of a barometer of this kind is, therefore, attained by joining together two tubes that have even but unequal bores, the longer and narrower one being uppermost. If the width of the upper tube were supposed to be to that of the under one as two to three, the scale would be enlarged three times, since, by descending three inches from the top, and consequently two at the bottom, the column would suffer a contraction of one inch in height.

This species of barometer is thus recommended by its simplicity and its ample range. But the bore of the tube being indispensably narrow, the mercury moves with difficulty, and resists the impression of minute changes of external action. When the conical-shaped tube is retained, the instrument is liable to some inaccuracy from the influence of the cohesion of the mercury, which varies with the diameter of the column in different parts of the tube.

Amontons likewise proposed another form of barometer, in which the mercurial column is subdivided among several short connected branches. (Plate I. fig. 10.) Suppose the instrument were to have only the third part of the usual height, the first, third, and fifth branches, enlarged above and below into very short cylinders, are filled with mercury; and the second, fourth, and sixth branches, which may have their bores narrower, are occupied with some light fluid, or simply with air. If the external pressure should suffer any diminution, the three mercurial columns which produce the counterpoise will each descend and push up the last fluid of the series by their combined effects. It is evident that, by multiplying those branches, the barometer will have its altitude proportionally reduced. But this construction, though specious in theory, is found to have no practical advantages. The instrument is, from its complication, very difficult to construct; its motions are sluggish, owing to the multiplicity of tubes, and the conjunction of fluids, and they are subject to derangements from the variable influence of temperature. It has therefore been generally abandoned.

These different forms of the instrument have been variously modified, and often brought forward with claims of novelty. We may notice, however, the *Sectoral Barometer*, proposed by Magellan, in which the mercury is always made to rise to the same high point of the tube, by drawing this less or more aside from the vertical position. The arc they describe will indicate the deviation from the perpendicular, and consequently the actual descent of the mercury. But the difference between the vertical and the oblique line is not measured by the inclination merely; it is proportioned to the versed sine of this angle, or nearly to the square of the arc. The advantage of this mode of observing is, therefore, best perceived in small variations of the mercurial column. In the hands of a skilful observer, the best and most accurate barometer, after all, is that of the original construction, with a tube rather wide, and a broad cistern. To apply minute divisions, is decidedly preferable to any enlargement of the scale. The measuring of such divisions has been since rendered extremely easy by the adaptation of the differential scale; a most valuable contrivance, proposed by Vernier early in the seventeenth century, but strangely neglected long afterwards. This delicate appendage being once adopted, it became the more desirable to improve the sensibility, and regulate the correctness, of the indications of the barometer.

The first object was carefully to cleanse the mercury, and to expel any portions of air or moisture adhering to the inside of the tube. The influence of aqueous vapour in depressing the mercurial column had been observed by Huygens; but other more evaporable fluids were afterwards found to occasion, by their presence, a still greater derangement. Homberg having, about the year 1705, washed a tube with alcohol to remove the impurities from its internal surface, remarked that the mercury introduced into it stood an inch and a half lower than usual; a depression which this ingenious chemist was disposed to attribute to the elasticity of the spiritous exhalations collected above the mercurial column; though other academicians, and Amontons among the rest, misled by their Cartesian prejudices, sought to ascribe the effect to the different sized pores of the glass. These anomalies were removed by heating, or rather boiling, the mercury in the tube till it was completely purged of air and moisture, and brought into close contact with the inside of the tube. But a new fact occurred which long puzzled the mechanical philosophers. The tube of a barometer, which had been filled with more than usual care, was observed to exhibit a luminous appearance when moved or slightly agitated in the dark. This curious phenomenon gave occasion to multiplied and prolonged controversies; it was attributed to the subtile matter of Descartes, or ascribed to a native phosphorescence, or a latent fire inherent in the mercury. Our countryman Hauksbee, in the year 1708, gave the first rational explanation of the fact, by referring it to electricity, which he had just begun to cultivate as a distinct science. It resembles exactly, indeed, the experiment of the exhausted flask, in which an electrical current flashes with a diffuse lambent flame, like the *aurora borealis*, or the *northern streamers*. The friction of the mercury against the inside of the tube excites electricity, while the vacuity, or rather the very attenuated vapour in which the supposed fluid plays, facilitates its expansion. When the vacuum is rendered very perfect, by the careful and accurate boiling of the mercury, the lambent flashing ceases for want of a fine medium to conduct and disperse the electrical influence.

The next point to which experimenters were led to direct their attention, was the effect of the width of the tube on the altitude of the mercurial column. Plantade, a lawyer of Montpellier, appears to have been one of the first who remarked that the mercury stands always lower in narrow tubes. This fact he communicated about the year 1730 to Cassini, who was then occupied in the south of France with carrying on the great trigonometrical survey. But the discrepancies observed by Plantade, being unfortunately confounded with other collateral circumstances, were for a time overlooked. In tubes having a narrow bore, the depression of the mercury, however, is very considerable, as may be readily perceived in a small glass syphon, of which the one branch is about half an inch in diameter, and that of the other branch less than the tenth of an inch. Thus, if the narrow tube had a width of only the thirteenth part of an inch, the depression of the mercury would amount to half an inch, which is about the third part of the elevation to which water in similar circumstances would be raised by capillary action. This effect has not been sufficiently examined, but it appears to result from the attraction of the particles of the mercury to each other, exceeding their attraction to the surface of the glass. Mercury, in contact with glass, therefore, tends to a spherical form, and always assumes a convex surface within a clean tube. Water and other liquids, again, manifest an opposite character, the mutual attraction of their particles being less than their adhesion to glass. Accordingly, they spread along a vitreous surface, instead of collecting into drops; and in narrow tubes they mount above the level, and invariably have a concave termination. If the bore be so small as to be reckoned capillary, the depression of mercury is, like the elevation of water, inversely as the diameter; but when the bore has a considerable width, the quantity of depression, depending on the curvature of the surface of the mercury, diminishes proportionally faster, and follows nearly the inverse duplicate ratio of the diameter. But on the subject of capillary action we refer our readers, with the utmost confidence, to a valuable paper communicated to the Royal Society of London, by Mr Ivory, late of the Military College at Sandhurst, one of the most original and profound mathematicians that our island has had the honour to produce.

The influence of the predominating attraction of the particles of mercury to themselves, above their adhesion to the sides of a glass tube, has not been considered with so much attention as it demands. Nothing is more common than to remark, that the mercury in the barometer is in the act of rising if it show a convex surface, but about to fall if it should appear concave. Now, the top of the mercurial column must always remain convex if the barometer be properly constructed, the tube perfectly clean, and the mercury purged of all impurities. But if the inside of the tube be anywise soiled, whether covered with humidity, or stained with mercurial oxyd, the metallic fluid will adhere so obstinately to the glass as to lose its convexity, and to subside into a flat surface, or even sink into a concavity, like water and other liquids. Hence the danger of boiling the mercury too long in the tube, as it becomes partially oxydated; and the thin crust so formed not only suspends the column higher, but obstructs the freedom of its motion. The same effect is produced by greasing the inside of the tube. Some respectable authors, from not attending to these facts, have hastily inferred that the convex appearance which mercury assumes in the barometer was merely accidental, and consequently removed by a more complete boiling and purification.

In the case of tubes having wide bores, the depression of the mercurial column may, without any sensible error, be disregarded. According to the accurate experiments made by Lord Charles Cavendish, and published by his son, the celebrated Mr Cavendish, the quantity of depression is only the 200th part of an inch in a tube of 6-10ths of an inch in diameter, the 28th part of an inch in a tube of 3-10ths diameter, and the 15th part of an inch in a tube of 2-10ths diameter. Wide tubes ought, therefore, to be preferred in the construction of barometers, both on account of the facility with which the mercury moves in them, and the smallness of its depression. The only circumstance to overbalance these advantages, would be the necessity and inconvenience of having a very large cistern. A quarter of an inch may be reckoned a good width of tube, and the corresponding depression is only the twentieth part of an inch.

In the syphon barometer, if both branches have the same diameter, the action is exerted on opposite sides, and therefore the effect of depression becomes entirely lost. For accurate purposes, this original form of the instrument has been resumed, and the inconvenience arising from the large variation of the lower level entirely obviated by an ingenious contrivance introduced about fifty years ago. This consists in the application of a leathern bag, instead of a wooden or ivory cistern, to hold the surplus mercury. Besides the barometric tube, there is placed adjacent to it another short one of the same width, communicating with the mercury contained in the bag, which being pressed by turning a screw below, is at each observation brought exactly to the same mark. The external atmosphere readily acts through the substance of the leather; but the mercury, from the powerful cohesion of its own particles, cannot be squeezed through the pores of that casing without violent compression. The addition of a bag within a cylindrical box, omitting the lower tube, likewise renders the barometer easily portable; since, for safe carriage, the mercury can be screwed up tight, to fill the whole cavity of its tube, but, on turning the screw again, the column will subside and rest on a broad base.

The last object which required nice observation, was to estimate the effect of heat in dilating the mercury, and consequently increasing the altitude of the equiponderant column. This correction could not be made with any sort of accuracy previous to the application of the thermometer, which, though invented half a century earlier than the barometer, was yet more than another half-century in arriving at perfection. Hero, a mechanical philosopher, who flourished at Alexandria about 130 years before Christ, has described in his *Spiritalia* a sort of huge weather-glass, in which water was made to rise and fall by the vicissitudes of day and night, or rather the changes of heat and cold. This machine had for ages been overlooked, or merely considered in the light of a curious contrivance. But Sanctorio, the inventor of the famous statical balance, a very learned and ingenious Italian physician, who was long professor of medicine in the university of Padua, and had laboured to improve his art by the application of experimental science, reduced the hydraulic machine of Hero into a more compendious form, and thus constructed, about the close of the sixteenth century, the instrument since known by the name of the *air thermometer*, which he employed with obvious advantage to examine the heat of the human body in fevers. Some years afterwards a similar instrument was contrived, perhaps without any communication, by Drebbel, a very clever and scheming Dutch artist, who visited London in the reign of James I., and introduced the knowledge of that instrument into England.

But this air thermometer was evidently of the same nature with what has been since called the *manometer;* it could measure only the dilatation or augmented elasticity of the air confined within its bulb, whether occasioned by heat or the diminution of external pressure. It was therefore considered merely as a weather-glass, indicating the state of the atmosphere; nor could its blended impressions, which might separately affect both the thermometer and barometer, be then distinguished. Had it been more closely studied, it must have led, by another path, to the discovery of the latter. But those irregularities to which the air thermometer was hence subject, appear to have created such doubts respecting the accuracy of the instrument, as occasioned its being neglected long afterwards.

The same country, however, which had given birth to the thermometer, began its improvement. After the principle of the barometer was established, the members of the Academy del Cimento, founded at Florence in 1657, and supplied with liberal funds by the Grand Duke of Tuscany, had, among other interesting physical researches, resumed the application of the thermometer; and, instead of air, they substituted alcohol or spirit of wine, another very expansible fluid not affected by atmospheric pressure, while they attached to the tube a scale graduated on a regular plan, though directed by no very precise measures. The instrument so constructed, but somewhat varied in its form, being copied by Italian artists, was dispersed over Europe under the name of the *Florence Glass.* From its careless execution, however, in the hands of itinerant venders, this thermometer, or rather thermoscope, appears never to have obtained an established reputation.

The great object was to bring thermometers to an exact correspondence. It was expedient, therefore, not only to select a proper fluid, but to adopt a uniform and consistent scale. Alcohol, linseed oil, and mercury, had been successively tried. The graduation was at first drawn from the temperature of cellars and deep caves, which, indicating the natural heat of the globe, had long been considered invariable ; but more enlarged experience discovered the inaccuracy of that supposition, and showed the mean temperature to be materially modified by the latitude of the place, and its elevation above the level of the sea. Congelation, or rather the inverted process, the thawing of ice, or the melting of snow, was then found to remain fixed ; a most important fact, which had been first noticed by Gürické, but overlooked till a considerable time afterwards. A stationary point was hence obtained, from which to commence the thermometer scale. But different modes were pursued for determining the divisions. Amontons, reverting to the air thermometer in spite of its acknowledged defects, found that the elasticity of air compressed in the bulb, and able at the temperature of melting snow to support a column of mercury fifty-four inches high, was capable of raising this to seventy-eight inches at the heat of boiling water; and he seemed contented in framing a rude standard, by merely dividing the intermediate space into inches and half-inches.

But about the same time, or nearly at the beginning of the eighteenth century, Newton himself cast a keen though rapid glance on the subject of heat, and proposed a thermometer of a much simpler and more elegant construction. Having adopted linseed oil as a fixed and uniform substance, capable of great dilatation, he discovered by experiment, that dividing the capacity of the bulb into ten thousand equal parts, the liquid expanded 256 parts from melting snow to blood-heat, and 725 parts to that of boiling water. These numbers, however, being inconveniently large, he reduced them somewhat more than twenty times, adopting 12 and 34 as the proportional divisions on his scale. But oil, being so viscid a substance, was found to trail and collect on the inside of the tube; and this thermometer, though constructed on a right principle, never came into general use.

Rōemer, the Danish astronomer who made the fine discovery of the progressive motion of light, was the first who proposed mercury as the fittest fluid for thermometers; and Halley and Amontons remarked about the same time, that it expands uniformly with heat, and remains nearly stationary at the point of boiling water. On this principle, Delisle of St Petersburg constructed, in 1733, a mercurial thermometer with a descending scale, the distance from freezing to boiling water occupying 153, or, in round numbers, 150 divisions, of which the bulb itself contains 10,000. A more ingenious method, but perhaps too refined, for graduating thermometers, was proposed by Renaldini, a distinguished Italian mathematician, in 1694. It consisted in adopting the scale, in the successive temperatures produced by mixtures in the different proportions of twelve parts of water at the moment of thawing and of ebullition. This suggestion led to a very important inference, since it proved that mercury expands uniformly with equal additions of heat, while alcohol swells constantly in a rising progression. But the capital improvement of the thermometer was effected by the skill and perseverance of Fahrenheit, whose name has remained justly attached to the instrument. This ingenious person, originally a merchant at Dantzic, who had the misfortune to fail in business, was induced, by his taste for mechanics and chemistry, to have recourse to the manufacture of thermometers, as the means of gaining a slender livelihood. But not meeting with sufficient encouragement at home, he removed, about the year 1720, to Holland, the great emporium of the arts, and fixed his future residence at Amsterdam. He now preferred mercury to alcohol for filling his thermometers; and, adopting the division of the bulb into 10,000 parts, he reckoned sixty-four of them as the expansion between freezing to blood heat, and thirty-two as the contraction from the same point to what he considered as extreme cold, or that produced by the mixture of one part of salt with three parts of snow. These numbers were extremely convenient, being found by a repeated bisection. With respect to the heat of boiling water, Fahrenheit discovered the important fact, that it varies with the state of atmospheric pressure. Taking the mean, however, he reckoned 180 degrees from freezing to ebullition, and therefore marked this point at 212 on his scale. The thermometer owes its improvement to Celsius, professor at Upsal, who in 1742 placed the commencement of the scale at congelation, and divided the interval thence to boiling water into an hundred degrees, extending such a portion downwards as might be wanted. This centesimal thermometer is exactly the same as what the French have since called the *centigrade*, which, from its fitness and simplicity, deserves to be universally adopted.

The thermometer having been thus carried by successive steps to perfection, it was found by delicate experiments, that, between the points of boiling and freezing, the expansion of mercury amounts to the fifty-fourth part of its bulk, or that it acquires, for each degree of heat, an increase of volume amounting to the 5412th part on the centesimal scale, or the 9742d part on the scale of Fahrenheit. A correction, therefore, on the height of the mercurial column in the barometer becomes necessary according to the changes of temperature which it undergoes. In this climate, the extreme variation arising from that cause will seldom exceed two tenths of an inch. But if the barometer be suspended in a room kept at an agreeable temperature, the error occasioned by the expansion of the mercury may, in ordinary cases, be disregarded, since it will scarcely amount to the twentieth part of an inch.

Since the barometer marks the condition of the distant atmosphere, and intimates those internal alterations which are generally connected with the change of the weather, it is particularly valuable at sea, by forewarning the mariner of the approach of a storm. But an instrument of the ordinary construction would not answer this purpose, the agitation of a vessel on a tempestuous ocean being such as will not only throw the ponderous mercurial column into violent oscillation, but communicate those sudden shocks which must infallibly break the tube. Various attempts have accordingly been made to obtain a *marine barometer* exempt from risk, and yet sufficiently sensible to the variations of the atmosphere. The conical or pendant barometer being, from the narrowness of its bore, rather sluggish, was first recommended for that purpose, though never adopted into practice. About the beginning of the eighteenth century, Dr Hooke and Amontons severally proposed to employ for a barometer on board ship, the manometer or air thermometer. To obviate the derangement arising from the influence of heat, there was to be placed beside it a spirit-of-wine thermometer, with a ball so large as to give expansions equal to those of the portion of air confined within the bulb. The difference between the two adjacent columns of liquid would therefore measure the variation of external pressure. But to procure such a nice adaptation would prove so extremely difficult in practice, that most probably this instrument was seldom or never actually constructed. Besides, the liquid column of the manometer, though light and narrow, would yet be much shaken by the rolling and pitching at sea. Notwithstanding these weighty objections, however, this compound manometer was tried in England, mercury being employed as the fluid both of expansion and pressure, and various adjustments applied by means of a complex machinery.

An ingenious and very substantial kind of marine barometer was, about twenty years ago, recommended by Blondeau, one of the professors of the naval academy at Brest. (Plate I. fig. 11.) It consisted of an iron tube, bent below into a syphon, and filled carefully with mercury, which carried a float. For this purpose, a musket-barrel, about three feet long, was chosen, having a very smooth and even bore, and an iron breech closely welded to it, instead of being soldered with brass, which might become corroded by the action of the mercury. The lower end of the tube had a collar of leather, to which was screwed a piece of iron, perforated through its whole length, and bent into an arch, having screwed likewise, with a collar of leather at its other extremity, a vertical cylinder of iron, four inches high, and of the same bore exactly as the tube. The contracted aperture at the end of the tube, not being exactly in the middle, was not always opposite that of the arch; and, therefore, by turning it occasionally aside, the communication could be contracted at pleasure, or even obstructed entirely. The cylindrical appendix was tapered at the top to a narrow orifice, through which an iron wire, attached to a small ivory float, had been introduced. To prepare this instrument for action, the mercury was first boiled in the tube; then the arch, filled with hot mercury, was screwed to the end, the cock opened, and the surplus mercury allowed to flow over; next the vertical piece, with its float, was screwed on, and a little mercury added, to give it due play. The origin of the scale was to be determined from the comparison with another good barometer of the ordinary construction; but, owing to the equality of the bores of the opposite tubes, the divisions were only half the usual size, or the inches were exhibited by half-inches.

This species of barometer is certainly free from all sort of risk, while the facility which, by means of turning the arch, it affords in checking the ascent and descent of the mercury, prevents in a great measure the oscillations of that fluid. If the instrument were properly suspended, therefore, its indications would be tolerably steady and regular. The chief objection to it consists in the diminutive range of its scale.

In every marine barometer, the main object is to give steadiness to the mercurial column, by retarding its motion in the tube; in short, to imitate the equalizing effect of the fly in mechanics. One form of construction was, instead of the cistern below, to annex a spiral tube composed of a number of horizontal convolutions. Passement, an ingenious Parisian artist, about the year 1758, improved on this idea. He twisted the barometer tube near the middle, at least twice round, and joined to its upper end a wide cylinder. But, more effectually to prevent all irregular oscillations, he took a tube with a very narrow or capillary bore, and contracted it below, while he annexed a wide cylindrical piece at its other extremity. The only thing wanted now to the perfection of this instrument, was to devise a mode of suspending it that should soften the jerks, and allow it generally to maintain a vertical position. Our English artists have, by repeated trials, at last succeeded in surmounting all the difficulties. The marine barometer, manufactured by Mr Cary of London, (Plate I. Fig. 12.), is one of the most approved kind. It consists of a capillary tube, about twenty-seven inches long, with a bore scarcely exceeding the thirtieth part of an inch, but terminated by a cylinder four or five inches high, and nearly three tenths of an inch in diameter. It is suspended by a spring and jimbols, near the top, at a certain point, which in each case is discovered by actual trial. By making the suspension lower, it is found that the agitation of the barometer will cause the mercury to rise a little ; while, by placing the suspension higher, the mercurial column suffers always some depression. The reason of this curious observation is not well explained. It probably results from the different centrifugal tendencies communicated to the opposite portions of the columns. The swinging of the instrument would evidently augment the pressure of the upper portion of the column, while it diminished that of the under portion. But this lower portion being longer than the other, its tendency to descend would be proportionally so much greater. About the point of suspension, however, the opposite effects of the centrifugal tendencies are balanced, since the superior force being employed to set in motion a narrower column, the reflux and efflux of the mercury in the upper cylinder must be preserved nearly equal.

Marine barometers, thus improved, are now very generally used, and with great benefit to the public service, on board ships of war and Indiamen. To facilitate the keeping of a register of barometrical observations, the meritorious and indefatigable Mr Horsburgh, hydrographer to the East India Company, published a set of engraved ruled sheets, adapted for the convenience of navigators. In these plates the height of the mercury, from twenty-seven to thirty-one inches, is represented in inches and tenth parts by horizontal lines, while each successive day has a space apportioned to it by vertical bars. The state of the barometer at every observation is marked with a dot, and these dots being afterwards connected together, exhibit an irregular waved line, stretching across the sheet, and indicating the series of the changes of the weather. At the lowest points, from which the curve again returns, a gale generally follows. From the observations made off the Cape of Good Hope during the month of May 1815, by an ingenious and active officer, Captain Basil Hall, then of his Majesty's sloop Victor, it appears, that whenever the mercury fell to 29.60 inches, a storm ensued; the column always rose when the gale abated, and when it reached near thirty inches, the weather became fair. Those gales often came on suddenly, without any visible change in the aspect of the sky, but the marine barometer never failed to give warning of their approach.

A very convenient substitute for the marine barometer, is the *sympesometer* constructed by Mr Adie, optician in Edinburgh. It is merely an improved *monometer*, partly filled with a fixed oil, and indicating the conjoined effect of heat and atmospheric pressure on the inclosed air; but the influence of temperature is corrected by means of a sliding scale, regulated by an attached mercurial thermometer. The sympesometer occupies little room, is easily rectified, and comparatively cheap.

To explain the cause of the variations of the barometer, has long perplexed philosophers. Many hypotheses have at different times been advanced on the subject; but it would be a mere waste of time to make any detailed recital of such crude and unsatisfactory attempts. The various and often imaginary effects of vapours, of heat, and of winds, have been employed in framing an explication of the changes of the atmosphere. The fact, that the mercurial column generally falls before rain, seemed at complete variance with the intimation of the senses, a notion having become universally prevalent, that the air is heavier when the sky appears lowering and overcast; another proof, were it wanted, how fallacious are all current opinions in matters of science.

Leibnitz, fancying he had discovered a new principle in hydrostatics, endeavoured, by a sort of metaphysical argument, to demonstrate that, though a body adds its own weight to the pressure of a fluid in which it is suspended, yet it will cease to be ponderous in the act of falling. This alleged principle will not, in the actual state of science, be thought to require any serious refutation; nor, were it even admitted, would it be found at all adequate to the explication of the phenomenon, since the weight of moisture precipitated from the whole body of atmosphere could never, by the absence of its pressure, occasion a diminution of the tenth part of an inch in the altitude of the mercurial column.

Dr Halley and Mairan sought to account for the depression of the barometer before a storm, by the withdrawing of the vertical pressure of the atmosphere, when borne swiftly along the surface of the globe by a horizontal motion. This hypothesis at first sight appears very plausible, and might seem further confirmed by a noted experiment which most authors have admitted without due examination. Hauksbee, a skilful and ingenious experimental philosopher, about the year 1704, placed two barometers, about three feet asunder, with their naked cisterns in two close square wooden boxes, connected by a horizontal brass pipe; one of these boxes had, inserted at right angles, an open pipe on the one side, and a second pipe, terminating in a screw, on the other side; to this end he adapted a strong globular receiver of about a foot in diameter, which had been charged, by injection from a syringe, with three or four atmospheres; then suddenly opening the stop-cock, and giving vent for the escape of the air through the box and over the surface of the included cistern, the mercury sunk equally in both the barometers more than two inches.

This elegant experiment might be deemed entirely conclusive, if a minute circumstance, on which the success really depends, had not unfortunately been overlooked. It will be perceived from the inspection of the figure which Hauksbee has given, that the exit pipe of the box was considerably wider than the pipe which conveyed into it the stream of air. This fluid, escaping from compression, would therefore be carried by its elasticity as much beyond the state of equilibrium; while the width of the orifice, by facilitating its emission, would allow the portion occupying the box and the connected reservoir to preserve its acquired expansion. If the pipe of discharge from the box had been much narrower than the other, an opposite effect must have taken place; for the air accumulated over the cistern, not finding a ready vent, would remain in a state of condensation. This curious fact is another of the many instances which might be cited, to show the great delicacy and circumspection required in performing philosophical experiments.

The same results, however, can be exhibited by a very simple apparatus. Let a small box, or rather a glass ball, have a short narrow tube inserted in the one side, and another wide tube opposite to this, with a cross slider of brass, for contracting the orifice at pleasure; and, to the under part of the ball, join a long perpendicular tube, bent back like a syphon to more than half its height, and containing a double column of water. Now, blow through the narrow tube into the cavity of the ball, while the orifice of emission is quite opened, and the liquid will rise several inches in the long stem; but, still continuing the blast, let the orifice be gradually contracted, and the column will first descend to its ordinary level, and then sink considerably below it.

The fall and rise of the mercury in the barometer must evidently be occasioned by some corresponding reduction or accumulation of the atmosphere at the place of observation. Whatever augments the elasticity of the air will cause part of the incumbent fluid to evade, and leave for the time a diminished vertical pressure. The efflux of wind might also produce a temporary reduction of the atmospheric column. But the real difficulty consists in explaining why the variations of the barometer should be greater in the high latitudes than between the tropics, and why they so much exceed in all cases the quantities which calculation might assign.

The influence of heat will account for the semi-diurnal variations of the barometer which are observed, especially within the torrid regions. From ten o'clock in the morning till four in the afternoon, the mercury generally falls; but, after that hour, it rises again, till ten o'clock at night, when it drops till four in the morning, and then mounts till ten o'clock in the forenoon. These regular changes, which amount to about the five hundredth part of the whole atmospheric pressure, depend on the prevalence of the alternating land and sea breezes, occasioned by the diversified action of the sun's rays upon the earth and water. The accumulation of air is greatest at four o'clock in the morning and evening, and the mercury then attains its highest point;

but it sinks lowest at ten o'clock in the morning and evening, when the incumbent mass has been the most reduced.

A similar reason will explain the effects of the northerly and easterly winds, in elevating the mercury of the barometer in our climate. A chill air, with enfeebled elasticity, is thus accumulated, and exerts a predominant pressure.

The augmented elasticity communicated to the air by the action of heat or the presence of humidity, and the reduction of the incumbent mass by the efflux of winds, have each their distinct influence in distributing the equilibrium of the atmospheric ocean. But the effects, particularly in the high latitudes, much surpass the regular operation of those causes. The only mode, perhaps, of removing the difficulty, is to take into consideration the comparative slowness with which any force is propagated through the vast body of atmosphere. An inequality may continue to accumulate in one spot, before the counterbalancing influence of the distant portions of the aerial fluid can arrive to modify the result. In the higher latitudes, the narrow circle of air may be considered as, in some measure, insulated from the expanded ocean of atmosphere, and hence, perhaps, the variations of the barometer are concentrated there, and swelled beyond the due proportion. We content ourselves with throwing out this hint at present, but hope to be able to resume and discuss the subject at some length under the article CLIMATE.

# BAROMETRICAL MEASUREMENTS.

It was remarked in the preceding article, that the decisive experiment by which Pascal established the reality of atmospheric pressure, had likewise suggested to this ingenious philosopher the method of determining the elevations of distant points on the surface of the globe. But the first attempts were very rude, proceeding on the inaccurate supposition that the lower mass of air is a fluid of uniform density. Different authors estimated variously from eighty to ninety feet as the altitude, which corresponds to a variation of the tenth part of an inch in the mercurial column. The Torricellian tube or cane, as it was then called, was, on its first introduction to England, carried accordingly to the tops of mountains, or conveyed to the bottom of pits and mines, or even let down to great depths in the sea.

Among those experimentalists who laboured most assiduously in the study and application of the barometer in this part of the island, we may mention George Sinclair. This ingenious person had been professor of philosophy in the university of Glasgow, but seems to have conscientiously resigned his office soon after the Restoration, rather than comply with that hated episcopacy which the minions of Charles II. had forced upon the people of Scotland. He then retired to the village of Tranent, not far from Edinburgh, and was employed as a practical engineer, in tracing the levels of the coal-pits, in directing the machinery employed in the mines at Leadhills, and afterwards in the great undertaking of conducting water from the heights of the Pentlands to supply the northern metropolis. Though not a profound mathematician, he was skilled in mechanics and hydrostatics, and possessed no small share of invention. Sinclair is said to have been the first who applied to the mercurial tube the name of *baroscope*, or *indicator of weight*; the more definite appellation of *barometer* or *measurer of weight*, not having been appropriated till many years afterwards. During his excursions in 1668 and 1670, he employed that instrument to measure the heights of Arthur's Seat, Leadhills, and Tinto, above the adjacent plains. He followed the original mode of using a tube sealed at the top, with a paper scale pasted against the side, which he carried to the top of the mountain, where he filled it with mercury; and, inverting it in a basin, he noted the altitude of the suspended column, and repeated the same experiment below; a very rude method, certainly, but no better was practised in England during the succeeding thirty years.

In a small scarce tract, printed in 1688, and bearing the quaint title of *Proteus bound with Chains*, Sinclair gives some judicious remarks on the variations of the barometer, considered as a weather-glass, and delivers very sound opinions, on the whole, respecting the causes of the chief meteorological phenomena. In a postscript to that piece, he proposes a most efficient and ingenious method of weighing up wrecks from the bottom of the sea. It consisted in employing two large arks, or square wooden boxes, fastened to the sides of the ship, and charged with air carried down to them by a succession of inverted casks, open at the lower end. An ark of a cubical shape, and twenty feet in every dimension, the smallest which he mentions, would, as he computes, have a buoyancy equivalent to 448,808 pounds Troy. It is remarkable that the celebrated Mr Watt always employed this very mode, using a large gazometer, floating in a pond dug in the court of his manufactory, and charged gradually by the action of bellows, for raising the ponderous engines constructed at Soho, and lifting them over his walls into the boats, which were stationed to receive them in the adjacent canal.*

---

* Sinclair was author of a well-known little book entitled *Satan's Invisible World Displayed*, which, at a former period, was sold at all the public fairs in the country, and devoured with eagerness and dismay by the Scottish peasantry. In a quarto volume, on *Hydrostatics, and the Working of Coal-mines*, printed in Holland, and published by subscription in 1672, he digressed so widely from his subject as to insert *A True Relation of the Witches of Glenluce*. But this was the folly of the age, from which several of the most learned men had not been able to escape. It is painful to observe, that James Gregory, the inventor of the reflecting telescope, who, although endowed with talents of the highest order, appears to have had a keen temper, and to have imbibed an hereditary attachment to royalism and episcopacy, should have stopped to attack an unoffending and less fortunate rival. He wrote a little tract against Sinclair's *Hydrostatics*, with the title of the *Art of Weighing Vanity*, and under the thin disguise of Patrick Mather, arch-beadle to the university of St Andrew's. It is a piece full of low scurrility, and memorable only for a very short Latin paper appended to it, containing the series first given to represent the motion of a pendulum in a circular arc. In the British Museum, there is a letter of Gregory to Collins, the Secretary of the Royal Society, boasting of his project, and soliciting information with which to overwhelm the poor author. But with all his eagerness to hunt down Sinclair, he never touches on the strange episode of the Witches of Glenluce. What a picture of times approaching so near our own!—Sinclair was restored to his chair at the Reformation, and lived a few years longer. He answered Gregory's attack in the same coarse style, charging him with total

In all the computations hitherto made from different altitudes of the barometer, the air was considered as a uniform fluid; no regard being had to the gradual diminution of density which must evidently take place in ascending the atmosphere. To estimate the effect of that gradation, it became requisite previously to determine the actual relation subsisting between the density of the fluid and its elasticity. This was first ascertained in England by Towley, who inferred, from some experiments of Boyle, that the elastic force which the air exerts is exactly proportional to its density. A similar conclusion was about the same time drawn by Mariotte, a French philosopher, from a still better series of experiments. Following out this very simple law, he thought of computing heights from barometrical observations, by the rules usually employed in constructing tables of logarithms; and had, therefore, obtained some glimpse, no doubt by a sort of conjectural process, of the remarkable result, that the density of the atmosphere decreases in a geometrical progression, corresponding to the elevations taken after an arithmetical one. But seemingly not aware of the importance of the principle at which he was pointing, Mariotte immediately deserted it; and calculating from a repeated bisection of the column of air between the two stations into successive horizontal strata, he contented himself with interpolating the densities according to a harmonic division, which he next abandoned for the simplicity of a series with equal differences. This able experimenter hence only sketched out a mode of investigating the problem of barometrical measurements, without arriving at any very definite or consistent rule of solution.

---

want of skill in the use of astronomical instruments, though by help of subscriptions he had erected a sort of observatory at St Andrew's. The manuscript of this reply is preserved in the College Library of Glasgow.

In 1686, the ingenious and active philosopher Dr Halley resumed the subject, and discovered the law that connects the elevation of the atmosphere with its density; of which he gave a clear demonstration, derived from the well-known properties of the hyperbole referred to its asymptotes.* Since the height of the mercury indicates the pressure, and consequently the elasticity of the external air, it must be proportioned likewise to the density. Wherefore the breadth of a given mass of air, or the thickness of a stratum which corresponds to a certain portion of the mercurial column, will be inversely as this altitude. Let O be the centre of a rectangular hyperbole, of which OA and OP are the asymptotes; and conceive the distances OA and OB to represent the

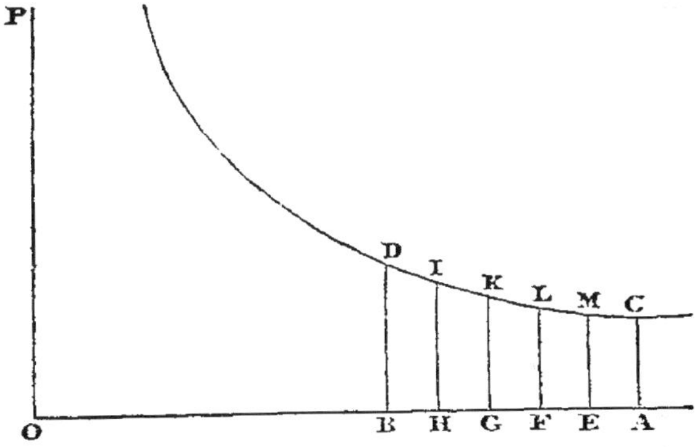


heights of the mercury at two stations. The perpendiculars AC and BD, which are reciprocally as OA and OB, must hence express the relative thickness of strata corresponding

* It appears, however, that he was anticipated by Huygens, who had found, as early as the year 1662, that the altitudes of successive strata of air being taken in arithmetical progression, the corresponding densities will form a geometrical series. This beautiful proposition he never published, indeed, but left it recorded upon one of the blank leaves of a Dutch almanac, containing various other speculations, and now preserved in the library of the university of Leyden.

to equal portions of the barometric scale. Divide AB into a multitude of equal segments, and erect the perpendiculars EM, FL, GK, and HI. The intercluded spaces, from AC to BD, will denote the successive thickness of the series of strata into which the whole mass of air between the two stations is subdivided. Consequently the aggregate or mixtilineal space DBAC, which is proportional to the logarithm of the ratio of OB to OA, will express the difference of atmospheric elevation when the mercurial column mounts from B to A. Taking equal ascents, therefore, in the atmosphere, the corresponding densities must form a decreasing geometrical series.

To apply this elegant theorem, Dr Halley availed himself of the best experiments which had been performed to determine the relative densities of air, water, and mercury. In different trials made near the earth's surface, it was found, when the barometer stood at $29\frac{3}{4}$ inches, that the air is 840, 852, or even 860 times lighter than water. Employing round numbers, therefore, and assuming the specific gravity of mercury to be $13\frac{1}{2}$, he reckoned $800 \times 13\frac{1}{2} \times 30 =$ 10,800 inches, or 90 feet, as the altitude of an atmospheric column which, near the surface, would exert a pressure equivalent to that of an inch of mercury. For the coefficient, which answers to the actual constitution of the atmosphere, Halley should have taken the thirtieth part of .4342945, the modulus of the common system of logarithms, or .0144748. But he proceeded less directly, having satisfied himself with taking the arithmetical mean between the differences of the logarithms of 29 and 30, and of those of 30 and 31; a compensation of errors which gives .0144765, hardly deviating from the former. Hence he gave this simple analogy for computing the heights of mountains by the barometer: *As the constant number* .0144765 *is to the difference between the logarithms of the barometric columns at the two stations, so is 900 feet to the elevation required.* The result of this operation is evidently the same as if the logarithmic difference had been multiplied by the number 62170; a very tolerable approximation at all seasons for a northern climate, and quite accurate, indeed, if the mass of intervening air had a medium temperature of 46° by Fahrenheit's scale. Dr Halley supposed that the observations themselves might, from the influence of heat, differ about a fifteenth part between summer and winter. But the thermometer was still so imperfect an instrument, it that could not be applied with confidence in correcting such variations.

The principle which Halley thus investigated might be derived from a simpler process. Conceive the atmosphere to be divided into a multitude of equally thin horizontal layers, it is evident that each successive stratum would, to the pressure of the superincumbent stratum, add its own weight, which being as its density or elasticity, is therefore proportioned to the collective pressure; and, consequently, those densities must continually increase in going downwards, exactly in the same way, and after a like progression, as money accumulates at compound interest, where a constant portion of the aggregate fund is regularly joined to the capital.* Such, in fact, is the distinguishing character of a geometrical progression, that the increase or decrease of each

* To illustrate still further this comparison; if we reckon the mean altitude of a uniform atmosphere to be 26000 feet, it will follow that, for every descent of 26 feet, the density of the air will increase one thousandth part. This rate of progression corresponds precisely with the weekly improvement of money invested at 5 per cent. per annum, supposing for the sake of simplicity, the year to contain only 50 weeks. Since money at the statute interest doubles in $14\frac{1}{5}$ years, so the air has its dilatation doubled at the height of $26 \times 50 \times 14.2$ or 18466 feet, answering to three miles and a half.

succeeding term is always proportional to the term itself. The logarithmic curve is hence the best adapted for exhibiting the relations which connect the densities with the elevations in the atmosphere; the axis of the curve expressing the elevation, while each ordinate represents the corresponding density of the stratum of air. It being a fundamental property of the logarithmic curve, that every subtangent applied to it has the same length, the exact determination of this in the case of our atmosphere is the only thing wanted for the final solution of the general problem.

Let AB, the absciss of a logarithmic curve, represent a line descending perpendicular through the atmosphere; the ordinates AC and BD will indicate the elasticity of the strata at A and B, and consequently, abating the influence of temperature, their densities. Draw the tangent DT, and form the elemental triangle $Dd\delta$; then BD : BT : : $\delta d$ : $\delta$D, or inversely BD : $\delta d$ : : BT : $\delta$D. But the thickness $\delta$D of each accrescent stratum being assumed to be the same, and the subtangent being constant, from the nature of the curve; the ratio of BT to $\delta$D is given; therefore the ratio of BD to $\delta d$ is likewise given, or the increment of the density in the successive strata is proportional to the density itself, the leading property of air. The pressure of the thin stratum at B is represented by BD$db$, which is equal to the rectangle under BT and $\delta d$. Hence, drawing the parallel DE, the collective area ABDC, which represents the weight of the atmosphere between the altitudes A and B, is equal to the rectangle under the subtangent and EC. But AB expresses the difference of the logarithms of the ordinates AC and BD, or the densities at A and B; and consequently this differ-

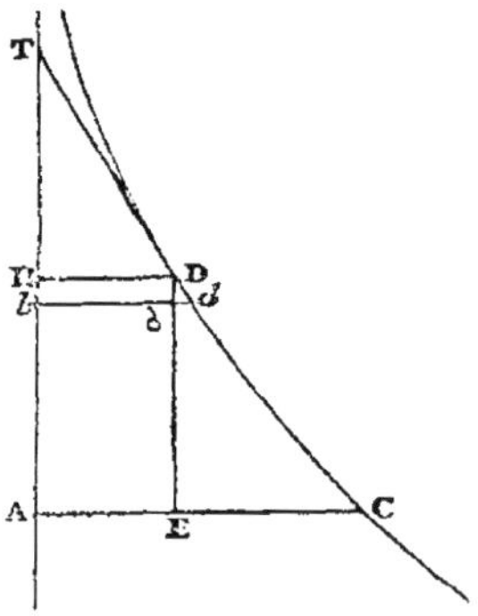

ence multiplied by the length of BT will indicate AB. In relation to our atmosphere at the point of congelation, the numerical value of the subtangent is about 26,000 feet, which is very nearly the product of 60,000 into the *modulus* of the ordinary logarithms.

Eleven years after Dr Halley had given his rule for barometrical measurements, this philosopher had an opportunity of applying it to discover the height of Snowdon in North Wales. He found that the barometer which stood at 29.9 inches on the sea-shore near Caernarvon, fell a few hours after, when planted on the summit of the mountain, to 26.1 inches, the altitude having been ascertained previously, by a trigonometrical observation, to be 1240 yards.

The year 1687 is memorable as being the date of the first publication of the *Principia*, which was drawn up chiefly at the urgent request of Halley, from disjointed materials that had lain a considerable time in the author's closet. In that immortal work, Newton resumed the problem of the gradation of atmospheric density, and solved it in that general way which suited his penetrating genius. He demonstrated that, supposing the particles of air, like other bodies, to have their weight or gravitating tendency diminished as the squares of their distances from the centre of the earth, if those distances be taken in harmonic progression, the corresponding densities of the atmosphere will form a geometrical one. But since the diminution of attraction at the greatest height we are able to reach amounts only to the two thousandth part of the whole, this difference is too minute to be adopted in practice; and the simpler law first established by the sagacity of Halley may be deemed sufficiently accurate for every real purpose.

Newton has given a sort of geometrical solution of the problem. But a more precise, and, in this case, a clearer investigation, is obtained by help of the symbols of the integral calculus. Let $x$ and $x'$ express the altitudes of two strata of atmosphere, $y$ and $y'$ the corresponding densities, $r$ the radius of the earth; and suppose further, that $e$ represents the altitude of the equiponderant column which measures the elasticity of the air. Since the density of the air depends on the incumbent pressure, its decrement must evidently be proportional to the weight of each superadded minute stratum, or to the density of this stratum multiplied into its thickness and power of gravitation. Whence $-edy = ydx\left(\frac{r}{r+x}\right)^2$, or $-\frac{edy}{y} = \frac{r^2dx}{(r+x)^2}$, of which the complete integral is $e$ Hyp. Log. $\frac{y'}{y} = \frac{r^2}{r+x} - \frac{r^2}{r+x} = \frac{r^2}{r+x} \cdot \frac{x'-x}{r+x'}$. If $r$ be regarded as indefinitely great in comparison of $x$, the expression will pass into $e$ Hyp. Log. $\frac{y'}{y} = x'-x$, which is only the common formula.

Little seemed wanting, therefore, to complete the practice of barometrical measurements, but the application of the thermometer, to correct the results. This instrument, however, advanced slowly to perfection, and more than forty years elapsed before it came into current use. Some of the continental philosophers likewise, biassed, perhaps, by a secret jealousy of the superiority which England had acquired in science, began to throw out doubts respecting the reality or accuracy of the law of geometrical progression in the atmosphere. Daniel Bernoulli, a man of candour on the whole as well as ingenuity, but who, with some proneness to speculative reasoning, had imbibed unfortunately many of the prejudices of the Cartesian and Leibnitzian schools, proposed in his capital work, the *Hydrodynamica*, which came out in 1736, certain vague hypotheses regarding the constitution of the atmosphere, as deduced from certain internal motions attributed to its component strata. The specious results of those calculations led him hastily to deviate from the principle of the geometrical progression of density in the upper regions. In this departure he was followed by Cassini and Horrebow, who concluded from some partial observations they had made, that the barometer, in its indications of atmospheric pressure, is subject to irregularity; and that, near the surface of the earth, it obeys a different law from that which obtains at great elevations. A strong light, however, was thrown upon the subject in 1753 by Bouguer, an able mathematician, and a very skilful and ingenious observer, who, with other academicians, had been employed for several years in measuring a degree of the meridian along the stupendous ridge of the Andes. From the comparison of more than thirty distinct observations, he deduced a simple and elegant rule for computing heights by means of the barometer. It is this, that the difference between the logarithms of the mercurial columns at the two stations being diminished by one thirtieth part, and the decimal point shifted four places to the right, will express the required elevation in *toises*. Since the English was to the French foot nearly as fifteen to sixteen, the rule would be accommodated to our measures, and the result expressed in feet, if the logarithmic difference were augmented by the thirtieth part, then multiplied by six, and the decimal point thrown back four places; or, which is the same thing, if that logarithmic difference were multiplied at once by 62,000. But Bouguer imagined, that this rule would not hold exactly in Europe, or in the lower regions of the torrid zone; and to explain the deviation, he had recourse to the forced supposition that the particles of air possess different degrees of elasticity. Lambert, a philosopher of great originality and penetration, afterwards published some excellent remarks on the comparison of barometrical measurements. But no material progress was made till 1755, when M. de Luc of Geneva resumed the subject, and carefully combined experiment with observation. For the space of fifteen years and upwards, he prosecuted his inquiries with diligence and perseverance, aided by the peculiar advantages of local situation, in a city abounding with skilful artists, and seated in the neighbourhood of lofty mountains. The discrepancies which had hitherto created so much embarrassment proceeded mostly from the inattention of observers to the disturbing influence of heat, and particularly its effect in expanding the air, and consequently augmenting the elevation due to a given difference of atmospheric pressure. De Luc's first object was to improve the thermometer of Reaumur, which, though greatly inferior to that of Fahrenheit, had been adopted in France and the adjacent parts of the Continent. Having ascertained that mercury has the valuable property of expanding equably with equal additions of heat, he substituted that metallic fluid for spirit of wine, but retained its arbitrary and inconvenient scale of eighty degrees between the points of freezing and boiling water. He next examined the dilatation of air at different temperatures, and corrected those results by numerous observations made on the mountains of Savoy, and the mines of the Hartz, in which the barometer was combined with the thermometer. The *formula* which he thence deduced for the computation of barometrical measurements was, in 1772, published in his *Recherches sur les Modifications de l'Atmosphère,* and seemed to attract, especially in England, a very considerable degree of notice. Dr Maskelyne, the astronomer-royal, adapted it to our system of measures, and De Horsley made annotations and comments on it. But, what was of more importance, other accurate observers, incited by De Luc's example, entered the same field of inquiry, provided with instruments of greater delicacy and much better construction. In 1775

Sir George Shuckburgh Evelyn visited the Alps, and combined trigonometrical operations with corresponding observations by barometers and thermometers from the hands of Ramsden; and about this time likewise, General Roy not only measured, with instruments made by that excellent artist, some of the principal mountains in Scotland and Wales, but instituted a series of manometrical experiments. It resulted from all these researches, that, for each degree on Fahrenheit's scale, mercury expands the 9700th part, and air the 435th part of their respective bulks. It further appeared that the atmosphere has its temperature almost uniformly diminished at equal ascents; and that the logarithmic difference, reckoning as integers the first four digits, expresses in English fathoms the height of an aerial column as cold as the point of congelation. General Roy proposed likewise another correction depending on the enfeebled gravity, and consequently the augmented altitude, of the equiponderant column of atmosphere in the lower latitudes, occasioned by the influence of centrifugal force arising from the earth's rotation. Several years afterwards, Professor Playfair, in a learned paper, printed in the first volume of the *Transactions of the Royal Society of Edinburgh*, examined all the circumstances which can affect barometrical measurements, and discussed each question with the correctness and perspicuity that we might expect from his distinguished abilities. At nearly an equal interval of time the celebrated Laplace resumed the subject in his *Mécanique Céleste*, and brought all the conditions together in a very complicated formula. Such an appearance of extreme accuracy, however, is perhaps to be regarded merely as a theoretical illusion, unsuited and inapplicable to any real state of practice. Biot has since attempted to arrive at a similar conclusion, by setting out *à priori* from some careful experiments on the relative density of air and mercury, performed by him in conjunction with Arago. He thence infers that, in the latitude of Paris, and at the point of congelation, air, under a mercurial pressure of 76 metres, or 29,992 English inches, is 10,463 times lighter than mercury at the temperature of water at its lowest contraction. This would give 26,090 feet for the height of a column of homogeneous fluid, whose pressure is equivalent to the elasticity of the atmosphere. The co-efficient adapted to common logarithms, and adjusted to the force of attraction at the level of the sea, would therefore be 60,148 feet, or 18,334 metres; scarcely differing sensibly from the quantity which Ramond had deduced from a very numerous set of experiments made by him on the Pyrenees. But Biot prefers, as the co-efficient, the number 18,393, answering for an elevation of 1200 metres, or about 4000 feet above the sea, which is not far from the general level of such observations. The formula is hence, in English feet, 60,346 $(1 + .002837 \cos. 2\psi)$

$$\left(1 + \frac{2\,(T + t)}{1000}\right) \log. \frac{H}{h};$$

where $\psi$ denotes the latitude of the place, T and $t$ the temperatures of the air at the two stations, as indicated by the centesimal thermometer, and H and $h$ the heights of mercurial columns corrected for the effects of heat.

This active writer has likewise given tables for expediting the calculation of barometrical measurements; in which he was anticipated, however, by Oltmans of Berlin, who published, in 1809, large *Hypsometrical Tables*, as they are called, accommodated to the complex formula of Laplace. Such tables might no doubt prove useful where very frequent computations are wanted, as in the case of the reduction of the numerous observations brought home by Baron Humboldt, for which, indeed, they were first designed. But still they contain a needless profusion of figures, and hold forth a show of extreme accuracy which the nature of the observations themselves can never justify. The mere calculation of barometrical measurements is a secondary object. The great difficulty is to procure good observations, and to combine tolerable accuracy with expedition. For this purpose, a very portable barometer is still wanted;—an instrument light and commodious, exempt from injury or derangement, and yet sensible to minute changes of atmospheric pressure. These properties, indeed, are seldom conjoined, and one advantage must generally be sacrificed to obtain another. A modification of the conical barometer already suggested would seem, on the whole, to be well adapted for the geological traveller. The two unequal tubes are cemented into a small cylinder of steel, with a fine stop-cock of the same metal confining the mercury. This renders the instrument so portable that it may be inclosed in a walking-stick; on turning the steel-cock, and inverting the staff, the height of the mercurial column is readily found.

A barometer of the most improved construction is represented in Plate II. fig. 6.; a portion of the tube is shown in fig. 7.; and a section of its cistern in fig. 8. By help of a screw pressing against the bottom of a leathern bag, inclosed within a cylindrical ivory box, the mercury is always brought up through a tabular aperture to the same precise level; or till its convex surface appears to touch a very thin line of light, which is admitted through a slip of ivory applied against the side of the chink or separation of this tube from a wider one immediately over it. The lower end of the mercurial column being thus adjusted, the length is easily measured by drawing gradually down a hollow brass tube, divided at intervals by wide slits, covered on one side by thin bits of ivory, till by that softened light a contact is observed with the edge of a slit and the convex top of the column. The fine Vernier which the moveable tube carries gives the altitude of the mercury in thousandth parts of an inch. A thermometer is likewise constantly attached to the instrument, for the purpose of indicating the temperature of the mercury, which, from the heat of the hand in carrying, or the influence of the solar beams, is commonly warmer than the external air.

This mountain barometer is suspended for observation by jimbols from a tripod, as exhibited in Plate II. fig. 5.; but its several parts can be folded up together into a convenient compass, tolerably well fitted for carriage, as represented in fig. 6. The whole apparatus may not exceed the weight of ten pounds; yet even this, moderate as it might seem, would be felt a serious encumbrance by a traveller who is engaged, day after day, in the labour of climbing mountains. The risk which the instrument incurs, besides, in transporting it perhaps over rough precipices, imposes a perpetual constraint, while to make correct observations with it must always require time and patient attention. A lighter and more compact, though less accurate, barometer will generally be preferred by the geological traveller, whose object is rather to extend our acquaintance with the altitudes of mountains, than to aim at a superfluous and often illusory precision. The portable instrument, invented by Sir Henry Englefield, and represented in Plate II. fig. 1. will on the whole, answer those views. Its cistern is formed of box-wood, sufficiently tight to hold the mercury, without preventing the access and impression of the external air. When this barometer is inverted, the mercury, therefore, subsides very slowly in the tube, which must be firmly suspended in a vertical position. For greater security, the mercury is now put into a leathern bag introduced within the cistern.

A very simple and convenient sort of portable barometer was lately invented in France by that celebrated chemical philosopher M. Gay-Lussac. (Plate II. fig. 2. and 3.) It consists of rather a wide syphon tube, filled with mercury, and sealed hermetically at the inverted end, having a very fine capillary hole formed about an inch under this, by nicely directing the flame of a blow-pipe against the side of the glass, and drawing a melted spot of it out to a point. The lower portion of the principal branch has its bore contracted to less than the tenth part of an inch, to prevent the mercurial column from dividing in the act of inverting it. The mercury is boiled as usual, and the tube may be concealed in a walking-stick, or lodged, like the complete mountain barometer, in a cylinder of brass, with movable sliders bearing the divisions of a Vernier at both ends. (Plate II. fig. 4.) For greater simplicity, however, the larger divisions might be engraved on the tube itself. This kind of barometer is of ready use, and very little exposed to hazard in carriage. It is commonly held in a reclined or inverted position; but, in making an observation, it must be gently turned back, and kept perpendicular till the mercury descends through the contracted bore, and slowly rises again in the opposite short branch; the scale is noticed at both ends of the incurved column, and the difference of those indications gives its correct altitude.

A modification of the conical barometer, which, in travelling, we have ourselves employed with great ease and advantage, should likewise be mentioned. The principal part of it consists of a small stop-cock made of steel, and represented in Plate I. fig. 13. A glass tube of 31 or 32 inches long, with a bore of the tenth part of an inch hermetically sealed at the top, and filled with quicksilver, is cemented into the one end of the stop-cock; and into the other end is cemented an open and wider tube, 16 inches or more in length, and having its diameter above the eighth part of an inch. This compound tube is lodged in a walking-stick, divided into inches and tenths through its whole extent, or only at the upper part, if uniform tubes be selected. In making an observation, the cock is turned, and the instrument inverted. The upper column then descends partly into the lower tube, till it becomes shortened to the proper altitude.

We have already stated the principles on which the calculation of barometrical measurements proceeds. But there still are some points, either assumed or overlooked, which may considerably modify the results. It is presumed, that, at equal successive heights, the temperature of the atmosphere decreases uniformly. This property, however, does not hold strictly, and it may be shown, from a comparison of the best observations, that the decrements of heat follow a quicker progression in the higher regions. But we shall soon have another opportunity to examine this subject, and trace out its various consequences.

The humidity of the air also materially affects its elasticity, and the hygrometer should therefore be conjoined with the thermometer in correcting barometrical observations. But nothing satisfactory has yet been done in regard to that subject. The ordinary hygrometers, or rather hygroscopes, are mere toys, and their application to science is altogether hypothetical. A most unphilosophical course has lately been pursued, by multiplying calculations grounded on very loose data, instead of instituting a nice and elaborate train of original experiments.

In the actual state of physical science, it is preposterous, therefore, to affect any high refinement in the formula for computing barometrical measurements. The whole operation may be reduced to a very short and easy process. But the simplicity of the calculation would be still greater, if the centesimal thermometer were generally adopted. It will be sufficiently accurate, till better data are obtained, to assume the expansion of mercury by heat as equal to the 5000th part of its bulk for every centesimal degree, while that of air is twenty times greater, being an expansion for each degree of the 250th part of the bulk of this fluid. 1. *Correct the length of the mercurial column at the upper station, adding to it the product of its multiplication into twice the difference between the degrees on the attached thermometers, the decimal point being shifted four places to the left.* 2. *Substract the logarithm of this corrected length from that of the lower column, multiply by six, and move the decimal point four places to the right; the result is the approximate elevation expressed in English feet.* 3. *Correct this approximate elevation by shifting the decimal point three places back to the right, and multiplying by twice the sum of the degrees on the detached thermometers; this product being now added, will give the true elevation.*

If it were judged worth while to make any allowance for the effect of centrifugal force, this will be easily done before the last multiplication takes place, by adding to twice the degrees on the detached thermometers the fifth part of the mean temperature corresponding to the latitude. The mean temperature itself is formed by multiplying the square of the cosine of the latitude by twenty-nine.

In illustration of these rules, we shall subjoin some real examples. General Roy, in the month of August 1775, observed the barometer on Caernarvon Quay, at 30.091 inches, the attached centesimal thermometer indicating 15°.7, and the detached 15°.6; while, on the peak of Snowdon, the barometer fell to 26.409 inches, and the attached and detached thermometers marked respectively 10°.0 and 8°.8. Here twice the difference of the attached thermometers is 11°.4, and twice the sum of the detached thermometers is 48°.8, which becomes 50.8 when augmented by the fifth part of the mean temperature on that parallel. Now, omitting the lower decimals, the first correction is .00264 × 11.4 = .030, to be added to 26.409. Wherefore,

| | |
|---|---|
| Log. 30.091 = | 1.4784866 |
| Log. 26.439 = | 1.4223450 |
| Difference = | .0561416 |
| Constant multiplier = | 60000 |
| Approximate height = | 3368.496 |

And, for the true height, the correction is 3.37 × 50.8 = 171.2, which gives 3340 for the final result.

We shall take another example from the observations made by Sir George Shuckburgh Evelyn, at the same period, among the mountains of Savoy. This accurate philosopher found the barometer, placed in a cabin near the base of the Mole, and only 672 feet above the surface of the lake of Geneva, to stand at 28.152 inches, while the attached and detached thermometers indicated 16°.3 and 17°.4; but, in another barometer carried to the summit of that lofty insulated mountain, the mercury sunk to 24.176 inches, the attached and detached thermometers marking 14°.4 and 13°.4. Wherefore, twice the difference of the degrees on the attached was 3°.8, and twice the sum of the degrees on the detached thermometers was 61°.6. Consequently the correction to be applied to the higher column was .0024 × 3.8 = .009, which makes it 4.185. Now,

| | |
|---|---|
| Log. 28.152 = | 1.4495092 |
| Log. 24.185 = | 1.3835461 |
| Difference = | .0659631 |
| Constant multiplier = | 60000 |
| Approximate elevation = | 3957.786 |

To correct this approximate elevation, remove the decimal point three places back, and multiply it by 61°.6, increased by 2°.9, the fifth part of the mean temperature, corresponding to the latitude; but 3.96 × 64.5 = 255.4, and 3957.8 + 255.4 = 4213. Hence the summit of the Mole is 4885 feet above the lake of Geneva, or 6083 feet above the level of the Mediterranean Sea.

The last example we shall give is drawn from the observation which Baron Humboldt made among the Andes, near the summit of Chimboraço, the highest spot ever approached by man. This celebrated traveller found there, that the barometer fell to 14.850 English inches; the attached thermometer in the tent being at 10°, and the detached in open air being 1°.6 under zero. But the same barometer, carried down to the shore of the Pacific Ocean, rose exactly to thirty inches, while both the attached and detached thermometers stood at 25°.3. Consequently the correction to be applied to the upper column is = .0015 + 30.6 = .045. Wherefore,

| | |
|---:|---:|
| Log. 30.000 = | 1.4771213 |
| Log. 14.895 = | 1.1730405 |
| Difference = | .3040808 |
| Constant multiplier = | 60000 |
| Approximate elevation = | 18244.848 |

Now, the difference of the detached thermometers, or 26°.9, being doubled and further increased by 5°.8, the fifth part of the mean temperature at the equator, makes 59°.6; and the final correction to be applied is therefore = 18.24 × 59°.6 = 1087, which gives 19,332 feet for the true elevation observed, or 2140 feet below the summit of Chimboraço.

These calculations are performed by the help of logarithms. It is desirable, however, to approximate at least to barometrical measurements without such aid. A very simple rule for this object has been given by Professor Leslie in his

*Elements of Geometry.* Since Log. $\frac{a}{b} = 2\text{ M}\left(\frac{a - b}{a + b} + \frac{1}{3}\left(\frac{a - b}{a + b}\right)^3 + \frac{1}{5}\left(\frac{a - b}{a + b}\right) \&\text{c.}\right)$, where M denotes the modulus of the logarithmic system; when $a$ approaches to $b$, the lower terms may be rejected without sensible error, or log. $\frac{a}{b} = 2\text{ M}\left(\frac{a - b}{a + b}\right)$, very nearly. Wherefore, in reference to our atmosphere, the modulus is expressed by the equiponderant column of homogeneous fluid, or 60,000 + .4342945 = 26,058 feet, or only 26,000 in round numbers; whence, *as the sum of the mercurial columns is to their difference, so is the constant number* 52,000 *feet to the approximate height.* Let General Roy's observation on Snowdon be resumed as an example: the analogy is 30 091 + 26.439 : 30.091 — 26.439, or 56.530 : 3.652 : : 52,000 : 3,359, the approximate elevation, differing very little from the logarithmic result.

This mode of calculation may be deemed sufficiently accurate for determining any altitude that exceeds not 5000 feet. But it will extend to greater elevations if the second term of the series be likewise taken, which is done by striking off three figures, and cubing the half of this number. Thus, resuming the mensuration of Chimboraço, 44.895 : 15.105 : : 52,000 : 17,496, and $(8.75)^3 = 678$, making together 18,166 for a nearer approximation.

The calculation of barometrical measurements, including the correction required, is rendered more easy and expeditious by means of a sliding rule made by Mr Cary, optician in London. This small instrument should always go along with mountain barometers; and it will be found a very agreeable companion to every geological traveller.

But portable barometers, in spite of every precaution, are yet so liable to be broken or deranged, that other auxiliary methods are desirable for ascertaining distant elevations. In this view, the variation of the boiling point of water was proposed by Fahrenheit as far back as the year 1724; the idea having occurred to him, as it had done before to Amontons, while engaged with experiments to perfect his thermometer. Little regard, however, seems to have been paid to the suggestion, till De Luc and Saussure made a series of observations on the heat of ebullition at different elevations above the surface. About thirty years since, Cavallo attempted to revive the scheme of Fahrenheit, but experienced much difficulty in preventing the irregular starts of the thermometer plunged in boiling water. The best and surest way of examining the heat of ebullition, is to suspend the bulb of the thermometer in the confined steam as it rises from the water.

The heat at which water boils, or passes into the form of steam, depends on the weight of the superincumbent atmosphere. By diminishing this pressure, the point of ebullition is always lowered. It appears that, while the boiling heat sinks by equal differences, the corresponding atmospheric pressure decreases exactly, or at least extremely nearly, in a geometrical progression, it being found that every time such pressure is reduced to one half, the temperature of boiling water suffers a regular diminution of about eighteen centesimal degrees. This beautiful relation assimilates with the law which connects the density and elevation of the successive strata of the atmosphere. The interval noticed between the boiling points at two distinct stations must be proportional to their difference of altitude above the level of the sea. We have, therefore, only to determine the co-efficient or constant multiplier, which may be discovered either from an experiment under the rarefied receiver of an air-pump, or from an actual observation performed at the bottom and on the top of some lofty mountain. We shall prefer at present the observation made by Saussure on the summit of Mont Blanc. This diligent philosopher found, by means of a very delicate thermometer constructed on purpose, that water which boiled at 101°.62 in the plain below when the barometer stood at 30.534 English inches, boiled at 86°.24 on the top of that mountain, while the barometer had sunk to 17.136. Wherefore the distance between the points of ebullition, or 15.38 centesimal degrees, must correspond to an approximate elevation of 15,050 feet, which gives 978½ feet of ascent for each degree, supposing the mean temperature of the atmospheric column to be that of congelation. But it will be more convenient to assume 1000 for the constant multiplier, which corresponds to the temperature of 5½°.

To reduce this very simple result into practice, it would be requisite to have a thermometer with a fine capillary bore, and nicely constructed, the stem six or eight inches long, and bearing ten or a few more degrees from the boiling point; these degrees to be divided into twenty or perhaps fifty equal parts engraved on the tube, which should be rather thick, and terminating in a bulb of about half an inch diameter. This thermometer, being fitted with a brass ring two inches above the bulb, should screw into the narrow neck of a small copper flask, which holds some water, but has a hole perforated near the top for allowing the steam to escape. The water may be made to boil by the application of a lamp. The difference between the indications of the thermometers at the two stations being multiplied by a thousand feet, will give the elevation corresponding to a temperature of 5½°. The correction for the actual mean temperature can easily be applied. If a more correct coefficient be afterwards determined, the same thousand, retained as a multiplier, may easily be adapted to another temperature.

This method of measuring elevations on the surface of the globe is, therefore, capable of great improvement, and might be employed with advantage in a variety of cases where observations with the barometer are not easily obtained. Its application would be most important to physical geography, in ascertaining the capital points for tracing the outline of the profile or vertical section of any country. The common maps, which exhibit mere superficial extension, are quite insufficient to represent the great features of nature, since the climate and productions of any place depend as much on its elevation above the sea as its latitude. Scientific travellers have accordingly turned their attention of late years to the framing of vertical sections. As a specimen, we give, (Plate II. fig. 9.) from Humboldt's *Geography of Plants*, a section across the American continent, one of the best and most interesting that has yet appeared. It consists, in fact, of four combined sections, traversing through an extent of 425 miles. The line begins at Acapulco, on the shore of the Pacific Ocean, and runs 195 miles, about a point of the compass towards the east of north, to the city of Mexico; then 80 miles, a point to the south of east, to La Puebla de los Angeles; again it holds a north-east direction of 70 miles, to the Cruz Blanca; and finally bends 80 miles east by south, to Vera Cruz, on the coast of the Atlantic. A scale of altitudes is annexed, which shows the vast elevation of the table-land of Mexico. An attempt is likewise made in this profile to give some idea of the geological structure of the external crust. *Limestone* is represented by straight lines slightly reclined from the horizontal position; *Basalt*, by straight lines reclined from the perpendicular; *Porphyry*, by waved lines somewhat reclined; *Granite*, by confused hatches; *Amygdaloid*, by confused points.

By this mode of distant levelling, a very interesting discovery, in another quarter of our globe, was made by Engelhardt and Parrot, two Prussian travellers. They proceeded, on the 13th July 1814, from the mouth of the Kuban, at the island of Taman, on the Black Sea; and, examining carefully every day the state of the barometer, they advanced with fifty-one observations, the distance of 990 wersts, or 711 English miles, to the mouth of the Terek, on the margin of the Caspian Sea. Similar observations were repeated and multiplied on their return. From a diligent comparison of the whole, it follows that the Caspian is 334 English feet below the level of the Black Sea. That the Caspian really occupies a lower level than the Ocean, had been suspected before, from a comparison of some registers of barometers kept at St Petersburg, and on the borders of that inland sea; but the last observation places the question beyond all doubt. It further appears, that within 250 wersts, or 189 miles, of the Caspian, the country is already depressed to the level of the Ocean, leaving, therefore, an immense basin, from which the waters are supposed to have retired by a subterranean percolation.

If the same plan of barometrical measurements had been followed by the adventurous explorers of the African continent, and the altitudes of the central lakes ascertained, the question regarding the course of the Niger would have been much sooner settled, and much vague and unsatisfactory discussion avoided. Even thermometrical observations of the temperature of springs, or of the ground at moderate depths, would have furnished an approximation to those elevations sufficiently near for solving that great problem in geography on scientific principles alone.

The spirit of modern enterprise has more lately carried barometers to the remotest and loftiest stations on the surface of our globe. It has been thus ascertained that some of the mountains in the equatorial parts of America attain the stupendous altitude of 25,000 feet, while the great chain of Upper India seems to rear its vast summit about 2000 feet still higher.—See Climate and Meteorology.

# CLIMATE.

---

THE word *Climate*, or κλιμα, being derived from the verb κλινεῖν, *to incline*, was applied by the ancients to signify that obliquity of the sphere with respect to the horizon from which results the inequality of day and night. The great astronomer and geographer Ptolemy distinguished the surface of our globe, from the equator to the arctic circle, into climates or parallel zones, corresponding to the successive increase of a quarter of an hour in the length of midsummer-day. Within the tropics, these zones are nearly of equal breadth; but, in the higher latitudes, they contract so much, that it was deemed enough to reckon them by their doubles, answering consequently to intervals of half an hour in the extension of the longest day. To compute them is an easy problem in spherical trigonometry. As the sine of the excess of the semidiurnal arc above a quadrant is to the radius, so is the tangent of the obliquity of the eclyptic, or of 23° 28′, to the cotangent of the latitude. The semidiurnal arcs are assumed to be 91° 52½′, 93° 45′, 95° 37½′, 97° 30′, &c.; and the following table, extracted from Ptolemy's great work, will give some general idea of the distribution of seasons over the surface of our globe :—

| Climate, or Parallel. | Latitude. | | Length of Midsummer Day. | | Breadth of Zone. | |
|---|---|---|---|---|---|---|
| I. | 0° | 0′ | 12h. | 00′ | 4° | 15′ |
| II. | 4 | 15 | 12 | 15 | 4 | 10 |
| III. | 8 | 25 | 12 | 30 | 4 | 5 |
| IV. | 12 | 30 | 12 | 45 | 3 | 57 |
| V. | 16 | 27 | 13 | 00 | 3 | 47 |
| VI. | 20 | 15 | 13 | 15 | 3 | 38 |
| VII. | 23 | 51 | 13 | 30 | 3 | 21 |
| VIII. | 27 | 12 | 13 | 45 | 3 | 10 |
| IX. | 30 | 22 | 14 | 00 | 2 | 56 |
| X. | 33 | 18 | 14 | 15 | 2 | 42 |
| XI. | 36 | 00 | 14 | 30 | 2 | 35 |
| XII. | 38 | 35 | 14 | 45 | 2 | 21 |
| XIII. | 40 | 56 | 15 | 00 | 2 | 9 |
| XIV. | 43 | 4 | 15 | 15 | 1 | 57 |
| XV. | 45 | 1 | 15 | 30 | 1 | 50 |
| XVI. | 46 | 51 | 15 | 45 | 1 | 41 |
| XVII. | 48 | 32 | 16 | 00 | 1 | 32 |
| XVIII. | 50 | 4 | 16 | 15 | 1 | 36 |
| XIX. | 51 | 40 | 16 | 30 | 1 | 10 |
| XX. | 52 | 50 | 16 | 45 | 1 | 40 |
| XXI. | 54 | 30 | 17 | 00 | 1 | 30 |
| XXII. | 55 | 00 | 17 | 15 | 1 | 00 |
| XXIII. | 56 | 00 | 17 | 30 | 1 | 00 |
| XXIV. | 57 | 00 | 17 | 45 | | 30 |
| XXV. | 58 | 00 | 18 | 00 | | |
| XXVI. | 59 | 30 | 18 | 30 | | |

These numbers are calculated on the supposition that the obliquity of the ecliptic was 23° 51′ 20″, to which, according to the theory of Laplace, it must have actually approached in the time of Ptolemy. They seem to be affected by some small errors, especially in the parallels beyond the seventeenth, as the irregular breadth of the zone abundantly shows; but they are, on the whole, more accurate than those given by Varenius.

Ptolemy describes the general appearances which the heavens will present on each parallel, and assigns the corresponding lengths of the shadow of the gnomon at both solstices. He justly maintains, in opposition to the more ancient opinion, that the equatorial region is habitable, since the action of the sun, not continuing long vertical, is there mitigated; but he will not venture to describe the inhabitants, because no person, he says, having yet penetrated so far south, the reports circulated respecting them appeared to be merely conjectural. He therefore passes over the first parallel to the second.

This *second* parallel, then, according to Ptolemy, runs through the isle of Taprobana, supposed to be Ceylon, in the latitude of 4° 15′. The *third* parallel, in the latitude of 8° 25′, traverses the gulf of *Adulitus*. The *fourth* parallel crosses the *Adulitic* gulf, in latitude 12° 45′. The *fifth* parallel passes through the isle of *Meröe*, in Upper Egypt, at latitude 16° 27′. The *sixth* parallel runs through the territory of the *Napati*, in latitude 20° 15′. All these climates or parallels lying below the tropic, the inhabitants are therefore *Amphiscians*, or see the sun pass twice over their heads in the course of the year. The *seventh* parallel, at the latitude 23° 51′, and consequently bordering the tropic, runs through Syene, in Upper Egypt. The *eighth* parallel, in latitude 27° 12′, traverses Ptolemais, in the Thebaid. The *ninth* zone, corresponding to a day of fourteen hours of length, passes through Lower Egypt, at the latitude of 36° 12′. The *tenth* parallel, in latitude 33° 18′, runs through the middle of Phœnicia. The *eleventh* parallel, at the thirty-sixth degree of latitude, passes through the isle of Rhodes. The *twelfth* parallel, in latitude 30° 35′, crosses

Smyrna. The *thirteenth* parallel traverses the Hellespont, in latitude 40° 56′. The *fourteenth* parallel, in latitude 43° 4′, runs through Marseilles. The *fifteenth* parallel passes through the middle of the Pontic Sea, in latitude 45° 1′. The *sixteenth* parallel runs through the sources of the Ister or Danube, in latitude 46° 51′. The *seventeenth* parallel, corresponding to a day of sixteen hours in length, traverses the mouths of the Borysthenes, in latitude 48° 32′. The *eighteenth* parallel, at the latitude of 50° 4′, crosses the Palus Mæotis. The *nineteenth* parallel passes through the most southern part of Britain, in latitude 51° 40′. The *twentieth* parallel crosses the mouths of the Rhine, in latitude 52° 50′. The *twenty-first* parallel passes through the mouths of the Tanaïs, in latitude 54° 30′. The *twenty-second* parallel, at the fifty-fifth degree of latitude, traverses the country of the *Brigantes*, in Great Britain, that is, the southern and larger portion of this island, reckoning from the Frith of Forth. The *twenty-third* parallel, in the fifty-sixth degree of latitude, passes through the middle of Great Britain. The *twenty-fourth* parallel, at the latitude of 57°, runs through *Caturactonium*, in Great Britain. The *twenty-fifth* parallel, corresponding to a day of eighteen hours, runs through the southern parts of Little Britain, in latitude 58°. The *twenty-sixth* parallel, corresponding to a day of 18½ hours in length, traverses the middle of Little Britain, in latitude 58° 30′. It should be observed, that the latitudes of the places in our own island are most inaccurately given by Ptolemy, and generally advanced about two or three degrees farther than their true position. By Little Britain he meant undoubtedly that part of Scotland which lies on the north side of the Friths of Forth and Clyde, and forms almost a peninsula.

The high zones become so narrow, that Ptolemy separates the *twenty-sixth* to an interval of half instead of a quarter of an hour in the length of the day; but he thinks it superfluous to extend this subdivision farther into such remote and inhospitable countries. Resuming the calculation, however, he places the parallel where midsummer-day is prolonged to nineteen hours, in the latitude of 61°, or the north of Little Britain. The parallel of 19½ hours would pass through the Ebudes, Hebrides, or Western Isles, in latitude 62°. The parallel of twenty hours runs through the island of Thule, in the latitude of 63°. The parallel of twenty-one hours would traverse the unknown Scythian nations, in latitude 64½°. The parallels of twenty-two and twenty-three hours would run through the latitudes of 65½° and 66°. He places in latitude 66° 8′ 40″ the arctic circle itself, where the sun does not set during the whole of midsummer-day. Within this circle the inhabitants are *Periscians*, or have the sun lingering above the horizon during part of the summer, and the shadow of the gnomon successively projected in every direction. In the latitude of 67°, the sun continues almost a whole month above the horizon; in the latitude of 69½°, he shines two months; in the latitudes of 73⅓°, 78⅓°, and 84°, that luminary displays his presence for three, four, and five months. At the pole itself, the sun appears, during the space of six months describing circles parallel to the horizon.

Climate, in its modern acceptation, signifies that peculiar condition of the atmosphere in regard to heat and moisture which prevails in any given place. The diversified character which it displays has been generally referred to the combined operation of several different causes, which are all reducible, however, to these two; *distance from the equator*, and *height above the level of the sea*. Latitude and local elevation form, indeed, the great bases of the law of Climate; and any other modifications have only a partial and very limited influence.

If we dig into the ground, we find the temperature to become gradually more steady, till we reach a depth of perhaps forty or fifty feet, below which it continues unchanged.

When this perforation is made during winter, the ground gets sensibly warmer till the limit is attained; but in summer, on the contrary, it grows always colder, till it has reached the same limit. At a certain depth, therefore, under the surface, the temperature of the ground remains quite permanent. Nor is there any indication whatever of the supposed existence of a central fire, since the alleged increase of heat near the bottom of the profoundest excavations is merely accidental, being occasioned by the multitude of burning tapers consumed in conducting the operations of mining. Accordingly, while the air of those confined chambers feels often oppressively warm, the water which flows along the floors seems comparatively cold, or rather preserves the medium heat.

It would be a hasty conclusion, however, to regard this limit of temperature as the natural and absolute heat of our globe. If we dig on the summit of a mountain, or any very elevated spot, we shall discover the ground to be considerably colder than in the plain below; or, if we make a similar perforation on the same level, but in a more southern latitude, we shall find greater warmth than before. The heat thus obtained at some moderate depth, is hence only the mean result of all the various impressions which the surface of the earth receives from the sun and the atmosphere.

The method employed hitherto for ascertaining the temperature at different depths under ground, consists in digging a hole, and burying a sluggish thermometer for several hours, or the space of a whole night. The celebrated naturalist and accurate observer, Saussure, in the month of October 1785, made an interesting set of observations on the banks of the Arve, near Geneva. By digging downwards on successive days, he reached at last the depth of 31 feet. While the surface of the ground had retained a heat of 60°.3 by Fahrenheit's scale, the temperature of the earth at the depth of

4 feet was 60°.8, at 16 feet 56°, at 21 feet 53°.6, and at 28 feet 51°.8. A thermometer buried 31 feet deep was found, when taken up in summer, to stand at 49°.5, and when raised in winter, to indicate 52°.2. Notwithstanding this great depth, therefore, it had still felt the vicissitude of the seasons, having varied 2°.7 in the course of the year. The extreme impressions must have taken six months to penetrate to the bulb, since the temperature was lowest in summer and highest in winter.

But this plan of observing is clumsy and imperfect, there not being sufficient time to allow the mass of earth to regain its proper degree of heat, and too much for the instrument to retain its impressions unaltered before it can be raised up and observed. In order to throw distinct light on a subject so curious and important, Robert Ferguson, Esq. of Raith, a gentleman whose elegant mind is imbued with the love of science, caused a series of large mercurial thermometers, with stems of unusual length, to be planted in his spacious garden at Abbotshall, about 50 feet above the level of the sea, and near a mile from the shore of Kirkaldy, in latitude 56° 10′. The main part of each stem having a very narrow bore, had a piece of wider tube joined above it; and to support the internal pressure of the column of mercury, the bulbs were formed of thick cylinders. The instruments, inclosed for protection in wooden cases, were then sunk beside each other to the depths of one, two, four, and eight feet, in a soft gravelly soil, which turns, at four feet below the surface, into quicksand, or a bed of sand and water. These thermometers were carefully observed from time to time by Mr Charles Norval, the very intelligent gardener at Raith; and we have had access to a register of their variations for nearly three years. It thence appears that, in this climate, and on naked soil, the frost seldom or never penetrates one foot into the ground. The thermometer at that depth fell to 33° of

Fahrenheit on the 30th December 1815, and remained at the same point till the 12th February 1816; but, in the ensuing year, it descended no lower than 34°, at which it continued stationary from the 23d December 1816 to 1st January 1817. At the same depth, of one foot, it reached the maximum 58° on the 13th July 1815, but, in the following year, it rose only to 54° on the 21st July; and, in the year 1817, it mounted to 56° about the 5th July. This thermometer, in the space of three years, travelled, therefore, over an interval of 25°, the medium being 45½, and attained its highest and lowest points about three weeks after the solstice of summer and of winter.

The thermometer planted at the depth of two feet sunk to 36° on the 4th February 1816; but it stood at 38° about the beginning of January 1817. It rose to 56° on the 1st of August 1815; but in the next year it reached only 53° on 24th July; and, in 1817, it again reached 56° on 10th July. At the depth of two feet, the extreme variation was therefore 20°; and the maxima and minima took place about four or five weeks after either solstice.

The thermometer of four feet depth had sunk to 39° about the 11th February 1816, and was stationary at 40° near the 3d February 1817. It rose to 54° on the 2d August 1815, and stood at 52° during the greater part of August and September in the years 1816 and 1817. It ranged, therefore, only 15°, the mean being 46½°, and the extreme points occurring near two months after either solstice.

The thermometer whose bulb was planted eight feet deep descended to 42° on the 16th February 1816, but stood at 42½° on the 11th February 1817. It rose to 51½° on the 11th September 1815, fell to 50° on the 14th September 1816, and mounted again to 51° on the 20th September 1817. This thermometer had, therefore, a range of only 9½°, the medium temperature being 46¾°, and the extremes of heat and cold occurring nearly three months after the solstice of summer and of winter.

These observations are quite satisfactory, and exhibit very clearly the slow progress by which the impressions of heat or cold penetrate into the ground. It will not be far from the truth to estimate the rate of this penetration at an inch every day. The thermometers hence attained their maximum at different periods, though in a tolerably regular succession. The mean temperature of the ground, however, seemed rather to increase with the depth; but this anomaly evidently proceeded from the coldness of the two successive summers, and particularly that of 1816, which occasioned such late harvests and scanty crops. Thus the thermometer of one foot indicated the medium heat of only 43°.8 during the whole of the year 1816. But it will be satisfactory to exhibit the leading facts in a tabular form. The following are the mean results for each month, only those for December 1817 are supplied from the corresponding month in 1815.

| | 1816. | | | | 1817. | | | |
|---|---|---|---|---|---|---|---|---|
| | 1 foot. | 2 feet. | 4 feet. | 8 feet. | 1 foot. | 2 feet. | 4 feet. | 8 feet. |
| January,..... | 33°.0 | 36°.3 | 40°.7 | 43°.0 | 35°.6 | 38°.7 | 40 .5 | 45°.1 |
| February,... | 33 .7 | 36 .0 | 39 .0 | 42 .0 | 37 .0 | 40 .0 | 41 .6 | 42 .7 |
| March,....... | 35 .0 | 36 .7 | 39 .6 | 42 .3 | 39 .4 | 40 .2 | 41 .7 | 42 .5 |
| April,......... | 39 .7 | 38 .4 | 41 .4 | 43 .8 | 45 .0 | 42 .4 | 42 .6 | 42 .6 |
| May,.......... | 44 .0 | 43 .3 | 43 .4 | 44 .0 | 46 .8 | 44 .7 | 44 .6 | 44 .2 |
| June,......... | 51 .6 | 50 .0 | 47 .1 | 45 .8 | 51 .1 | 49 .4 | 47 .6 | 47 .8 |
| July,.......... | 54 .0 | 52 .5 | 50 .4 | 47 .7 | 55 .2 | 55 .0 | 51 .4 | 49 .6 |
| August,....... | 50 .0 | 52 .5 | 50 .6 | 49 .4 | 53 .4 | 53 .9 | 52 .0 | 50 .0 |
| September,... | 51 .6 | 51 .3 | 51 .8 | 50 .0 | 53 .0 | 52 .7 | 52 .0 | 50 .7 |
| October,...... | 47 .0 | 49 .3 | 49 .7 | 49 .6 | 45 .7 | 49 .4 | 49 .4 | 49 .8 |
| November,... | 40 .8 | 43 .8 | 46 .3 | 45 .6 | 41 .0 | 44 .7 | 47 .0 | 47 .6 |
| December,.... | 35 .7 | 40 .0 | 43 .0 | 46 .0 | 37 .9 | 40 .8 | 44 .9 | 46 .4 |
| Mean of whole year, | 43 .8 | 44 .1 | 45 .1 | 46 .0 | 44 .9 | 45 .9 | 46 .2 | 46 .6 |

If the thermometers had been sunk considerably deeper, they would no doubt have indicated a mean temperature of 47°.7. Such is the permanent temperature of a copious spring which flows at a short distance, and about the same elevation, from the side of a basaltic or greenstone rock. Profuse fountains and deep wells, which are fed by percolation through the crevices of the strata, furnish the surest and easiest mensuration of the temperature of the earth's crust. The body of water which bursts from the caverns of Vaucluse, and forms almost immediately a respectable and translucid river, has been observed not to vary in its temperature, by the tenth part of a degree, through all the seasons of the year. It is, therefore, an object highly important for scientific travellers to notice the precise heat of springs in favourable situations, as they issue from their rocky beds. Such choice observations would accurately fix the medium temperature of any climate. It is only requisite to exclude the superficial and the thermal springs, which are not difficult to distinguish.

From a comparison of meteorological observations made at distant points on the surface of our globe, the celebrated astronomer Professor Mayer of Gottingen, was enabled to discover an empirical law which connects most harmoniously the various results. Round the pole, the mean temperature may be assumed at the precise limit of freezing, since the fields of ice accumulated in that forlorn region seem at this present period neither to increase nor diminish. But under the equator the medium heat on the level of the sea is found to be 84½° of Fahrenheit, or 29 centesimal degrees, the division of the thermometric scale which is the best suited to philosophical purposes. At the middle point, or the latitude of 45°, the temperature is likewise the exact mean, or 14½° centigrade. From that centre, the heat diminishes rapidly northwards, and increases with equal rapidity towards the south. Hence the mean temperature of any place, at the level of the sea, is calculated in centesimal degrees, by multiplying the square of the cosine of the latitude into the constant number 29; or it is found by multiplying the supplemental versed sine of double the latitude into $14\frac{1}{2}°$. The variation of temperature for each degree of latitude is hence denoted centesimally, with very great precision, by half the sine of double the latitude; being, in fact, this quantity diminished in the ratio of 58, the double of 29, to 57.3, the length of an arch equal to the radius. From these data the following table is computed; in which are likewise annexed the corresponding degrees of Fahrenheit's, with the successive differences.

| Lat. | Centesimal. | Dif. | Fahrenheit. | Dif. | Lat. | Centesimal | Dif. | Fahrenheit. | Dif. |
|---|---|---|---|---|---|---|---|---|---|
| 0 | 29°.00 | .00 | 84°.2 | .00 | 26 | 23°.43 | .39 | 74°.2 | .70 |
| 1 | 28 .99 | .01 | 84 .2 | .02 | 27 | 23 .02 | .41 | 73 .5 | .72 |
| 2 | 28 .96 | .03 | 84 .1 | .05 | 28 | 22 .61 | .42 | 72 .7 | .74 |
| 3 | 28 .92 | .04 | 84 .0 | .07 | 29 | 22 .18 | .43 | 71 .9 | .76 |
| 4 | 28 .86 | .06 | 83 .9 | .11 | 30 | 21 .75 | .43 | 71 .4 | .77 |
| 5 | 28 .78 | .08 | 83 .8 | .13 | 31 | 21 .31 | .44 | 70 .3 | .79 |
| 6 | 28 .68 | .10 | 83 .6 | .18 | 32 | 20 .86 | .45 | 69 .5 | .81 |
| 7 | 28 .57 | .11 | 83 .4 | .20 | 33 | 20 .40 | .46 | 68 .7 | .83 |
| 8 | 28 .44 | .13 | 83 .2 | .23 | 34 | 19 .93 | .47 | 67 .9 | .84 |
| 9 | 28 .29 | .15 | 82 .9 | .27 | 35 | 19 .46 | .47 | 67 .0 | .85 |
| 10 | 28 .13 | .16 | 82 .6 | .30 | 36 | 18 .98 | .48 | 66 .2 | .86 |
| 11 | 27 .95 | .18 | 82 .3 | .32 | 37 | 18 .50 | .48 | 65 .3 | .87 |
| 12 | 27 .75 | .20 | 82 .0 | .36 | 38 | 18 .01 | .49 | 64 .4 | .88 |
| 13 | 27 .53 | .22 | 81 .6 | .40 | 39 | 17 .50 | .49 | 63 .5 | .88 |
| 14 | 27 .30 | .23 | 81 .1 | .42 | 40 | 17 .01 | .49 | 62 .6 | .89 |
| 15 | 27 .06 | .24 | 80 .7 | .44 | 41 | 16 .52 | .49 | 61 .7 | .90 |
| 16 | 26 .80 | .26 | 80 .2 | .47 | 42 | 16 .02 | .50 | 60 .8 | .90 |
| 17 | 26 .52 | .28 | 79 .7 | .50 | 43 | 15 .52 | .50 | 59 .9 | .91 |
| 18 | 26 .23 | .29 | 79 .2 | .52 | 44 | 15 .01 | .51 | 59 .0 | .91 |
| 19 | 25 .93 | .30 | 78 .7 | .54 | 45 | 14 .50 | .51 | 58 .1 | .92 |
| 20 | 25 .61 | .32 | 78 .1 | .57 | 46 | 13 .99 | .51 | 57 .2 | .92 |
| 21 | 25 .28 | .33 | 77 .5 | .60 | 47 | 13 .49 | .50 | 56 .3 | .91 |
| 22 | 24 .93 | .35 | 76 .9 | .63 | 48 | 12 .98 | .51 | 55 .4 | .91 |
| 23 | 24 .57 | .36 | 76 .2 | .65 | 49 | 12 .48 | .50 | 54 .5 | .90 |
| 24 | 24 .20 | .37 | 75 .6 | .67 | 50 | 11 .98 | .50 | 53 .6 | .90 |
| 25 | 23 .82 | .38 | 74 .9 | .68 | 51 | 11 .49 | .49 | 52 .7 | .89 |

| Lat. | Centesimal. | Dif. | Fahrenheit. | Dif. | Lat. | Centesimal. | Dif. | Fahrenheit. | Dif. |
|---|---|---|---|---|---|---|---|---|---|
| 52 | 10°.99 | .50 | 51°.8 | .90 | 72 | 2°.77 | .30 | 37°.0 | .54 |
| 53 | 10 .50 | .49 | 50 .9 | .88 | 73 | 2 .48 | .29 | 36 .5 | .52 |
| 54 | 10 .02 | .48 | 50 .0 | .87 | 74 | 2 .20 | .28 | 36 .0 | .50 |
| 55 | 9 .54 | .48 | 49 .2 | .86 | 75 | 1 .94 | .26 | 35 .5 | .47 |
| 56 | 9 .07 | .47 | 48 .3 | .85 | 76 | 1 .70 | .24 | 35 .1 | .43 |
| 57 | 8 .60 | .47 | 47 .5 | .84 | 77 | 1 .47 | .23 | 34 .6 | .41 |
| 58 | 8 .14 | .46 | 46 .6 | .83 | 78 | 1 .25 | .22 | 34 .2 | .40 |
| 59 | 7 .69 | .45 | 45 .8 | .81 | 79 | 1 .05 | .20 | 33 .9 | .36 |
| 60 | 7 .25 | .44 | 45 .0 | .79 | 80 | .86 | .19 | 33 .6 | .34 |
| 61 | 6 .82 | .43 | 44 .3 | .78 | 81 | .71 | .17 | 33 .3 | .31 |
| 62 | 6 .39 | .43 | 43 .5 | .77 | 82 | .56 | .15 | 33 .0 | .27 |
| 63 | 5 .98 | .41 | 42 .8 | .76 | 83 | .43 | .13 | 32 .8 | .23 |
| 64 | 5 .57 | .41 | 42 .0 | .74 | 84 | .32 | .11 | 32 .6 | .20 |
| 65 | 5 .18 | .39 | 41 .3 | .71 | 85 | .22 | .10 | 32 .4 | .18 |
| 66 | 4 .80 | .38 | 40 .6 | .68 | 86 | .14 | .08 | 32 .3 | .15 |
| 67 | 4 .43 | .37 | 40 .0 | .67 | 87 | .08 | .06 | 32 .2 | .11 |
| 68 | 4 .07 | .36 | 39 .3 | .65 | 88 | .04 | .04 | 32 .1 | .07 |
| 69 | 3 .72 | .35 | 38 .7 | .63 | 89 | .01 | .03 | 32 .0 | .05 |
| 70 | 3 .39 | .33 | 38 .1 | .60 | 90 | .00 | .00 | 32 .0 | .01 |
| 71 | 3 .07 | .32 | 37 .5 | .57 | | | | | |

It hence appears, that, near the extremities of the quadrant, or towards the pole and the equator, there is scarcely any sensible variation of the mean temperature, and that the whole change within the arctic circle, or between the tropics, amounts only to eight degrees on Fahrenheit's scale. Very little increase of heat is, therefore, observed in advancing through the torrid zone to the equator; and the intensity of the cold would not be sensibly augmented in penetrating from the arctic circle to the pole. The existence of an open sea towards the extreme north is hence not improbable.

On the other hand, the character of the climate changes rapidly in the temperate zone. Hence likewise the variety of vegetable productions with which those happier regions abound. Such a country as France, for example, stretching from about the 40th to the 50th degree of latitude, and through a difference of five centesimal degrees of mean temperature, yields not only plentiful crops of wheat, barley, and oats, but raises olives, fig-trees, and vines.

The gradation of temperature in different latitudes may be clearly shown by a geometrical diagram. Let the figure below represent a quadrant, 90 the pole, and 50 the latitude of any place; on the radius as an axis and a parameter, describe a parabola, which will consequently pass through the pole; draw the perpendicular 50 B, and the portion of it AB, intercepted within the parabola, will express the mean temperature of the given place at the level of the sea, which in the present case should amount to 12 centesimal degrees.

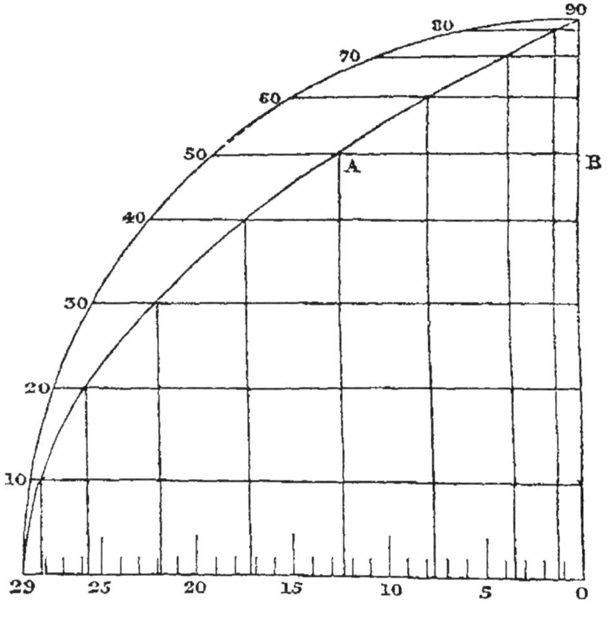


Since each elementary zone of the sphere is equal to the corresponding belt of a circumscribed cylinder, the whole heat accumulated on its surface must be proportional to the area of the annexed parabola, and, consequently, the mean temperature is two-thirds of what obtains at the equator, and therefore 19½°, or 66° 8′ on Fahrenheit's scale. Such must be the temperature of the great mass of the earth, if it has derived all its heat from external impressions. But, at the very small depths to which we can ever penetrate, the influence of the immediate vicinity only is felt; nor, in the profoundest mines, has any tendency yet been perceived towards increase of temperature in the higher latitudes, or decrease in the lower.

These superficial impressions are all produced, either directly, or through the intervention of the atmosphere, by the action of the solar rays. It may be calculated from experiment, that the entire and unimpaired light of a vertical sun will communicate one centesimal degree of heat every hour to a sheet of water of a foot in thickness. Consequently, since the surface of a sphere is four times that of its generating circle, such a sheet of water, spread over the whole of the globe, would receive six degrees of heat every day. But the very inferior capacity of the atmosphere for heat, being estimated as equal to that of a body of water about twelve feet in depth, if the aerial mass finally received and retained all the calorific impressions, it would every day have its temperature raised half a degree, and, therefore, augmented to 182½° in the course of a whole year. This annual accession of heat, however, is quickly dispersed by the mobility of the fluid medium, and gradually absorbed into the earth, or more quickly diffused through the waters of the ocean, which, besides, occupy at least three-fourths of the whole surface of our globe. The luminous matter communicated by the incessant shining of the sun, whether received on the ground, or intercepted in its passage through the air, would hence be capable of communicating one centesimal degree of heat to the body of the earth in the space of 1323 years; a quantity too small, perhaps, to be yet perceived, though its influence may be afterwards detected by very delicate observations.

It is easy to demonstrate, from the laws of optics, that the quantity of light which falls on a horizontal surface must be proportional to the sine of its obliquity. Hence the aggregate light received under the equator at either equinox, is to what would accumulate during twenty-four hours, if maintained at its highest intensity, as the diameter to the circumference of a circle. This daily accession of heat, confined to the mass of atmosphere, would therefore, in that climate and season, amount to 633,000 parts of a degree. At the pole itself, during the complete circuit of the sun in midsummer's day, the measure of heat would be about a fourth part greater, or 797,000 parts; the continued endurance of the sun above the horizon more than compensating for the feebleness of his oblique rays.

In general, the quantity of light received at any place from the sun, in the space of one day, is denoted by the product of the sine of the semidiurnal arc, or the distance from noon to the time of sunset, into the cosines of the latitude and declination, joined to the product of that arc itself into the sines of the latitude and declination; the latter part of the expression being considered as additive or subtractive, according as the declination lies on the same or on the opposite side of the latitude. Hence, at Edinburgh, in the latitude of 56°, the heat collected during one day at the summer solstice is 307,000 parts, but at the solstice of winter only 166,000.

If a calculation be instituted for the quantities of heat during the half-yearly periods, from the equinox of spring to that of autumn, and from the autumnal equinox again to the vernal, the following table will be formed.

| | Summer. | Winter. | Whole Year. |
|---|---|---|---|
| Equator, . . . | 116° | 116° | 232° |
| Tropic, . . . | 127 | 87 | 214 |
| Latitude 45°, . . | 120 | 42 | 162 |
| Arctic circle, . . | 102 | 12 | 114 |
| Pole, . . . | 84 | 00 | 84 |

The annual accumulation at the latitude of 45° is thus 162°, which differs very little from 158°, the mean between the calorific effects at the equator and at the pole. It may be observed likewise, that the effects vary more slowly at the extremes, than near the middle of the quadrant. Thus, from the equator to the tropic, and from the arctic circle to the pole, the differences are 30° and 28°; but in the narrower intervals, from the tropic to the latitude of 45°, and thence to the arctic circle, the differences are 52° and 48°. The property now stated corresponds with the changes of mean temperature in different latitudes.

If a current of air from the equator, having the ordinary temperature of 29°, were supposed to travel to the pole, from which an equal and contrary current would consequently flow towards the equator, each journey would transport 58° of heat. Between two and three such journeys, performed every year, would therefore be sufficient to disperse the whole accumulation of 148°. This only requires the existence of a wind advancing northwards at the rate of forty-six miles every day. It is not necessary even that the wind should either continue permanent, or blow directly north. The same effect would be produced if it were to blow indifferently to every point of the compass, and only at the rate of three miles an hour; a supposition which agrees perfectly with actual observation.

The circulation excited in the body of our atmosphere

thus prevents the heat shed by the sun on different parts of the earth's surface from an excessive accumulation. In proportion as the equatorial regions grew warmer from the predominance of illumination, the polar wind would rush with more rapidity, till it had tempered the excess. This balance of the accession, and the consequent dispersion of heat, has probably been long attained, and it now regulates the gradation of climates in successive latitudes. But if air had possessed the capacity for heat which belongs to hydrogen gas, it would have produced a more equable diffusion of temperature, insomuch that the temperature of the equator could not have become ten degrees warmer than that of the poles. On the contrary, had our atmosphere been less fluid, or less capable of containing heat, the inequality of different climates would have risen to a higher pitch. That variety of temperature which occurs at present on the surface of the globe, was requisite for the development of the different vegetable tribes which clothe it. The same harmony connects the system of this lower world, which irradiates the expanse of the celestial regions.

The equilibrium of temperature preserved over the globe by the circulation of the atmosphere, is not, however, very quickly produced. Hence the remarkable increase of heat which takes place during the summer months in the higher latitudes. But within the arctic circle, another powerful agent of nature is constantly tempering the inequality of the seasons. The vast beds of snow, or fields of ice, which cover the land and the sea in those dreary retreats, absorb, in the act of thawing or passing again to the liquid form, all the surplus heat collected during the continuance of a nightless summer. The rigour of winter, when darkness resumes her tedious reign, is likewise mitigated by the warmth evolved as congelation spreads over the watery surface.

Of the light received from the sun, which, by its union with other bodies, constitutes heat, a considerable portion is always detained and absorbed in its passage through the atmosphere. Even a vertical ray shot through the clearest air will lose more than the fifth part of its intensity before it reaches the surface of the earth. In most cases, the loss which light will suffer in the shortest transit through the atmosphere, may be estimated at one-fourth of the whole. But the oblique rays must undergo a much greater absorption. If, from their slanting course, they have to encounter twice the number of aerial particles, their intensity must be reduced to nine-sixteenths, or the square of three-fourths; and if they describe triple the vertical tract, only twenty-seven sixty-fourth parts, or the cube of three-fourths, will reach the ground. In general, if the tracts of light follow an arithmetical progression, the diminished force with which it escapes and arrives at the ground, will form a decreasing geometrical progression. To determine the train of aerial particles which the oblique rays of the sun must traverse in their passage through the atmosphere, is a nice problem, which requires a skilful application of the integral calculus. Without stopping to engage at present in the details of this intricate investigation, it may suffice to remark, that, in general, the length of the tract is nearly in the inverse ratio of the sine of the sun's altitude. But the following table, to every five degrees, is calculated from rigorous formulæ, the length of oblique tract being reduced to the standard of air uniformly dense. These quantities again are diminished in the ratio of the sine of obliquity, to express the calorific action which those enfeebled and slanting rays finally exert at the surface of the earth.

| Sun's Altitude. | Measure of Atmospheric Tract. | Intensity of the Light transmitted. | Calorific Action at the Surface. |
|---|---|---|---|
| 90° | 1.000 | .750 | .740 |
| 85 | 1.004 | .749 | .747 |
| 80 | 1.015 | .747 | .735 |
| 75 | 1.035 | .742 | .717 |
| 70 | 1.064 | .736 | .691 |
| 65 | 1.103 | .728 | .660 |
| 60 | 1.154 | .718 | .609 |
| 55 | 1.220 | .704 | .577 |
| 50 | 1.305 | .687 | .526 |
| 45 | 1.413 | .666 | .454 |
| 40 | 1.554 | .640 | .411 |
| 35 | 1.740 | .606 | .348 |
| 30 | 1.995 | .563 | .282 |
| 25 | 2.359 | .507 | .214 |
| 20 | 2.905 | .434 | .148 |
| 15 | 3.841 | .331 | .086 |
| 10 | 5.610 | .199 | .035 |
| 5 | 10.450 | .050 | .004 |
| 0 | 37.850 | .00002 | ... |

It hence appears that, even when the sky is most serene, only one half of the sun's light can reach the surface of the earth from an altitude of 25°, or one-third from that of 15°, and that, if the obliquity of the rays were increased to 5°, no more than the twentieth part of them would actually be transmitted. The annual quantity of light which falls may be computed as equivalent at the equator to an uniform illumination from an altitude of 17° 46′; and, in the mean latitude of 45°, and at the pole, the effects are the same as if the rays had respectively the constant obliquities of 13° 2′ and 7° 17′. Wherefore, under the most favourable circumstances, of 1000 parts of light transmitted from the sun, only 378 can, at a medium estimate, penetrate to the surface at the equator, 228 in the latitude of 45°, and 110 at the pole, of their oblique rays; but the shades received by a given portion of the surface are still less, being only 115, 51, and 14. In cloudy weather, the portion of light that can finally reach the ground, will seldom amount to the third of those quantities; and when the sky becomes darkened with accumulated vapours, almost every shining ray is intercepted in its passage.

The light which at last gains the surface being there absorbed and converted into heat, is in this form profusely delivered to the ambient air, or more feebly conducted downwards into the body of the earth. But the rays which fall on seas or lakes are not immediately arrested in their course; they penetrate always with diminishing energy, till, at a certain depth, they are no longer visible. This depth depends, without doubt, on the clearness of the medium, though probably not one tenth part of the incident light can advance five fathoms in most translucid water. The surface of the ocean is not, therefore, like that of the land, heated by the direct action of the sun during the day, since his rays are not intercepted at their entrance, but suffered partially to descend into the mass, and to waste their calorific power on a liquid stratum of ten or twelve feet in thickness.

But the surface of deep collections of water is kept always warmer than the ordinary standard of the place, by the operation of another cause, arising from the peculiar constitution of fluids. Although these are capable, like solids, of conducting heat slowly through their mass, yet they transfer it principally in a copious flow by their internal mobility. The heated portions of a fluid being dilated, must continue to float on the surface; while the portions which are cooled, becoming consequently denser, will sink downwards by their superior gravity. Hence the bed of a very deep pool is always excessively cold, since the atmospheric influences are modified in their effects by the laws of statics. The mean temperature of the climate is not communicated by these variable impressions; every change to warmth being spent on the upper stratum, while every transition to cold penetrates to the bottom, which thus experiences all the rigours of winter, without receiving any share of the summer's heat. But, if the beds of profound bodies of water remain perpetually cold, their surface undergoes some variety of temperature, and is generally warmer than the average weekly or monthly heat of the air.

These principles are confirmed by observations made on our own lakes, and strikingly exemplified in those of Switzerland, which have a breadth proportioned to the stupendous altitude of their encircling mountains. It appears, from the careful observations of Saussure, that the bottoms of those majestic basins, whether situated in the lower plains, or embosomed in the regions of the upper Alps, are almost all of them equally cold, being only a few degrees above the point of congelation. That accurate observer found the temperature of the Lake of Geneva, at the depth of 1000 feet, to be 42°, and could discover no monthly variation under 160 feet from the surface. In the course of July, he examined the Lakes of Thun and Lucerne: the former at the depth of 370, and the latter at that of 640 feet, had both the temperature of 41°, while the superficial waters indicated respectively 64° and 68½° by Fahrenheit's scale. The bottom of the Lago Maggiore, on the Italian side of the Alps, was a little warmer, being 44° at the depth of 360 feet, while the surface was almost as high as 78°. Barlocci has since found that the Lago Sabatino, near Rome, at the depth of 490 feet, was only 44½, while the thermometer, dipped at the surface, marked 77°.

Through the friendship of Mr James Jardine, civil engineer, we are enabled to give the results of his observations on some of the principal Scottish lakes, which, as might be expected from him, were conducted with the most scrupulous accuracy. The instrument which he employed for exploring the temperature at different depths, was free from the ordinary objections; being a register thermometer, let down in a horizontal position, which could acquire the impression in not many seconds, and might be drawn up leisurely, without risk of subsequent alteration. It would appear that the variable impressions of the seasons do not penetrate more than fifteen or twenty fathoms; that below this depth, an almost uniform coldness prevails. Thus, in the deepest part of Loch Lomond, on the 8th September 1812, the temperature of the surface was 59°.3 of Fahrenheit; at the depth of 15 fathoms, 43°.7; at that of 40 fathoms, 41°.3; and from that point to about 3 feet from the bottom, at 100 fathoms, it decreased only the fifth part of a degree. Again, on the preceding day, the superficial water of Loch Katrine being at 57°.3, the thermometer, let down 10 fathoms, indicated 50°.6; at the depth of 20 fathoms it marked 43°.1; at the depth of 35 fathoms it fell to 41°.1; and on the verge of the bottom, at 80 fathoms, it had only varied to 41°. At the same place, on the 3d September 1814, the heat of the surface was 56°.4; at the depth of 10 fathoms, 49°.2; at that of 20 fathoms, 44°; at that of 30 fathoms, 41°.9; and at that of 80 fathoms, 41°.8.

Hence it is that, even in the northern latitudes, the deep lakes are never, during the hardest winters, completely frozen over. But if the same water be let into a shallow basin, it will, in a rigorous season, be chilled thoroughly, and converted into ice. This may even happen when spread above the surface of salt water, which is always considerably denser. Thus, frost takes no effect on Loch Ness, nor on the river of that name, which, in a rapid course of a few miles, discharges the surplus water into the sea. But in very severe winters, a sheet of ice appears formed along the shore; the impressions of cold being almost wholly expended on the accumulation of fresh water, since the chilled portions of this which continually descend, are stopped in their progress by the greater density of the recumbent sea water.

The seas and the ocean itself obey the same law of the distribution of heat, only the difference of temperature experienced by sounding in the Mediterranean is less conspicuous than in the fresh-water lakes. Saussure found that the temperature at the bottom, in the Gulfs of Nice and of Genoa, at the depths of 925 and of 1920 feet, was the same, or 55°.8, the heat of the superficial water being about 69°. But the mean temperature, or that of the body of the land on the same parallel of latitude, is 59°. The smallness of the diminution here observed may perhaps be attributed to the effect of evaporation in such hot confined bays, the water at the surface being thus rendered salter, and consequently disposed, by its acquired density, to sink into the colder mass below.

In open seas, and in damper climates, the depression of temperature is greater in the inferior strata. This difference becomes augmented in proportion of the extreme variation of the seasons. Lord Mulgrave, on the 4th September 1773, in the latitude of 65° north, drew up water from the depth of 4100 feet, which he found to have the temperature of 40°, while the thermometer, dipped at the surface, stood, on the 19th June, at 55°. In the latitude of 66°, a register thermometer, let down 4680 feet, marked 26°, while the air was at 48½°; and on the 31st August, in the latitude of 69°, while the exterior thermometer indicated 59½°, the temperature of the water at the depth of 4040 feet was only 32°. In shallow seas the two extremes are brought closer together, and therefore a similar difference of temperature now occurs at moderate depths. The water lying on the surface, which, in the vicissitude of the season, comes to be chilled, is not precipitated as before to a fathomless abyss. The increase of cold below is hence considered as an indication of the proximity of banks, if not of the approach to land itself.

A like gradation of temperature is produced by the alternating influence of the seasons in deep and stagnant masses of air. When this active fluid is confined in profound caverns, opening to the sky without being much exposed, and either perpendicularly or gently inclined, its lower strata become intensely and permanently cold. The mild air of summer floats motionless at the mouth of the pit; but, in winter, the superior air, cooled many degrees perhaps below the freezing point, and therefore greatly condensed, precipitates itself continually to the bottom.

This fact takes place in most caverns, and in draw-wells which are left uncovered. Saussure found, on the 1st of July, when the thermometer in the shade stood at 78° of Fahrenheit, that a cave in the Monte Testacco, a small hill in the vicinity of Rome, formed entirely by the enormous accumulation of broken pottery, had the temperature of 50°; and two other caves in the same porous mass were cooled to 44°. On the 9th July, when the external air was at 61°, the cave of St Marino, at the foot of a sandstone hill, about 2080 feet above the level of the sea, indicated only 44½, which is 8° below the mean temperature of the soil in that situation. In the grotto of Ischia, and in the caves of Cesi and of Chiavenna, the thermometer marked likewise 44½°; but in the caves of Caprino, on the borders of the Lake of Lugano, it stood at different times of the year at 37° and at 42°; and in those of Hergisweil, near Lucerne, the heat of the interior, on the 31st of July, was only 39½°.

But this phenomenon is still more striking in certain peculiar circumstances. The famous Swedish mine of Dannemora, which yields the richest iron ore in the world, presents an immense excavation, probably two or three hundred feet in depth. On the occasion of some repairs which suspended the usual labours, the basin appeared some years since full of water, with huge blocks of ice floating in it. The silver mine of Köngsberg, in Norway, has for its main shaft a frightful open cavern, perhaps three hundred feet deep and thirty feet wide, of which the bottom is covered with perpetual snow. Hence, likewise, on the sides of Ætna and of the mountains in Spain, the collected snows are preserved all the year in caves and crevices of the rocks, from which natural stores the muleteers carry down, during summer, to the villages and the cities of the plain, a material so necessary to comfort in those parched climates.

Such is the disposition induced in a confined column of air; but in a free atmosphere the gradation of temperature is exactly reversed, the lower strata being invariably warmer than the upper. This most important fact in meteorology and physical geography was thought sufficiently explained, in the infancy of physical science, from the proximity of the heat supposed to be reflected by the surface of the earth. But it were idle to attempt any serious confutation of such crude ideas. The true cause of the cold that prevails in the higher regions of the atmosphere is undoubtedly the enlarged capacity which air acquires by rarefaction. From the unequal action of the sun's rays, and the vicissitudes of day and night, a quick and perpetual circulation is maintained between the lower and upper strata; and it is obvious that, for each portion of air which rises from the surface, an equal and corresponding portion must likewise descend. But that which mounts up, acquiring an augmented attraction from heat, has its temperature proportionally diminished; while the correlative mass falling down, carries its share of heat along with it, and, again relaxing its attraction, seems to diffuse warmth below. A stratum at any given height in the atmosphere is hence affected both by the passage of air from below and by the return of air from above, the former absorbing a portion of heat, and the latter evolving it. But the mean temperature at every elevation is on the whole still permanent, and consequently those disturbing causes must be exactly balanced, or the absolute measure of heat is the same at all heights, suffering merely some external modification from the difference of capacity in the several portions of the fluid with which it has combined. That temperature is hence inversely as the capacity of air having the rarity due to the given altitude.

It only remains, therefore, to discover the capacity of air, or its attraction for heat under successive pressures, or at different degrees of rarity. But this problem requires a very nice investigation, and appears incapable of being resolved by any direct procedure. If the elaborate experiments of Dr Crawford and others, on the capacity of air in its ordinary state, gave such erroneous results, what hope could be formed of ascertaining even its minute shadings, by any similar plan of operation? But, independently almost of any theory, a simple method occurs for conducting this research. A delicate thermometer, suspended within the receiver of an air-pump, indicates a decrease of temperature as the process of rarefaction advances; and, on stopping this operation at any stage, the thermometer will slowly regain its former state. If now, when the equilibrium is restored, the air be suddenly re-admitted, the dilated portion which had remained in the receiver, liberates the heat absorbed by it during the progress of rarefaction. The thermometer accordingly rises quickly through a certain space, then becomes for a short while stationary, and afterwards slowly subsides. But the instrument does evidently not measure the whole of the heat thus evolved, a great part of it being spent in warming up to the same point the internal surface of the receiver. This action, however, is merely superficial, since its effect appears to be momentary. Consequently, the internal surface of the receiver, with that of the plate on which it stands, as penetrated by the sudden impression to a certain very minute depth, forms a constant film of matter, which, as well as the body of air itself, draws its supply from the extricated heat. Under the same receiver, therefore, although the air will not seize the whole of the heat disengaged in the act of admission, it must always retain a proportional share of it. A series of experiments, at successive degrees of rarefaction, must hence discover, if not the absolute, yet the relative changes of the air's capacity for heat.

To institute this inquiry with the desired success, an excellent and powerful air-pump was used, having a receiver of the very largest dimensions, of an oblong spheroidal form, approaching, however, nearly to the globular, and with a narrow bottom. The apparatus being placed in the middle of a close room, which had a steady temperature, a thermometer, with a slender stem, open at top, and a small bulb, of extreme sensibility, was fixed in a vertical position, a few inches above the centre of the plate. Having replaced the receiver, and allowed it to stand some time, one-fifth of the air was now extracted from under it; and, after a considerable interval, the cock was suddenly opened to restore the equilibrium, when the mercury of the thermometer, which had been stationary, mounted up very rapidly 3.0 centesimal degrees, from which point it afterwards slowly descended.

The temperature of the room having been regained, two-fifths of the air in the receiver were then extracted; and, after some lapse of time, the external communication being repeated, the thermometer rose instantly 5.3 centesimal degrees. On extracting three-fifths of the internal air, the corresponding ascent of the thermometer, at the restoration of the equilibrium, was 7.0 of those degrees; and, when the contents of the receiver had been rarefied five times, the heat evolved, on the re-admission of the air, amounted to 8.0 degrees. The rate of progressive effect was thus evidently diminishing. On pushing the rarefaction as far as it was really practicable, or till the residual air had become rarefied about 300 times, the change indicated by the thermometer did not reach to more than 8.3.

But, to determine the absolute quantity of heat which is disengaged in the transition of air from a rarer to a denser state, it becomes necessary to ascertain what part of it was consumed on the sides of the receiver. By varying the size of the receiver, and consequently altering the proportion between its surface and its contents, some light may be thrown on this question. Another similar receiver was therefore provided, having half the former dimensions; and with this the same set of experiments was repeated. Its included air being reduced successively to the density of four-fifths, three-fifths, two-fifths, and one-fifth, and then rarefied as much as possible, the thermometer mounted each time through the shorter spaces of 1.8, 3.2, 4.2, 4.8, and finally 5.0 centesimal degrees. These quantities evidently follow the same proportion as the former, of which, indeed, they are only three-fifths. But the smaller receiver having, under the fourth part of the surface of the larger, only the eighth part of its contents, exposes comparatively twice the extent of surface. The rise of temperature which its included air exhibits, must consequently be the same as what would have obtained within the larger receiver, if, while its capacity remained the same, its surface had been actually doubled. If we suppose the air to hold one part of the heat, while two parts and four parts are respectively expended on the insides of the receivers, the results would correspond with observation; for the whole quantity evolved being in both cases the same, the air under the larger receiver would retain one-third, and under the smaller receiver, only one-fifth; the impressions being thus in the ratio of five to three. The same conclusion may be obtained somewhat differently. If the heat spent on the inside of the large receiver had been spread over twice the surface, it would have raised the temperature only 1°.5; but this mounted really to 1°.8, and therefore the difference .3 was the effects of 1°.8 derived from the contained air. Of the heat thus shared between the air and the doubled surface, one part was hence retained, and five communicated. Consequently, to obtain the true results, it is only necessary to multiply the second set of quantities by five, or the first set by three. If no waste, therefore, took place against the inside of the receiver, the heat evolved in the passage of air from the densities of four-fifths, three-fifths, two-fifths, one-fifth, and extreme rarefaction, to its ordinary state, would be 9, 16, 21, 24, and 25 centesimal degrees.

It is not difficult to discover the law of this progression. They are obviously formed by the successive addition of the odd numbers 9, 7, 5, 3, and 1; and are, consequently, the excesses of the square of 5 above the squares of 4, 3, 2, and 1. Wherefore, if the square of the density be taken from unit, the remainder, multiplied by 25, will express, in centesimal degrees, the rise of temperature which accompanies the return of the air to its ordinary state.

The numbers thus obtained, however, do not still express the final results. If the restoration of four parts of the air included under the receiver, to their usual density, disengage heat sufficient to raise the temperature of the whole five parts 9 degrees, its real measure must have been $11\frac{1}{4}$ degrees, or the former augmented in the ratio of 4 to 5. For the same reason, if three-fifths, two-fifths, and one-fifth of the air in the transition of density, evolve portions of heat which would elevate the temperature of the mass 16, 21, and 24 degrees, the actual quantities are $26\frac{2}{3}°$, $52\frac{1}{2}°$, and $120°$, or those numbers multiplied by 5, and divided by 3, 2, and 1.

These conclusions are easily reduced to formulæ. Let $\theta$ denote the density of the air, and $25\ (1-\theta^2)$ will express, in centesimal degrees, the elevation of the thermometer which would follow the re-admission of the air, if none of the heat were spent on the inside of the receiver. Consequently, $25\left(\frac{1-\theta^2}{\theta}\right)$, or $25\left(\frac{1}{\theta}-\theta\right)$, will exhibit on the same scale the whole quantity of heat evolved in the restoration of density. The last formula is extremely simple, implying that 25, multiplied into the difference between the density of air and its reciprocal, will represent the measure of heat due to the change of condition. This result may be either additive or subtractive; it may express the heat emitted in the condensation of air, or the heat absorbed during its opposite rarefaction.

Thus, the heat extricated from air which has its density doubled, is $25°\ (2-\frac{1}{2})$, or $37\frac{1}{2}°$; and the same quantity is withdrawn, either when this air recovers its former density, or when air of the ordinary state expands into double its volume. Hence the copious heat extricated by the sudden compression of air. If it were condensed thirty times, the heat discharged would amount to $25\left(30-\frac{1}{30}\right)$, or $749°$, which is more than sufficient for the inflammation of fungous or soft substances. On this principle are constructed the pneumatic matches invented by Mollet of Lyons, which produce their effect by the momentary action of a small syringe.

But, to discover the relative capacity or attraction which

air of a given density has for heat, it would be necessary to know the extent of the natural scale, or the position of the absolute zero. The conclusions, however, from different data, are not very constant; yet several experiments appear to fix nearly the point from which the infusion of heat commences at 750 centesimal degrees below congelation. On this supposition, therefore, air which is rarefied thirty times has its capacity doubled, the heat contained in it being dilated only fifteen times. For the same reason, air sixty times rarer than ordinary acquires a triple attraction for heat, which, in this union, becomes attenuated only twenty times. But these inferences are merely speculative, and the law of the gradation of temperature in the atmosphere is quite independent of the existence of an absolute term of heat.

The last formula now investigated has been already laid before the public, without any explication, however, or indeed indication, of the process by which it was discovered. The experiments on which it rests were begun many years since, and have been repeated with every precaution. But the mean results only are retained; and, for the sake of simplicity, a few slight modifications have been introduced, to adapt the apparatus to more convenient proportions. Though it was impossible to blow a receiver that should have *exactly* half the dimensions of another, nothing seemed easier, from the general mode of investigation, than to apply the minute corrections which any small deviations of size or form required. The mixture of obscure and intricate computations has been thus avoided.

The gradation of the effects disclosed by this experimental research is more easily and clearly traced in geometrical diagrams. Having divided the absciss AF into five equal portions, erect the several perpendiculars EG, DH, CI, BK, and AL, equally respectively to 9, 16, 21, 24, and 25 parts on any scale, and connect the points F, G, H, I, K, and L, by a curve line. It is readily perceived that the curve now traced must be a parabola, formed on the axis LA. Wherefore, if AF represent the height of the barometric gage of the air-pump, and AD its altitude, after a partial exhaustion, the parallel DH will express the rise of the included thermometer on the re-admission of the air which had been extracted. The axis AL itself denotes the extreme effect, or what would take place if the rarefaction were pushed to the utmost. While the mercurial column, therefore, descends by equal intervals to E, D, C, B, and approximates to A, the elevation of the thermometer through the spaces EG, DH, CI, BK, and nearly AL, advances at first uniformly, and afterwards continually more slowly, till it becomes stationary.

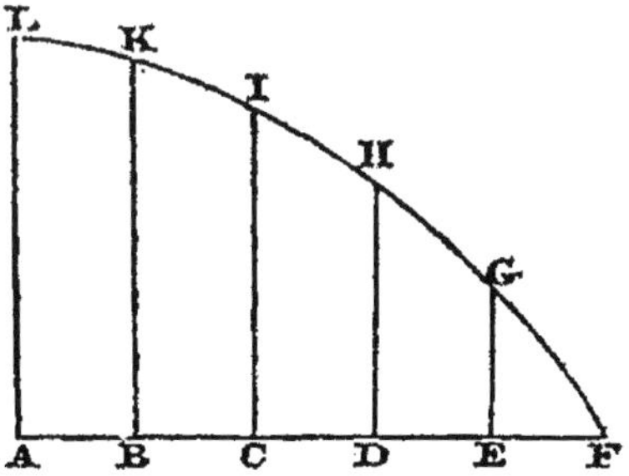


But the parallels to the axis of a parabola do not express the whole of the heat disengaged by the attenuated air during the resumption of its density. The line DH, for instance, marks only the rise of temperature communicated to the entire mass AF, by the heat evolved from the portion AD; and to represent the true measure of this heat, it must, therefore, be increased in

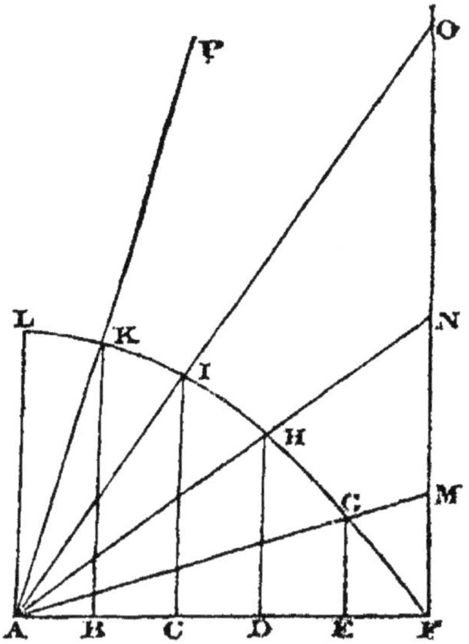

the ratio of AD to AF. Hence, from the point A, and through the extremities of the parallels EG, DH, CI, and BK, draw the oblique lines AG, AH, AI, and AK, to meet the extended perpendicular FO; the intercepted segments, or parabolic tangents, FM, FN, FO, and FP, (if this last were completed,) will exhibit the real portions of heat liberated from any of the corresponding densities AE, AD, AC, AB.

Let the part AL of the axis be taken equal to the parameter, and the ordinate AF must likewise be equal to it: draw the vertical tangent LQ meeting the parallel DH in Q; join the oblique line AQ, and produce it to meet the perpendicular in R. From the property of the curve, QH : AD : : AD : AF; but from the mutual relation of the diverging lines AN, AR, and the parallels DQ and FR, AD : AF : : AH : AN : : QH : NR; whence QH : NR : : QH : AD, and, therefore, NR is equal to AD. Consequently, FN, which expresses the whole evolution of heat corresponding to the density AD, is equal to the difference between AD and FR. But since AD : AF : : DQ or AF : FR, the line FR is the reciprocal of AD, the parameter AF being considered as unit. The measure of heat evolved, or the parabolic tangent FN, is, therefore, as before

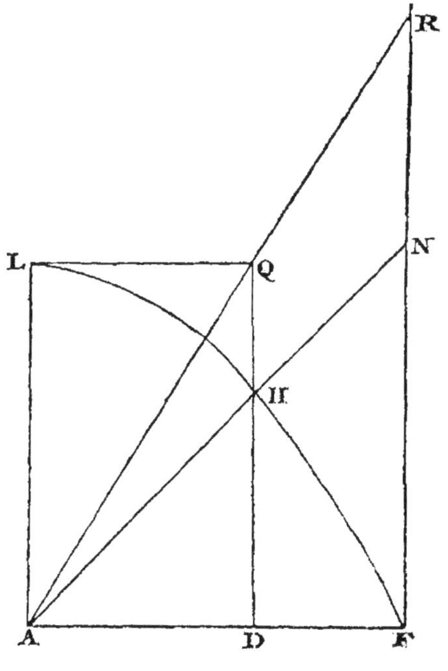

expressed, by the difference of the density of the air and its reciprocal.

The same result may be represented geometrically in another way. Let LIG be a rectangular hyperbola referred to its asymptotes AE and AF, the axis being AO. If the perpendicular IC, or the absciss AC, express the ordinary density of the air, and the ordinate DH denote any other density, the intercepted segment HN, or the difference between DH and its reciprocal AD or DN, will exhibit the additional share of heat required for its constitution. On the contrary, if the air acquires the higher density BK, the quantity of heat which it must evolve in its transition will

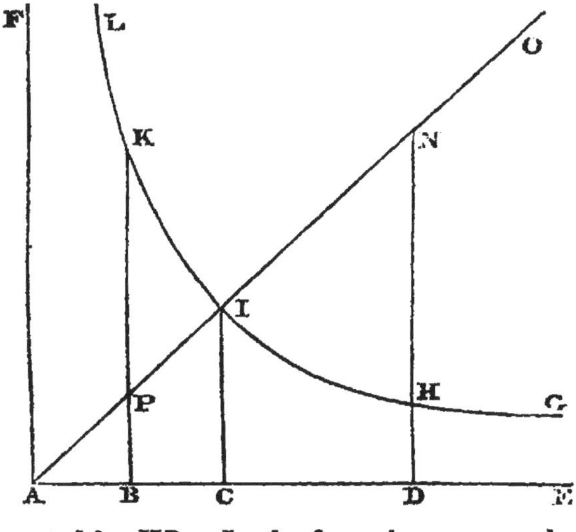


be represented by KP. In the figure here annexed, the air is presumed to have its density reduced to the half in one case, and doubled in the other; the quantities of heat, HN and KP, which are evolved and absorbed, being then equal and opposite.

Since the absolute quantity of heat contained in every part of any vertical column of the atmosphere has been shown to remain unchanged, these diagrams must likewise represent the diminution of temperature in the higher strata, corresponding to the decreased density of the air at different elevations. The same formula will determine the measure and gradation of this effect. Reckoning the density of the air at the surface of the earth unit, the difference between the density at any given altitude and its reciprocal, being multiplied by 25, will express the mean diminution of temperature in centesimal degrees; or if 45 be employed as the multiplier, the product will exhibit the same result in degrees of Fahrenheit's scale.

This very simple deduction from theory is amply confirmed by numerous and extensive observations. But a few leading facts will perhaps be deemed sufficient for exemplification.

According to Lasius, the same barometer which, at Goslar, an ancient town seated in the bosom of the Hartz Forest, stands at 29,500 inches, would fall to 26,444 on the top of the Brocken, in that mining district. This gives .896 for the density of the air on the summit, the reciprocal of which is 1.116; but $1.116 - .896 = .22$, and $.22 \times 25° = 5°.5$, the calculated difference of temperature. The actual difference is very nearly the same, being only 5°.2; as we had once an opportunity ourselves of observing, having found the temperature of a copious spring at Goslar to be eight centesimal degrees, while that of the noted Hecken-Brunnen, or *Witch-Well*, on the summit of the Brocken, was only 2°.8.

Saussure, whose accuracy always inspires confidence, found that, while at his villa of Conche, near Geneva, the barometer stood at 28.500, another similar instrument fell to 25.165 on the top of the mountain of Nant Bourant. The diminished density of the air at this elevation was, therefore, .890; the difference between which and its reciprocal 1.123 being multiplied by 25°, gives 5°.82. But a thermometer buried a whole night at two feet deep in that lofty station, marked only 12.75, while it indicated 6.25 more, or 19°, a few days afterwards, when sunk to the same depth at Conche. The discrepancy here is thus less than half a degree.

On the top of a higher mountain, the Chapieu, the same observer found the ground, at a depth of two feet, to be colder by 6°.44 than at Conche; but the corresponding density of the air and its reciprocal were .872 and 1.147; consequently, 25° × .275 = 6°.87.

While the barometer at Conche stood at 28.500 inches, the mercurial column was only 19.836 inches on the summit of Mont Cervin, a still loftier mountain. The density of the air at this elevation was therefore .696, which being taken from its reciprocal 1.437, leaves .741 to be multiplied by 25°, indicating 18°.52 as the diminution of temperature. The actual medium difference ascertained from corresponding thermometrical observations, made at depths in the ground from one to three feet, on the top of Mont Cervin and at Conche, was 18°.25, almost exactly the same.

Such is the nice agreement on the whole, between theory and observation, with regard to the decrease of the mean temperature in the higher regions of the atmosphere. This gradation of cold varies, however, to a certain extent with the seasons. Since the heat derived from the sun is chiefly accumulated at the surface of the earth, the changes of temperature which take place through the year in the elevated strata of our atmosphere, must evidently be less than what are experienced below. The lofty tracts of air, remote from the primary scene of action, preserve nearly an equable temperature, and scarcely feel the extreme heat of summer or winter's frost. In ascending the atmosphere, the decrease of warmth is hence more rapid in the fine season, and more slow in the darkened period of the year. In many places, it will not be far from the truth, perhaps, to assume 30° for the multiplier during the summer months, and only 20° during those of winter.

Thus General Roy, a diligent and experienced observer, found, in the month of August, the air at the top of Snowdon was, in the course of a whole day, at an average, 7.2 centesimal degrees colder than on Carnarvon Quay; but the difference between the density at that elevation, and its reciprocal, or between .878 and 1.139, being only .261, would require nearly 28 for the multiplier.

In the early part of September, the same observer noticed the centesimal thermometer to stand at 10 degrees lower on the top of Ben Lawers than at Weem, the relative density of the air at that height being .868. The difference from its reciprocal is .284, which would hence require to be multiplied by 35 to give the actual diminution of temperature.

Again, Saussure found, on his visit to Mont Blanc, the air on its summit to be 31 centesimal degrees colder than at Geneva. The relative density was .592, which being taken from its reciprocal 1.689, leaves 1.097; consequently the multiplier required is 28.

This ingenious philosopher passed several days encamped on the Col du Géant, where he found, from the mean of eighty-five observations, the temperature of the air to be only 4°.54, or 20°.3 colder than at Geneva; but the relative density of that elevated stratum and its reciprocal were .704 and 1.420; the difference of which, or .716, would require to be multiplied by 28½, to indicate the diminished temperature.

The observations made on the decreased temperature of the higher regions of the atmosphere by the ascension of balloons, appear generally to indicate rather a slow rate of diminution; but it should be recollected that those daring aerial flights have seldom been performed except in the fine season of the year. Besides, the car, the balloon, and its cordage, will not immediately acquire the temperature of the elevated strata, but continue for a considerable time to diffuse a sensible portion of heat. A memorable example, however, is entirely conformable to the general principle. Charles, the first who ascended the atmosphere by means of a balloon filled with hydrogen gas, found, on the 1st of December 1783, the thermometer depressed 11 centesimal degrees at the greatest elevation, the column of the barometer having sunk from 29.24 to 20.05 inches. This would require a multiplier less than 20.

In ascending through equal heights, the density of the atmosphere, as derived from the incumbent pressure, diminishes in a continued proportion. This density is hence represented by the ordinates of a logarithmic curve, of which the absciss denotes the altitude above the surface. The cold which prevails in the upper strata would no doubt modify the dilatation of the air; but since it follows nearly the same progression, it cannot materially affect the general results. In the figure annexed, the vertical spaces expressing the elevations of one, two, three, and four English miles, the horizontal lines 1D, 2F, 3H, 4K, represent the corresponding mean temperature in the temperate latitudes; but, from the nature of the logarithmic curve, if it were supposed to be continued in like manner below the surface, the horizontal lines at the same equal distances would be the reciprocals of the former. Let this extension of the curve be, therefore, folded back, and placed over its first portion, rising from OA, and the difference between the horizontal lines, or the distance between the two branches of the curve, as, from the point A, they constantly spread from each other, will exhibit the diminution of temperature corresponding to the elevation. These branches of the curve at first diverge at the same angle; but the wider branch afterwards spread-with a larger sweep, and their mutual distance slowly increases.

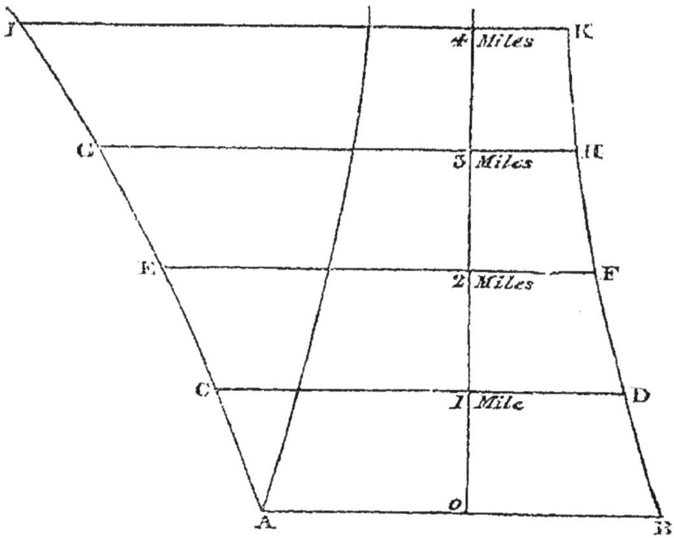


The gradually increasing progression, which marks the diminution of temperature in ascending the atmosphere, is still more apparent from another property of the logarithmic curve. The difference between any two ordinates is constantly proportional to the intercepted area. Consequently, the successive diminution of temperature corresponding to the elevations of one, two, three, or four miles, are expressed by the areas ABDC, CDFE, EFHG, and GHKI. But, since the component spaces 0C, 1E, 2G, and 3I, augment evidently faster than 0D, 1F, 2H, and 3K, diminish, the decrements of heat encountered in ascending the atmosphere must on the whole increase. It is further evident, that if those spaces were reduced to extremely narrow belts, they might be considered as proportional merely to their several lengths; and hence the momentary decrements, or the rate of the diminution of temperature at the heights of one, two, three, and four miles, are expressed by the transverse lines

CD, EF, GH, and IK. If a calculation be instituted for the temperate climates, while the mean temperature at the level of the sea decreases one centesimal degree during an ascent of 540 feet above the surface, it suffers a similar diminution in 529 feet at the altitude of a mile; but at two, three, four, and five miles of elevation, the same difference will obtain at the contracting intervals of 498, 454, 401, and 346 feet. Should Fahrenheit's scale be preferred, those numbers multiplied by five, and divided by nine, will give respectively 300, 295, 277, 252, 223, and 192, for the ascents due to a decrement of one degree at the surface, and at the heights of one, two, three, four, and five miles.

Hence the altitude of any place above the surface of the ocean may be nearly ascertained from an observation of its mean temperature. In the milder climates it will be sufficiently accurate, in moderate elevations, to reckon an ascent of 540 feet for each centesimal degree, or 100 yards for each degree on Fahrenheit's scale, of diminished temperature. Thus, the Black Spring, a copious perennial source which bursts forth on the ridge of the Pentland Hills, in the vicinity of Edinburgh, is found to indicate only 7.2 centesimal degrees, or 1°.6 less than the medium due to that parallel. But 1°.6 multiplied by 540 gives 868 feet; which differs by a few feet only from the altitude as determined by actual levelling.

Again, we had ascertained the temperature of a plentiful and deep-seated spring in the meadow below Schweitz, near the margin of the branching lake of the Forest Cantons, to be only ten centesimal degrees, or $3\frac{1}{2}°$ lower than the standard temperature of that latitude. This gives $3\frac{1}{2} \times 540$, or 1890 feet, for the elevation of the central valley of Switzerland above the level of the sea.

As a final example, we may cite an observation made in Upper India by Dr Francis Buchanan, who found a spring at Chitlong, in the Lesser Valley of Nepaul, to indicate the temperature of 14.7 centesimal degrees; which is 8°.1 below that of the standard of its parallel, or 27° 38′. Wherefore 8.1 × 540, or 4374 feet, is the elevation of that valley.

When the altitude above the sea is very considerable, however, a multiplier somewhat decreasing from 540 must be taken. But it is easy to show, that the same result very nearly will be obtained by applying a small correction. If the product of 540 into the diminution of temperature be again multiplied by the square of this difference in centesimal degrees, and then divided by the constant number 15,000, the quotient will express the quantity to be subtracted. Thus, in the last example, the square of 8.1 or 65.61 multiplied into 4374 makes 286,978, and this divided again by 15,000 gives 18, to be deducted from 4356 for the corrected altitude.

This correction, therefore, amounts only to 33½ feet at the elevation of one mile; but for two, three, four, and five miles, it would rise to 266, 904½, 2128, and 3350 feet. Though it augments thus rapidly, it may in ordinary cases be totally disregarded, and the allowance of one degree by Fahrenheit's scale for every hundred yards of ascent is a rule of most easy recollection.

That the decrements of heat corresponding to equal ascents in the atmosphere are not uniform, however, but augment gradually with an accelerating progression, is confirmed by the appeal to actual observation. Thus, in proceeding from Geneva to the vale of Chamouni, through an elevation of 2094 feet, Saussure found the temperature of the air to decrease only 3¼ centesimal degrees, or at the rate of one degree in the rise of 644 feet. Between Chamouni and the Col du Géant, an interval of 7940 feet, the diminution of temperature was 17°, being a degree for each 465 feet of ascent. Again, the summit of Mont Blanc, though only 4400 feet higher than the Col du Géant, was 10°.8 colder, which gives a degree for every rise of 407 feet. The acceleration of cold is here very perceptible; but the gradation from Geneva to Chamouni was unusually slow, owing to the position of this sequestered spot, where the action of the sun-beams in summer, being sheltered from the influence of dispersing winds, accumulates above the standard. Near the level of the sea, the ascent, corresponding to the decrement of a centesimal degree, very seldom exceeds 600 feet.

Since, in ascending from the surface, the temperature constantly diminishes, there must in every latitude exist a certain limit of elevation at which the air will attain the term of congelation. The mountains, likewise, which rear their heads above that boundary, are covered with eternal snow. In the higher regions of the atmosphere, especially within the tropics, the temperature varies but little throughout the whole year. Hence, in those brilliant climates, the line of perpetual congelation is strongly and distinctly marked. But in countries remote from the equator, the boundary of frost rises after the heat of summer as the influence of winter prevails, thus varying its position over a belt of some considerable breadth.

The height of the limit of perpetual congelation for any latitude, is easily determined from the principles already established. Let $t$ denote the mean temperature in centesimal degrees at the surface of the ocean, and $x$ the density of the upper atmosphere at the line where the reign of frost begins. It is evident from the formula, that $\left(\frac{1}{x} - x\right) 25 = t$; whence $x^2 + .04\, tx = 1$; and this quadratic equation being solved, gives $x = \sqrt{1 + .0004t^2} - .02\, t$. From the density of the air thus found, the corresponding latitude is discovered by the application of logarithms, as in the barometrical measurements. Hence the following table is computed:

| Latitude. | Height of Curve of Congelation English Feet. | Latitude. | Height of Curve of Congelation English Feet. | Latitude. | Height of Curve of Congelation English Feet. |
|---|---|---|---|---|---|
| 0° | 15207 | 31° | 11253 | 62° | 3365 |
| 1 | 15203 | 32 | 11018 | 63 | 3145 |
| 2 | 15189 | 33 | 10778 | 64 | 2930 |
| 3 | 15167 | 34 | 10534 | 65 | 2722 |
| 4 | 15135 | 35 | 10287 | 66 | 2520 |
| 5 | 15095 | 36 | 10036 | 67 | 2325 |
| 6 | 15047 | 37 | 9781 | 68 | 2136 |
| 7 | 14989 | 38 | 9523 | 69 | 1953 |
| 8 | 14923 | 39 | 9263 | 70 | 1778 |
| 9 | 14848 | 40 | 9001 | 71 | 1611 |
| 10 | 14764 | 41 | 8738 | 72 | 1451 |
| 11 | 14672 | 42 | 8473 | 73 | 1298 |
| 12 | 14571 | 43 | 8206 | 74 | 1153 |
| 13 | 14463 | 44 | 7939 | 75 | 1016 |
| 14 | 14345 | 45 | 7671 | 76 | 887 |
| 15 | 14220 | 46 | 7402 | 77 | 767 |
| 16 | 14087 | 47 | 7133 | 78 | 656 |
| 17 | 13947 | 48 | 6865 | 79 | 552 |
| 18 | 13798 | 49 | 6599 | 80 | 457 |
| 19 | 13642 | 50 | 6334 | 81 | 371 |
| 20 | 13478 | 51 | 6070 | 82 | 294 |
| 21 | 13308 | 52 | 5808 | 83 | 226 |
| 22 | 13131 | 53 | 5548 | 84 | 167 |
| 23 | 12946 | 54 | 5290 | 85 | 117 |
| 24 | 12755 | 55 | 5034 | 86 | 76 |
| 25 | 12557 | 56 | 4782 | 87 | 44 |
| 26 | 12435 | 57 | 4534 | 88 | 20 |
| 27 | 12145 | 58 | 4291 | 89 | 05 |
| 28 | 11930 | 59 | 4052 | 90 | 0 |
| 29 | 11710 | 60 | 3818 | | |
| 30 | 11484 | 61 | 3589 | | |

This table, though calculated from theoretical data, will be found to coincide with actual observation. It suggests a variety of important results; and the position of the snowy boundary may direct the intelligent traveller to estimate the elevation of any country. Thus, Ben Nevis, the highest mountain in Scotland, is generally seen covered with snow through the whole year, except during two or three weeks in the month of July. Its summit, therefore, does not reach the limit of congelation, which, at the latitude of 57°, has an altitude of 4534 feet. Accordingly, the height of that mountain is only 4380 feet. Again, we learn from the relations of travellers, that, though perpetual snow covers the stupendous ridge of the Himalaya Mountains in Upper India, it descends but a short way along their sides, and leaves, for the *ghauts* or passages below, a grassy plain. But the boundary of congelation traverses the parallel of 30° at the altitude of 11,484 feet, and consequently that towering range may rise probably 4000 or 5000 feet higher, and thus surpass the elevation of Mont Blanc. The pretended altitudes of 23,000 or even 27,000 feet, so recently assigned to those mountains, are hence scarcely credible.

The gradation of altitude which marks the snowy boundary in different latitudes is most clearly perceived from the inspection of a diagram. Thus, if the horizontal line repre-

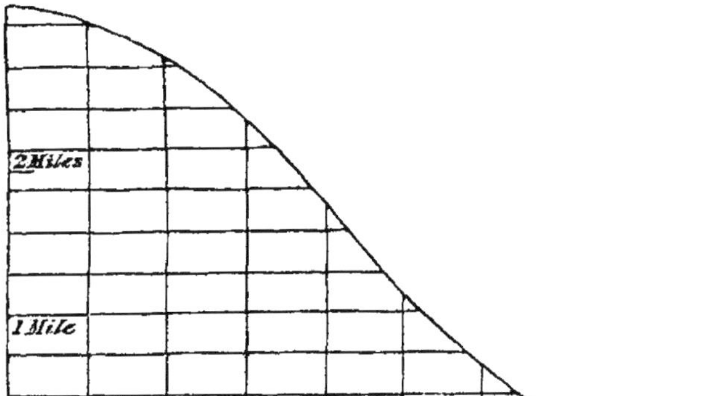


sent the distance of the pole from the equator, divided into 90°, and each vertical line corresponding to the latitude denote the height of the limit of perpetual congelation in English miles, the extremities of these perpendiculars being connected, will form a curve of contrary flexure, of which the one half is nearly the reverse of the other. This curve, therefore, bends slowly at first from the equator, most rapidly about the middle latitudes, but again slants near the surface of the pole.

The ordinates of this curve of congelation, or the altitudes of the limit of eternal frost, might also be found, with sufficient precision, from the corrected gradation of cold, if the primary multiplier were in each parallel of latitude adapted to the mean temperature. In the high latitudes, the correction to be applied is small, but it becomes very considerable within the tropics. Thus, in the middle latitude of 45°, it is only 110 feet; but under the equator it amounts to 877 feet. But, except in this extreme case, the altitude of the snowy boundary will, on the whole, differ very little from the product of the mean temperature by 540. That line must, therefore, approximate to a coincidence with the curve formed, by referring the series of temperatures to the successive latitudes. Let the tangent 0 90 be applied equal

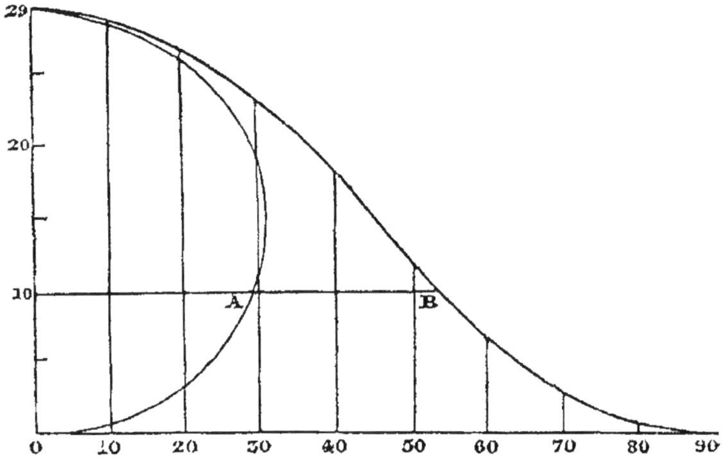


to the semicircumference 0 A 29 of a circle, and every intermediate point B be determined, by making any parallel 10 AB equal to the intercepted arc A 29. If the absciss then represent the meridian extending from the equator to the pole, the several ordinates will express the corresponding mean temperatures at the level of the sea. The reason of this construction is easily perceived; for the arc O A of the generating circle being equal to the complement of the corresponding latitude, its chord is equal to the cosine of this latitude, which belongs to a circle having twice the diameter of the former; but the ordinate from B, or the perpendicular segment O 10, is proportional to the square of the chord O A, and consequently to the square of the cosine of the latitude.

The curve now constructed is what has been called the *Companion of the Cycloid*. In fact, it differs from the cycloid only by the defect of the generating circle. If each parallel AB were extended as far beyond B as 10 A, the cycloid would be traced. This deficient cycloid has obviously the same aspect as the curve of perpetual congelation. It is only less flattened towards the equator, where the correction of altitude chiefly operates. A close inspection, however, is requisite to distinguish their difference.

The companion of the cycloid, or curve of temperatures might also be formed, by unrolling half the convex surface of a cylindrical wedge or *ungula*. The curve already delineated was only the expansion of the annexed figure. If the semicircumference of the base of the cylinder were divided into 90 equal parts, to denote the successive latitudes from the equator to the pole, the perpendicular

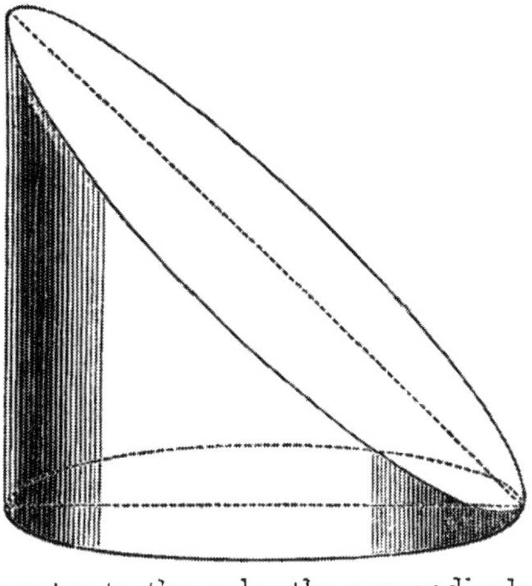

lines terminating in the oblique section would express the corresponding mean temperatures. These lines, again, might be easily measured by parallels of temperature, or arcs of circles, dividing the altitude of the wedge into twenty-nine equal parts, and running parallel to the plane of its base. If the whole wedge or ungula were taken, the combination of vertical lines and of horizontal arcs, would exhibit all the gradations of mean temperature in its progress and subsequent decline from pole to pole. It is easy to perceive that, on this plan, an elegant instrument could be constructed, equally calculated for illustration and practical use.

The curve of perpetual congelation must evidently rise higher during the tide of summer, and again descend in the winter months, thus oscillating between certain limits of elevation. The intervening belt is narrow under the equator, but it enlarges to a very considerable extent in the higher latitudes. On the breadth of this zone, where frost holds a doubtful reign, depends the formation of *glaciers* along the flanks of the snowy mountains. The fields of snow which are alternately melted and congealed, become at last changed by this process into solid ice, often grouped and fashioned by such irregular action into the most fantastic shapes. In its native seat, this icy belt acquires continual additions to its height, till the accumulating pressure at last tears the mass from its base, and precipitates its dissevered fragments to a lower level. In its new position, below the inferior boundary of congelation, the enormous pile suffers, on the whole, a very gradual thaw, which is sometimes protracted for several centuries. Meanwhile, in the higher magazine, another icy belt is again slowly collecting, which will, in due time, repeat the succession, and maintain the eternal circle of production and decay.

Within the tropics, the zone of undecided frost is so very narrow, that scarcely any trace of a glacier has been ever observed. But as that zone enlarges in the higher latitudes, the appearance of vast glaciers constitutes a very striking feature in the aspect of the lofty mountains. They occur frequently along the sides of the Pyrenees, but they are still more conspicuous in the recesses of the central chain of the Swiss Alps. Glaciers are likewise seen as far north as the verge of the arctic circle. Along the western shore of Norway, and the coast of Lapland, stretching onwards to the promontory of the North Cape, huge masses of columnar ice descend from the cliffs, or against the precipitous sides of the mountains, almost to the surface of the ocean.

Through the whole train of investigation now pursued, the warmth which vivifies our globe has been viewed as flowing from the fountain of the sun. The easiest mode of conceiving the subject, is to consider the heat which permeates all bodies, and unites with them in various proportions, as merely the subtle fluid of light in a state of combination. When forcibly discharged, or suddenly elicited from any substance, it again resumes its radiant splendour. Such changes are exhibited in the phenomena of electricity, and in many of the operations of chemistry. But the production of light is not less striking in the ordinary business of life. The glowing spark struck from flint, and the consuming fire, which even the rudest tribes had learned to excite, by the rapid friction of two pieces of dry wood, were by the ancient philosophers regarded as only detached portions of the divine and celestial flame which illumines the empyreal vault. The same notion was embraced by the poets, and gives sublimity to their finest odes. Heavenly Light, Invisible Light, and the Light of Life,* are lofty and refined expressions, which gleam through the mystic hymn of Orpheus.

* ΟΥΡΑΝΙΟΝ ΦΩΣ—ΑΦΑΝΤΟΝ ΦΩΣ—ΖΩΗΣ ΦΩΣ.

Those poetical images which have descended to our own times, were hence founded on a close observation of nature. Modern philosophy need not disdain to adopt them, and has only to expand and reduce to precision the original conceptions. If any body be exposed to the sun's incident rays, it will experience a rise of temperature exactly proportioned to the quantity of light intercepted or absorbed. When those rays are either concentrated or attenuated, the effect produced by their detention is proportionally augmented or diminished. The calorific action is occasioned solely by that portion of the light which remains united to the opposing substance. The rays which are either transmitted or reflected, leave no trace of their energy. Neither a very bright reflecting surface, nor a fine pellucid medium, is sensibly heated on being exposed even to a meridian sun. But since reflection and transmission cannot be perfect, a certain calorific effect, however small, and varied by circumstances, is always the result. A polished surface of silver may absorb from the tenth to the fifth part of the perpendicular rays; and we have seen that the thin body of atmosphere itself will detain one fourth part of the vertical light shot down through it, and one half of the beams that slant at an inclination of 25°. The absorption of light in its passage through water is proportionally greater; the perpendicular rays lose half their force by a descent of seventeen feet in that medium; and become reduced to one-fourth by traversing a path of thirty-four feet, which corresponds to the mass of an atmosphere. It hence follows, that only the hundred thousandth part of the vertical rays can penetrate below forty-seven fathoms. But this faint remnant is scarcely equal to the glimmer of the closing twilight. The depths of the ocean are never visited, therefore, by the cheering influence of the great luminary of day.

The rise of temperature produced by the same measure of the absorption of light, is in different substances inversely as their capacity or attraction for heat. Thus, if two very thin glass balls of the same diameter, and coated with China ink, be filled, the one with water and the other with mercury, and exposed to the sun, the latter will mark an accession of five degrees of heat, while the former will indicate only two. Or, if two equal and similar pieces of glass and lead were both of them blackened and presented to the incident solar rays, the lead will acquire a higher temperature than the glass, in the ratio of twelve to seven.

This theory agrees, therefore, in all its bearings, exactly with observation. But facts have been alleged of a contrary tendency; and some ingenious yet incorrect experiments, performed by a celebrated astronomer, appear to have, for several years, deceived the public into a belief, that, besides the ordinary mixture of coloured rays emanating from the sun, there is emitted also a cluster of invisible and less refrangible rays, which excite or constitute heat.

And as if such an hypothesis were not already sufficiently complex, it was afterwards assumed, that there exists likewise another series of dark rays, more refrangible than usual, whose office it is merely to abstract oxygen from the substances on which they fall. But a closer examination of the circumstances has proved, that, in forming these opinions, philosophers have proceeded to generalise too fast. If, while the sky is perfectly clear, which seldom happens indeed in this climate, the solar spectrum formed behind a prism be observed by a differential thermometer in a darkened room, the intensity of the heat will be found to augment most rapidly in passing from the one extremity to the other, from the limit of the violet to that of the red. When a prism of flint-glass was used, we found that the spectrum being divided into four equal spaces, the effects indicated by the thermometer, and corresponding to the range of the violet, the green, the yellow, and the red, were as the square numbers 1, 4, 8, and 16. If the spectrum were formed by another refracting medium, as the intermediate coloured spaces would alter their proportions, so likewise would the calorific action vary. But, in every case, the violet and the red rays which occupy the extremes must differ very widely in their relative energies. Without the spectrum, however, and beyond the absolute limit of illumination, no sort of action whatever is betrayed.* The distinct effects produced by

---

* It may be worth while to state here a very simple yet decisive experiment, which we have repeated more than once. If a circle of black paper ten inches in diameter be pasted over the middle of a burning-glass of the diameter of one foot, it will leave around the margin of the lens a sort of circular prism, which bends the coloured rays into a hollow double cone, whose intermediate apex occupies the focus. On directing the lens, thus partly shaded, to a bright sun, in a clear sky, a card held perpendicular to the beam, and drawn either forwards or backwards from the focus, will exhibit a narrow coloured ring, opening with intense brilliancy. Between the focus and the lens, the red rays will mark the outer edge of the vivid ring; but beyond that limit they will cross over and tint the inside. In either case it was easy to mark their extreme boundary. To examine the graduating calorific action of this condensed spectrum, a broad and flat piece of black sealing wax, having the gloss removed from its surface by a file, and a thin slip of white paper laid across it, was then held in the cone of light, two or three inches before the focus; and in the space of a few seconds a very narrow ring, shading inwards, however, was distinctly impressed in the wax, the roughened surface being melted and left hollow and glossy at the exact limit of the red on the outside, and smoothing slightly away on the inside. Beyond the apex of the double cone of light, a similar glazed narrow ring was imprinted, softening from the inside. In both cases, the circular impression on the wax corresponded exactly with the trace of the red rays, an interval of the fiftieth part of an inch being easily distinguishable.

the extreme coloured spaces may possibly depend on the peculiar qualities of the several rays, but perhaps proceed from the character of the diluted or concentrated energy which they display. The violet or blue rays will probably hasten the process of darkening a solution of the nitrate of silver; yet have we found that white light, attenuated to a certain degree, has likewise a power to accelerate this change. It is probable that the collected beams only retard the effect, by exciting heat, which may dispose the oxygen to adhere more strongly to its substratum. In like manner, the intense action of the red rays on the bulb of a thermometer, may proceed more from their condensation, than from some exclusive quality which they possess. What the nature is of that peculiar constitution which fits a ray of light to excite in the organ of vision any peculiar sensation of colour, it seems impossible to decide. Perhaps the whole distinction consists in the difference of the celerity with which it moves. The peculiar energy of any coloured pencil of light may depend chiefly, if not entirely, on its density. From the violet to the red, the successive tints of the spectrum rise with augmented force. The blue rays appear feeble, the green cause a milder impression, the yellow and the orange begin to glow with intensity, and the red dazzle the eye. The language of painters, being grounded on nice observation, marks appositely the gradation of strength. Blue is by them termed a cold colour, green a soft colour, yellow a rich colour, and red a warm colour.

For measuring the intensity, or at least the calorific action of light, no instrument is so finely adapted, by its peculiar delicacy, as the *Photometer*, which consists of a *Differential Thermometer* inclosed in a thin pellucid case, and having one ball made of black, and the other of clear glass. It will besides admit of some variety in its form and construction, and may be rendered on the whole very commodious and portable. Yet, owing to a combination of circumstances, this elegant instrument has only been partially and reluctantly admitted; and the philosophic world has still to discharge an act of justice, by receiving it into the favour and distinction which it so well deserves. Some, indeed, affecting to display superior sagacity, have taken the trouble to remark that it was only a species of thermometer, and not strictly a photometer, since it measures heat and not light. But what does the thermometer itself indicate, except *expansion?* As heat is measured by the expansion it occasions, so light is determined by the intensity of the heat which, in every supposition, invariably accompanies it. What other mode, after all, could be imagined for detecting the presence of light? How can an unknown quantity be expounded, but in terms of one already known?

The photometer is adapted for a variety of important meteorological researches. If such instruments, in the hands of skilful observers, had been dispersed to the remote regions of the globe, we should ere now have obtained a body of precise facts, highly instructive in themselves, and calculated to illustrate the nature of different climates. Meanwhile, we shall endeavour to state the general consequences which may be drawn from even a scanty range of photometrical observations.

The direct and absolute action of the sun's rays on the photometer, at the elevation of 30°, may be reckoned in this climate at 120 millesimal degrees. The effect is produced by the incidence of a pencil of light, which has for its base a circle of the same diameter as the black ball, but modified and regulated in its amount by the subsequent dispersion of the accumulated heat from the whole surface of the sphere, which is four times greater than that of the generating circle. If a thin disc were, therefore, substituted instead of the ball, and presented to the perpendicular rays of the sun, the impression would be doubled, or raised to 240°; or, if the emission of heat from the posterior surface of the disc were prevented, the calorific effect would amount by this accumulation to 480°, or 86 degrees on Fahrenheit's scale. Such is the rise of temperature which a dark surface of dry mould, sloping at an angle of 30°, yet exactly fronting the sun, might acquire under a diaphanous shell of glass, if scarcely any portion of the heat were supposed to be conducted downwards into the mass of the earth. But since the rays of light which traverse the atmosphere under an obliquity of 30° have, in comparison with perpendicular beams, their force diminished in the ratio of 750 to 563; the action of a vertical sun, through a thin capsule of glass, might heat up a dark horizontal surface 113° by Fahrenheit's scale. On removing the glass cover, this effect, in a calm still air, would be reduced about two thirds, or to 75½°.

The colour of the ground or substratum has very little influence in modifying the calorific action of the solar beams. The different shades of brown or gray have almost the same absorbent quality as black itself; yellow and orange begin to cause a partial reflection; but a surface even of pure chalk will not reflect perhaps the fifth part of the incident light. Hence the arid sands of Africa are often heated by a vertical sun to near the point of boiling water, insomuch as to scorch the feet of the wretched traveller, and to accumulate warmth sufficient for the slow roasting of an egg. Even within the arctic circle, the calorific action of a bright sun, in the height of summer, may exceed 49°. In those dreary regions, accordingly, amidst a dissolving world of ice, the pitch is yet sometimes softened or melted in the seams of the sides or the decks of ships.

It is obvious that the accumulated effect of the incident rays must increase in proportion as the conducting power of the medium is diminished. Hence, at an elevation of three miles and a half, where the density of the atmosphere is reduced to one half, the heat communicated would, on this account alone, augment from 75½ to 83 degrees. But the effect would be further increased, from the smaller absorption of heat in its passage to the surface. Under the equator, the whole accumulated action would, therefore, amount to 96 degrees. All travellers, accordingly, complain of the scorching rays which the sun darts from a dark azure sky on the summits of lofty mountains. Yet the contrast is more striking in the higher latitudes. Thus, in the middle parallel of 45°, the action of the sun at the summer solstice would excite a heat of 69° at the level of the sea, and of 90° at an elevation of three miles and a half; but at the winter solstice it would communicate only 17° below and 46° at the altitude assumed. Saussure was accordingly very much struck with the force and brilliancy of the sun-beams on the top of Mont Blanc.

It might easily be computed that, on the supposition of a perfect calm, the surface of the earth, under the equator, will, at the medium of a year, have its temperature raised 12°; and that, in the latitude of 45°, the mean annual impression would be only 5°. But, of the whole of the light received, the calorific action on a black mould, whether emitted from the sun or shed indirectly by the sky, may be deduced from the indication of the photometer. It is only required to diminish the power of the sun's rays in the ratio of the sine of their obliquity, and to reduce the action of the light reflected from the canopy of clouds to one half, or what is due to the medium inclination of 30°, then to multiply the sum of these quantities by eight, and divide by three, or to take two thirds of the quadrupled effect. Thus, suppose, while the sun's altitude is 40°, that the photometer marks 155°, which it very seldom ever reaches in this climate, and that it indicates only 20° if merely screened from the direct action of the sun. Now, 155° multiplied by the sine of 40°, makes 87°.8, which is augmented to 97°.8 by the addition of the half of 20°; and this number again being increased in the ratio of 3 to 8, gives finally 261 millesimal degrees, or 47° on Fahrenheit's scale. When the sun is obscured in clouds, the reflected light, from a dappled sky, will sometimes in summer affect the photometer to the extent of 50°. This corresponds to a heat of 16° of Fahrenheit communicated to the ground. During the fine season the photometer seldom in cloudy weather indicates less than 15°, which is equivalent to an impression of 6° on the embrowned surface of the earth. While the sun is enveloped in clouds, if the rest of the sky assumes a fine azure hue, the photometer will only mark 10°. But in the gloomy days of winter, the minute portion of light which pierces through the congregated mass of clouds will scarcely affect the photometer 5°, or excite a heat of 2° by Fahrenheit on the ground.

These augments of temperature are communicated to the ground unimpaired, only in the case, however, of a perfect calm. The agitation of the atmosphere will scatter the heat before it has accumulated. When the wind creeps along the surface of the earth at the rate of eight miles in the hour, it diminishes the calorific action of the light from the sun and from the sky one half; but if it sweeps with a velocity of 16, 24, or 32 miles in the hour, it will reduce the whole effect successively to the third, the fourth, or the fifth of its standard. The impression made on the ground seldom, therefore, exceeds the third part of the computed measure, and often will not amount to one fifth.

The simplest and most accurate method of examining the temperature acquired during the day on the surface of the earth, is to employ a differential thermometer of the pendant kind, about one or two feet in length, and having its lower ball surmounted by a small cylindrical cavity supporting the coloured liquor. This instrument being suspended or held in a vertical position, the lower ball resting on the ground, will evidently mark, by its moveable column, the difference between the temperature of the surface and that of the ambient air. In this way, we have found, during the summer months in this climate, that the ground was, by Fahrenheit's scale, generally two or three degrees warmer than the air near it in cloudy weather, and perhaps ten or fifteen degrees warmer when the sun shone powerfully upon it. But, under similar circumstances, the effect varies very considerably, according to the nature of the surface. While fresh ploughed land, for instance, indicates an increased temperature of perhaps 8°, a grass plot close beside it will scarcely show a difference of 3°. Nor is this distinction owing to any greater absorption of light by the black mould; the reflection from the surface, in both cases, being extremely small. A thin layer of hay, whether spread on the naked soil or on the green turf, will betray the same diminished effect. The fibres of the grass exposing a multiplied surface to the contact of the air, the greater portion of the heat is hence dissipated before accumulation. A corresponding effect has been remarked with respect to the impressions of cold. Thus, in the neighbourhood of Edinburgh, after a long tract of rigorous weather, the frost was found to have penetrated thirteen inches into the ground in a ploughed field, but only eight inches in one piece of pasture ground, and four inches in another. But, in some of the streets of that city, the frost had descended even below two feet, so as to begin to affect the water-pipes. The greater density and solidity of the pavement had no doubt conducted the frigorific impressions more copiously downwards, while the loose and spongy blades of grass had mostly scattered and wasted those impressions in the open field. This consideration, it is obvious, might lead to very important practical results.

The unequal action of light at the surface of the earth, whether produced by the various obliquity of the sun's rays, the different inclination of the horizon, or the alternating succession of day and night, is attempered, we have seen, by the actual flow of the heated portions of the atmosphere. Between the poles and the equator, a perpetual circulation of air is maintained, which confines the accumulating effects of heat within narrow limits. The prevalence, on the whole, of northerly winds in this hemisphere during summer, and of southerly winds in winter, tends likewise to mitigate the extreme impressions of hot and cold. But a current of warm air excited at first by the presence of the sun, continues to rise from the ground, and occasions the descent, therefore, of an opposite current of cold air, which, as the equilibrium of temperature is not soon restored, may be protracted through a great part of the night. The combined influence of these currents is hence continually exerted in cooling down the surface of the earth; but their activity being the greatest while the solar beams fall most copiously, the accumulation of heat is checked in little more than an hour after mid-day, while its further dissipation is prolonged through the whole of the night, sun-rise being generally the moment when the ground is coldest.

Such a concatenated system of aerial currents might hence appear sufficient to explain the gradation and general balance of temperature which prevails on the surface of our globe. An horizontal stream of air must evidently cause the flow of an opposite one, since the action must be the same on every part of the same parallel of latitude. The difference of the temperature of the surface from that of the ambient air will maintain the constant play of an ascending and a descending current. In clear and calm weather, this interchange between the higher and lower strata of the atmosphere will be most vigorous, owing then to the concentrated impression of the sun-beams. The perpetual commerce maintained in our atmosphere by the medium of these combined horizontal and vertical currents, forms no doubt an essential part of the system which attempers the constitution of this globe. But it is not the only mode by which nature seeks to preserve the harmony of her productions; and recent discovery has detected the existence of another auxiliary principle, extremely active, of most rapid and extensive influence, and continually at work, though subject to various modifications. To understand rightly, however, the operation of this principle, it will be necessary to recall the chief facts which have been disclosed relative to the propagation of heat.

It is well known, that, though partial causes may disturb the equilibrium of temperature among bodies, there is a constant tendency to restore it again. Yet heat still remains in the state of combination, without ever assuming a distinct form. Its balance is, therefore, maintained by a very different process from that which establishes the equilibrium between the several communicating parts of a liquid. The substratum of heat is not passive, nor do the calorific particles themselves merely flow from their redundance towards another situation where they happen to be deficient. But since the presence of heat is invariably accompanied by corpuscular distention, that portion of the substance which gains it will in the same degree expand. The actual transfer of heat through any mass will hence give occasion to a connected series of minute internal contractions and expansions. To consider the subject more fully, we shall suppose the conducting substance to be, 1. a solid; 2. a liquid; and, 3. a gaseous fluid.

1. When the surplus heat is conducted through a solid substance, a sort of alternating vermicular motion is excit-

ed along the whole train of communication. If heat were left to the energy of its own repulsion, it would, like light, dart with a speed almost instantaneous. But the time consumed by those interior oscillatory movements retards immensely the rate of transmission. The quickness of the oscillations themselves depends on the elasticity of the conducting substance; but their energy and extent are proportioned to the extreme difference of temperature, and the shortness of the tract, modified essentially by the peculiar nature of the conducting substance. In equal circumstances, glass transmits heat faster than wood, and metal faster than glass. But, even in the same class of conductors, the effects are very different; thus, box delivers the impressions of heat more quickly than cork, and silver conveys them with greater rapidity than platinum.

2. When the conducting substance is a liquid. The ordinary transmission of heat through a solid is now greatly augmented, from the diffusion occasioned by the mobility of the affected portions of the medium. Below the freezing point, ice will conduct heat through its substance; but after it has melted into water, a new and powerful agency is brought into play. The liquid particles, as they become successively warmer, acquire a corresponding expansion, and, therefore, rise upwards and spread through the mass, *carrying* with them and dispersing the heat which they have received. This diffuse buoyancy will depend evidently on the dilateable quality of the liquid. It is greater in alcohol than in water, and in water than in mercury; it is even more active in hot than in cold water. Near the point of congelation, indeed, the joint conducting power o water is scarcely superior to that of mere ice. The actual flow of a liquid, by whatever cause it is produced, must evidently accelerate the dissipation of heat.

3. When the medium of transmission is a gaseous sub-

stance, the heat is partly still conducted through the substance of the communicating mass, as if this were solid, and partly transferred by the streaming of the corpuscles, which come to be successively affected. But a new principle seems here to combine its influence, and the rate of dispersion in aeriform media is found to depend chiefly on the nature of the mere heated surface. From a metallic surface the heat is feebly emitted; but from a surface of glass, or still better, from one of paper, it is discharged with profusion. If two equal hollow balls of thin bright silver, one of them entirely uncovered, and the other closely enveloped in a coat of cambric, be filled with water slightly warmed, and then suspended in a close room, the former will only lose 11 parts of heat in the same time that the latter will dissipate 20 parts. Of this expenditure 10 parts from each of the balls are communicated in the ordinary way, by the slow recession of the proximate particles of air as they come to be successively heated. The rest of the heat, consisting of 1 part from the naked metallic surface, and of 10 from the cased surface, is propagated through the same medium, but with a certain diffusive rapidity, which in a moment shoots its influence to a distance, after a mode altogether peculiar to the gaseous fluids.

But those effects are modified by the different proximity of the air to the metallic surface. If the silver ball be covered with the thinnest film of gold-beater's skin, which exceeds not the 3000th part of an inch in thickness, the power of dispersion will be augmented from 1 to 7; if another pellicle be added, there will be a farther increase of this power, from 7 to 9; and so repeatedly growing, till after the application of five coats, when the repellent energy will reach its extreme limit, or the measure of 10.

The approximation of the metallic substratum thus evidently diminishes the power of the external pellicle in darting heat. No absolute contact exists in nature; but air must approach to a boundary of pellicle, or cambric, much nearer than to a surface of metal, from which it is always divided by an interval of more than the 500th part of an inch. A vitreous surface has very nearly the same property as one of the cambric or paper; from its closer proximity to the recipient medium, it imparts its heat more copiously and energetically than a surface of metal in the same condition.

By what process the several portions of heat thus delivered to the atmosphere shoot through the fluid mass, it seems more difficult to conceive. They are not transported by the streaming of the heated air, for they suffer no derangement from the most violent agitation of their medium. The air must, therefore, without changing its place, disseminate the impressions it receives of heat, by a sort of undulatory commotion, or a series of alternating pulsations, like those by which it transmits the impulse of sound. The portion of air next the hot surface, suddenly acquiring heat from its vicinity, expands proportionally, and begins the chain of pulsations. In again contracting, this aerial shell surrenders its surplus heat to the one immediately before it, now in the act of expansion; and thus the tide of heat rolls onwards, and spreads itself on all sides.

But these pulsations are not propagated with equal intensity in all directions. They are most powerful in the perpendicular to the projecting surface, and diminish as they deviate from that axis in the ratio of the sine of the angle of obliquity.

Nor are the vibratory impressions strictly darted in radiating lines, but each successive pulse, as in the case of sound, presses to gain an equal diffusion. Different obstructions may hence cause the undulations of heat to deflect considerably from their course. Thus, if a cornucopia, formed of pasteboard, present its wide mouth to a fire, a strong heat will, in spite of the gradual inflection of the tube, be concentrated at its narrow end; in the same way, probably, as waves flowing from an open bay into a narrow harbour, now contracted and bent aside, yet without being reflected, rise into furious billows.

But the same pulsatory system will enable the atmosphere to transmit likewise the impressions of cold. The shell of air adjacent to a frigid surface, becoming suddenly chilled, suffers a corresponding contraction, which must excite a concatenated train of pulsations. This contraction is followed by an immediate expansion, which withdraws a portion of heat from the next succeeding shell, itself now in the act of contracting; and the tide of apparent cold, or rather of deficient heat, shoots forwards with diffusive sweep.

That quality which enables a surface to propel the hot or cold pulses, likewise fits it under circumstances to receive their impressions. If a vitreous or varnished surface emits heat most copiously, it will also, when opposed to the tide, arrest with entire efficacy the affluent wave; and if, on the other hand, a surface of metal sparingly parts with its own heat, it detains only a small share of each warm appulse, and reflects all the rest.

Hence the construction of the *Pyroscope*, a delicate and valuable instrument, adapted to measure the warm pulses of air, or the intensity of the heat that darts continually from a fire into a room, which has been vaguely and inaccurately termed *radiant heat*. It is in fact only a modification of the *differential thermometer*, one of the balls being completely gilt with a thick gold or silver leaf. The *pyroscope* being placed at some distance from the fire, the hot pulses are mostly thrown back from the bright metallic surface; but on the naked glass ball they produce their full impression; and the same instrument will serve equally to indicate the pulsations excited from a cold surface. Thus, in a warm apartment, the pyroscope placed before a mass of snow, a block of ice, or even a pitcher of water from the fountain, will quickly intimate the chilling impressions propagated through the ambient medium. Nor has the brightness of the fire, or the glare of the snow, any sensible influence to affect the result; for, of the small portion of light transmitted, what falls on the diaphanous ball passes almost without obstruction, and what strikes the gilt ball, especially if this be covered with silver leaf, is nearly all reflected.

But the pyroscope, in its simple form, is scarcely calculated for making nice observations, when exposed out of doors to the agitation of winds, and the effulgence of light. A greater share of this light will generally be detained by the gilt surface, than what is absorbed in its passage through the diaphanous ball. On the other hand, all the effects on the instrument will be diminished by the rapidity of the circulation of the air.

The application of the pyroscope incontestibly shows, that no pulsation, whether calorific or frigorific, ever takes place, except from a conterminous surface, which is either hotter or colder than the surrounding gaseous medium. If the general equilibrium of temperature be interrupted, such aerial pulses will arise, whether the boundary of excitation be solid or liquid. They are produced feebly from a surface of metal, though evidently stronger from mercury than from silver; but they are projected with most energy from a surface of glass, and still more powerfully from one of varnish, of paper, of ice, and of water.

But since those hot or cold pulses are darted from such various surfaces, and since the softness of the external coat, and its tendency to fluidity, seem vastly to augment their power; may they not likewise be excited from a boundary of air itself? This extension of a great principle in the economy of nature has never yet been surmised; nor can it be readily brought to the test of direct experiment. A body of air, whether hotter or colder than the general medium, it is evident, could not remain for a moment in the same detached situation, but would continue to rise or to fall. In a confined place, however, a stratum of warm air may float incumbent over a mass of colder fluid. The difference of temperature between the conterminous surfaces would no doubt be constantly diminishing, from the slow intermixture and subsequent diffusion of the adjacent portions of the air. This tendency to an equilibrium might be counteracted, by causing either a constant stream of warm air to enter by the edges of the ceiling of the apartment, or one of cold air to flow from the side of the floor. Such would be the most favourable arrangement for performing the experiment. Both the ceiling and the floor would obviously soon acquire the same temperature as the current spreading over them, and could therefore exert no influence whatever in projecting the aerial pulses. If under such circumstances those pulses be hence found to exist, they must necessarily proceed from the action excited at the bounding surface which divides the warm from the cold stratum of air.

But an arrangement, less perfect indeed, yet sufficient for ascertaining the main fact, is entirely within our reach. In a close apartment, where a good fire is constantly kept up, the ceiling and the floor may be discovered by the pendant differential thermometer to have exactly the same temperature with its adjacent stratum. Yet the upper portions of the confined air of the room will be found several degrees warmer than the lower. Instead of being divided only into opposite ranges, the whole mass from the floor to the ceiling will, in consequence of the expansion and buoyancy of its heated particles, form a series of intermediate strata, not distinguished however by any very precise boundaries. But the intensity of action being proportional to the difference of temperature, the effect on the pyroscope must evidently be the same, whether it is produced by a single set of large pulses or by several sets of smaller ones. If, instead of one bounding surface, for example, above which the air is six degrees warmer than immediately below it, we suppose six such boundaries, each having an excess of temperature of only one degree, the pulses excited at the first of these intermediate surfaces, and successively augmented as they reach the second, third, and fourth, &c. surfaces, will at last acquire the same energy as if the aggregate difference of six degrees had been all exerted at once.

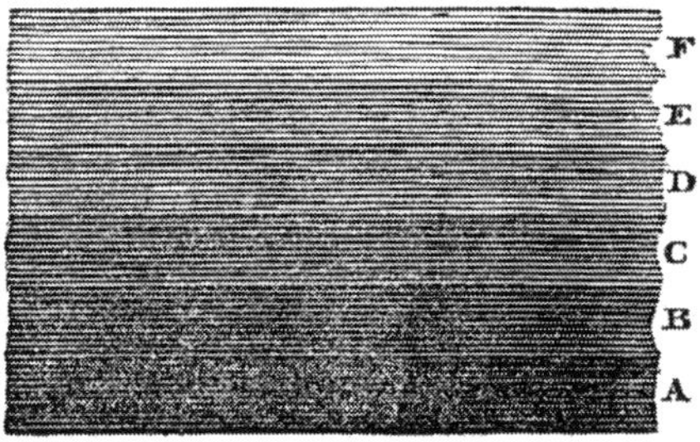


Thus, the under surface of the stratum F darts pulses downwards, which being augmented in succession at the under surfaces of the strata E, D, C, B, and A, may have finally the same intensity as if they had originated from the apposition of the extreme strata F and A. Accordingly, having planted a large screen immediately before the fire, if a delicate pyroscope be placed about the middle of the room, and a broad circular piece of metal suspended a few inches above it; on withdrawing this canopy after some time, the instrument will indicate a small impression of heat, seldom exceeding, however, one degree. But the effect may be rendered more sensible by a moderate concentration of the power excited. Suppose a hemispherical shell of thin brass, or even planished tin, having a diameter about four times greater, adapted under the naked or sentient ball of the pyroscope, which might occupy by its surface the place of the diffuse focus, or the middle space between the centre and the bottom of the cup. Thus mounted, the instrument will now, in the same situation, mark a very sensible calorific impression, amounting, at least in ordinary cases, to three or four degrees. Hot pulses are, therefore, actually shot downwards from all the upper strata of the confined air of a room in which a fire is kept steadily burning.

The experiment can likewise be reversed. Let an inverted pyroscope, composed of a pendant differential thermometer, have its sentient ball fitted with a small hemispherical cup which is turned downwards. This instrument being set on the floor, will remain at zero; but if lifted only a few feet, it will indicate a visible impression of cold received from below, which will increase to three or four degrees when the pyroscope is suspended near the top of the room. Wherefore, the upper surfaces of the successive decumbent strata, being comparatively colder, send upwards a series of chilling pulsations. Each of the conterminous boundaries appears thus to perform a double operation, shooting downwards impressions of heat, and darting upwards equal impressions of cold. Such a mutual exchange of influence must evidently tend to accelerate that progress to an equilibrium which the gradual intermixture of the different strata, if left quite undisturbed, would in time produce. The air of a close apartment exposed to the action of a steady fire is hence kept agitated through its whole mass by a series of opposite tremors, which continually disperse in all directions the irregularities of temperature.

If the action of the pulses excited in the air of a small room be made thus apparent, how much more striking should we expect to find the effect produced by the mingled tide of commotion collected in free space from the vast body of the atmosphere itself? Taking even the lowest range of strata, to the height perhaps of two miles, including scarcely one third part of the whole aerial mass, the difference of temperature between its extreme boundary will amount to twenty centesimal degrees, or 36 on Fahrenheit's scale. The order of the series, however, is exactly the reverse of what takes place in a close room, the air of the superior regions being invariably colder than at the surface of the earth. Accordingly, the simple pyroscope, exposed in calm weather to a clear and open sky, will at all times, if not disturbed by the influence of a strong light, indicate large impressions of cold, amounting to five, or perhaps even ten degrees. In most cases it may be sufficient to screen this instrument from the direct action of the sun's rays. But the action of light will be almost neutralized, by opposing a diaphanous ball to one gilt with silver, or contrasting a ball of the different shades of green or blue to another coated with pure gold leaf. But, to procure consistent results, it is still more necessary to guard against the deranging influence of winds.

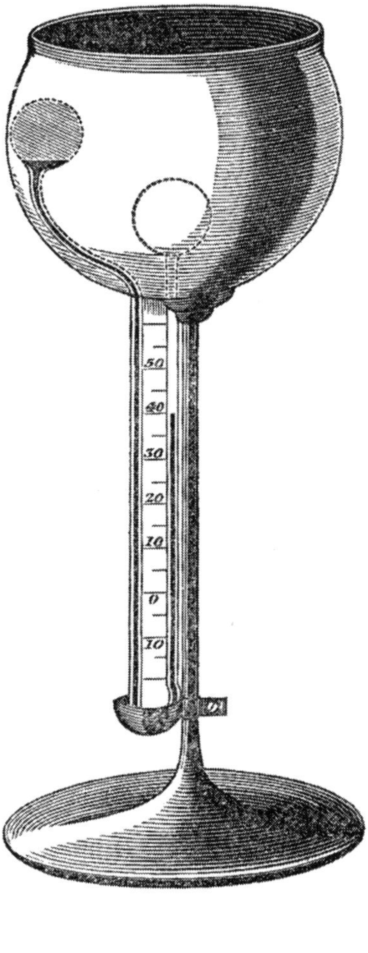


All these requisites are attained by adapting the pyroscope to the cavity of a polished metallic cup, of rather an oblong spheroidal shape, the axis, having a vertical position, being occupied by the sentient ball, while the section of a horizontal plane passing through the upper forms the orifice. The cup may be made of thin brass or silver, either hammered or cast, and then turned and polished on a lathe; the diameter being from two to four inches, and the eccentricity of the elliptical figure varied within certain limits according to circumstances. The most convenient proportion, however, is to have this eccentricity equal to half the transverse axis, and, consequently, to place the focus at the third part of the whole height of the cavity, the diameter of the sentient ball being likewise nearly the third part of that of the orifice of the cup. In order to separate more the balls of the pyroscope, the gilt one may be carried somewhat higher than the other, and lodged in the swell of the cavity, its stem being bent to the curve, and the neck partially widened, to prevent the risk of dividing the coloured liquor in carriage. A lid of the same thin unpolished metal as the cup itself is fitted to the mouth of the æthrioscope, and only removed when an observation is to be made. The scale may extend to 60 or 70 millesimal degrees above the zero, and about 15 degrees below it.

This instrument, exposed to the open air in clear weather, will at all times, both during the day and the night, indicate an impression of cold shot downwards from the higher regions. Yet the effect varies exceedingly. It is greatest while the sky has the pure azure hue; it diminishes fast as the atmosphere becomes loaded with spreading clouds; and it is almost extinguished when low fogs settle upon the surface. The name ÆTHRIOSCOPE (from Αιθριος, *serenus, sudus, frigidus,)* may, therefore, be justly appropriate to this new combination of the pyroscope. The sensibility of the instrument is very striking, for the liquor incessantly falls and rises in the stem with every passing cloud. But the cause of its variations does not always appear so obvious. Under a fine blue sky, the *æthrioscope* will sometimes indicate a cold of 50 millesimal degrees; yet on other days, when the air seems equally bright, the effect is hardly 30°.

Particular winds at different altitudes seem to modify the result, and so perhaps may the transition from summer to winter. Four months are scarcely elapsed since this instrument was first contrived, and during that time very few opportunities have occurred for making nice observations.

The æthrioscope might be reduced to a smaller and more compact form, by conjoining with it a pendant differential thermometer. Neither of the glass balls in this case requires to be gilt; but the lower one is encased by a hollow sphere of brass, composed of two pieces which screw together and the upper ball occupies the focus of the cup, which needs scarcely be more than two inches wide. This variety of the instrument is equally accurate, but, under any change of temperature, rather slower in its action than the former.

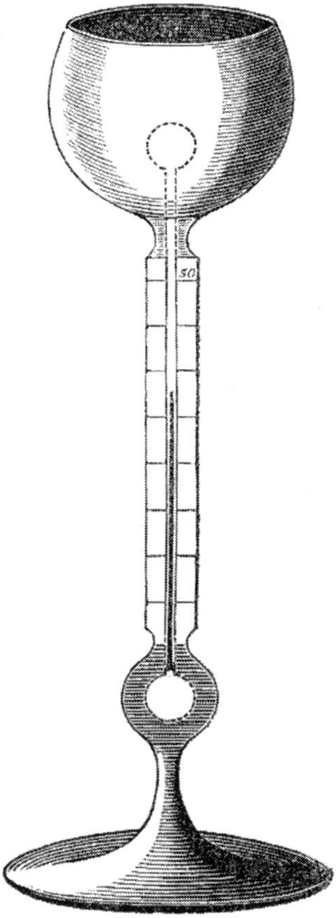


On replacing the metallic lid, the effect is entirely extinguished, and the fluid in the stem of the differential thermometer sinks to zero. A cover of pasteboard has at first precisely the same influence; but after it has itself become chilled by this exposure, it produces a small secondary action on the sentient ball, scarcely exceeding, however, the tenth part of the naked impression. A lid of glass or of mica intercepts the impression like one of paper; for the admission of light has no deranging effect if the æthrioscope be rightly constructed, and highly polished. The minute secondary action is almost extinguished, if screens of paper, glass, or mica, be held at some distance above the mouth of the instrument.

In the first form of the æthrioscope, it is not essential (though it augments the effect) that one of the balls should be gilt, since the other, being placed naked in the focus, receives the pulses much concentrated. If both the balls were formed of black glass, the æthrioscopic impressions would evidently be blended with opposite photometrical ones. As long as the action of the light from the sky predominates, the instrument indicates an excess of heat; but after the pulses darted from the higher atmosphere come to prevail, it will mark the excess of cold. The liquor is stationary when these contending influences are exactly balanced.

To ascertain whether the cold pulses are shot obliquely as well as in the vertical direction, the æthrioscope may be constructed to turn towards any portion of the sky. To effect this, the form best adapted for the reflection would be that of an hyperbola, whose asymptotes have an inclination equal to the visual angle of the space to be explored. But to obtain accurate results, the focal ball must be small, and the hyperbolic conoid wide and much extended. It will answer nearly the same purpose, however, to adopt a truncated spheroid of great eccentricity. Let the height of the focus, for instance, be one inch, that of the entire cavity nine inches, and consequently the widest diameter six inches. The figure on the other side is rather more distended, its extreme width being equal to double the eccentricity, and the focal ball dividing the height of the orifice in the ratio of 1 to $3 + \sqrt{8}$, or 6 to 35 nearly. While that sentient ball remains always in the same position, the axis of the instrument can, by means of a screw acting on the limb of a quadrant, be depressed or elevated to any given angle. But the effect will chiefly be produced by the direct impressions;

for the lateral pulses, striking less obliquely against the cavity of the spheroid, will be feebly reflected. This moveable æthrioscope was placed in a convenient situation out of doors, when the sky appeared free from clouds, and had assumed a clear blue tint. The spheroid being turned first upright, the effect was noted; but this continued still unchanged on depressing the axis successively till it had approached the limit of energetic range, or within twenty degrees of the horizon.

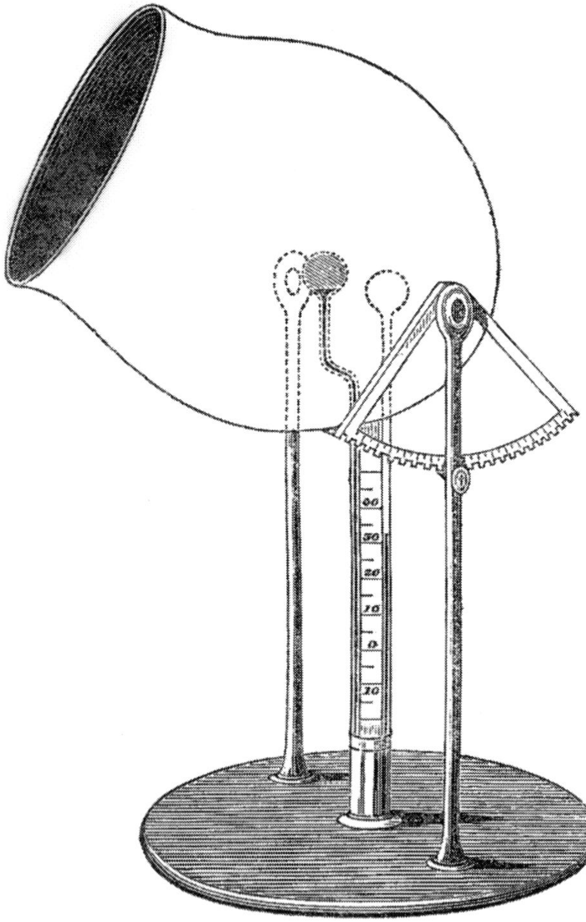

From every portion of the sky that subtends a given visual angle, there is hence received the same quantity of the frigorific pulses. But such would likewise be the result if they were showered from the horizontal surfaces of the successive strata which divide the atmosphere; since, although the intensity diminishes in the ratio of the sine of obliquity, a projecting space proportionally broader is for each elemental angle brought into action, as evidently appears from the annexed figure. This entire agreement between theory and observation is most satisfactory.

The same sectorial form of the æthrioscope discloses also the peculiar influence of clouds in obscuring the frigorific pulses excited in the atmosphere. When the sky was completely obscured by a dense canopy of clouds, the instrument being pointed to the zenith, marked only five millesimal degrees; but on lowering it successively to the angle of thirty degrees above the horizon, it continued to indicate still the same effect. Water almost completely absorbs the pulsatory impressions of heat or cold; and may not clouds, consisting of diffuse aqueous particles, produce a similar effect? But the feeble action of five degrees, amounting scarcely to the eighth part of what is observed in clear weather, could not be any remnant of the pulses from the higher celestial regions which had penetrated through the mass of vapours, because if the vertical transit, through the obstructing range, allowed only an eighth part to escape, the oblique passage of thirty degrees, redoubling the extent of absorption, would have reduced the final discharge to five-eighths of a degree. The impression measured by the æthrioscope in this case must therefore have originated wholly in the strata of air between the under surface of the clouds and the ground. But in that narrow space, the extreme difference of temperature would be comparatively small. Hence the frigorific action is found always to diminish as the clouds descend.

Nor does their variable denseness appear materially to affect the result, which is often the least, when a very thin, whitish, but low vapour, gathers in the atmosphere. Hence the æthrioscope might, with great facility, be employed to estimate the altitude of clouds.

As the higher strata of the atmosphere thus dart cold pulses downwards, so the lower strata must evidently project equal pulses of heat upwards. But to measure these, it would require the æthrioscope to be inverted, and furnished with a pendant pyroscope. The instrument, now carried to the top of a lofty mountain, and directed to the plain below, would indicate a considerable impression of heat, nearly proportional to the quantity of ascent; and therefore amounting, for example on the summit of Chimboraço, to perhaps twenty millesimal degrees. But in the same situation the common æthrioscope might be expected to mark an impression of cold from above as just so much diminished. No opportunity, however, has yet occurred, on a large scale, for making these interesting observations. A balloon would afford the readiest mode of verifying the theory.

The inverted æthrioscope likewise discovers the quality and measure of the pulses projected from the ground. These in general are very feeble, seldom in this climate exceeding three or four degrees. In the progress of a bright day, as the ground grows warmer than the incumbent air, it excites hot pulses, but as the sun declines the effect gradually diminishes, till this again returns, increasing with a contrary character, when the surface of the earth has become relatively colder.

The same instrument being suspended a few feet above the ground while the sky appeared clear and blue, a silver tray was laid under it, and the reflected impression of cold amounted to twenty-five degrees; but on interposing a plate of glass, it was reduced to two degrees; and on removing this and pouring a sheet of water over the silver, the effect was absolutely extinguished. The absorbent influence of water, and consequently of clouds, was thus distinctly shown.

The nature and intensity of the cold and hot pulses excited in the several strata of the atmosphere, may be easily understood from the annexed figure. Let two equal and opposite circles touch the straight line AB, which divides a stratum of cold from another of warm air. While the opposite diameters CD and *cd* represent the force of the perpendicular pulses of cold darted downwards, and of heat upwards, the chords CF, CE, CH, and CG, and C*f*, C*e*, C*h*, and C*g*, will likewise exhibit the strength of the pulses which are shot obliquely.

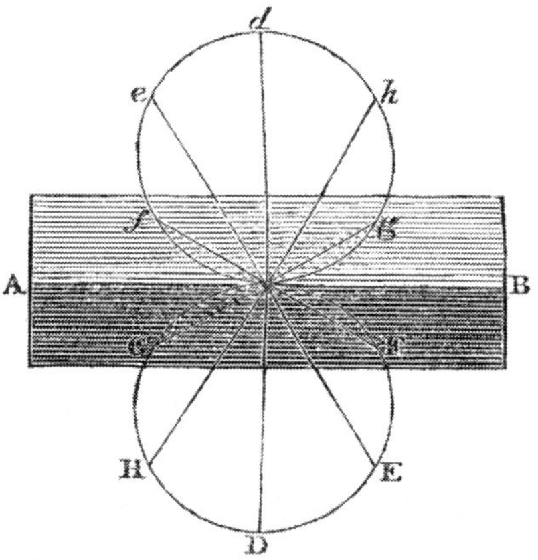


The æthrioscope thus opens new scenes to our view. It extends its sensation through indefinite space, and reveals the condition of the remotest atmosphere. Constructed with still greater delicacy, it may perhaps scent the distant winds, and detect the actual temperature of every quarter of the heavens. The impressions of cold which arrive from the north will probably be found stronger than those received from the south. But the instrument has yet been scarcely tried. We are anxious to compare its indications for the course of a whole year, and still more solicitous to receive its reports from other climates and brighter skies.

But the facts discovered by the æthrioscope are nowise at variance with the theory already advanced on the gradation of heat from the equator to the pole, and from the level of the sea to the highest atmosphere. The internal motion of the air, by the agency of opposite currents, still tempers the disparity of the solar impressions ; but this effect is likewise accelerated by the vibrations excited from the unequal distribution of heat, and darted through the atmospheric medium with the celerity of sound. Any surface which sends a hot pulse in one direction, must evidently propel a cold pulse of the same intensity in an opposite direction. The existence of such pulsations, therefore, is in perfect unison with the balanced system of aerial currents.

The most recondite principles of harmony are thus disclosed in the constitution of this nether world. But we have left no room for pursuing the details. In clear weather, the cold pulses then showered entire from the heavens will, even during the progress of the day, prevail over the influence of the reflex light, received on the ground, in places which are screened from the direct action of the sun. Hence, at all times, the coolness of a northern exposure. Hence, likewise, the freshness which tempers the night in the sultriest climates, under an expanse of an almost constant azure sky. In our northern latitudes, a canopy of clouds generally screens the ground from the impressions of cold. But, within the arctic circle, the surface of the earth is more effectually protected by the perpetual fogs which deform those dreary regions, and yet admit the light of day, while they absorb the frigorific pulses, vibrated from the higher atmosphere. Even the ancients had remarked that our clear nights are generally likewise cold. During the absence of the sun, the celestial impressions continue to accumulate ; and the ground becomes chilled to the utmost in the morning, at the very moment when that luminary again resumes his powerful sway. But neither cold nor heat has the same effect on a green sward as on a ploughed field, the action being nearly dissipated, before it reaches the ground, among the multiplied surfaces of the blades of grass. The lowest stratum of air, being chilled by contact with the exposed surface, deposits its moisture, which is either absorbed into the earth, or attracted to the projected fibres of the plants, on which it settles in the form of dew or hoar frost. Hence the utility in this country of spreading awnings at night, to screen the tender blossoms and the delicate fruits from the influence of a gelid sky; and hence, likewise, the advantage of covering walled trees with netting, of which the meshes not only detain the frigorific pulses, but intercept the minute icicles that, in their formation, rob the air of its cold.

The novelty and importance of the principles which regulate the distribution of heat in the atmosphere seemed to require a full discussion; but we have already outrun the space allotted for this article, and must for the present content ourselves with subjoining a few corollaries and general remarks.

From what has been explained, it is easy to perceive why the climate of an island should have a distinct character from that of a wide continent. The latter will experience a much greater range of heat and cold than the former, though the mean temperature, in like circumstances, will be the same in both. The proximity of the ocean quickly absorbs the inequality of the solar impressions. The passage from summer to winter is hence but feebly marked in detached islands. The prevalence, likewise, in such situations, of a land breeze during the day, and of a sea breeze during the night, especially near the tropics, reduces to very moderate limits the effects produced by the vicissitude of light and darkness.

The coldness of particular situations has very generally been attributed to the influence of piercing winds which blow over elevated tracts of land. This explication, however, is not well founded. It is the altitude of the place itself above the level of the sea, and not that of the general surface of the country, which will mould its temperature. A cold wind, as it descends from the high grounds into the valleys, has its capacity for heat diminished, and, consequently, becomes apparently warmer. The prevalence of northerly above southerly winds may, however, have some slight influence in depressing the temperature of any climate.

It has often been assumed as an incontrovertible fact, that the clearing of the ground, and the extension of agriculture, have a material tendency to ameliorate the character of any climate. But whether the sun's rays be spent on the foliage of the trees, or admitted to the surface of the earth, their accumulated effects in the course of a year, on the incumbent atmosphere, must continue still the same. The direct action of the light would no doubt more powerfully warm the ground during the day, if this superior efficacy were not likewise nearly counterbalanced by exposure to the closer sweep of the winds; and the influence of night must again re-establish the general equilibrium of temperature. The drainage of the surface will evidently improve the salubrity of any climate, by removing the stagnant and putrefying water; but it can have no effect whatever in rendering the air milder, since the ground will be left still sufficiently moist for maintaining a continual evaporation, to the consequent dissipation of heat.

Much has been said of the comparatively low temperature of the American continent. Its majestic forests, impervious to the solar beams, and its immense lakes, diffusing their vapours, are supposed sufficient to make the climate cold and damp; and the most splendid prospects have been delineated of those changes, which an active, free, and rapidly extending population is destined to produce in reclaiming that vast region from the rankness of nature, and opening its soil to the genial influence of heaven. Much inaccuracy and exaggeration, however, has prevailed on this subject. The extremes of summer and winter probably differ more in America than in the old world; but the mere temperature on any parallel appears, when carefully taken, to be nearly the same. Thus, Mr Warden (*Chorographical Description of the District of Columbia*, 1816) found a spring in the vicinity of the capital of the United States, and therefore a little above the level of the sea, to mark 58° in Fahrenheit's scale; but two chalybeate springs in the same quarter, though perhaps deeper seated, showed a heat of 62° and 64°. Chalybeates are not deemed warm springs, or affected by their slight mineral ingredient. But on the same parallel of 39°, the mean temperature of the ancient continent is 63½°.

After the above had been written, the third volume of the *Mémoires de Physique et de Chimie* of the Society of Arcueil was published, containing an elaborate paper of the celebrated and accomplished traveller Baron Humboldt, *On Isothermal Lines, and the Distribution of Heat over the Globe.* Laying aside every theoretical consideration, this distinguished philosopher has here endeavoured to class the results of his own very numerous thermometrical observations, with all those which he could collect from different quarters. The first difficulty that occurs is the mode of ascertaining the mean temperature of any place. It was found that half the sum of the maximum and minimum observed during the day and night, or through the summer and the winter, does not represent the true quantity. M. Humboldt has remarked that, between the parallels of 46° and 48°, the thermometer at the moment of sun-setting indicates in every season very nearly the medium temperature of the day. If this fact could be extended to other latitudes, it would greatly abridge the labour of meteorological observations.

From the data employed by M. Humboldt, he calculates

that, corresponding to the latitudes of 30°, 40°, 50°, and 60°, the mean temperatures on the west side of the Old Continent are 70°.5, 63°.1, 50°.9, and 40°.6; but, on the east side of the New Continent, only 66°.9, 54°.5, 38°, and 42°.2. The isothermal lines, or lines of equal temperature, under the tropics run across the Atlantic nearly parallel to the equator; but in the middle latitudes, they bend southerly towards the coast of America. These lines run nearly parallel into the interior of the New Continent, till they pass the Rocky Mountains, after which they bend again northwards. The climate is comparatively mild on the western shore of North America.

It is well known that the difference between the temperature of the winter and of the summer months increases in advancing from the equator to the pole. On the isothermal lines of 68°, 59°, 50°, and 41°, by Fahrenheit's scale, M. Humboldt estimates the extreme range of the year at 22°, 29°, 32½°, and 36°, on the Old Continent, and at 27°, 42°, 41½°, and 52°, on the New Continent

# COLD.

It is often disputed whether cold has any actual existence, or should be considered as merely the privation of heat. Nor is that question of a modern date. Plutarch attempted to discuss it, in his tract *De primo frigido*; and the reasonings which he there employs, though abundantly vague, are yet curious. Cold, he says, affects the senses as well as heat; and it is not less active, since it condenses and consolidates bodies. He therefore inclines to the opinion, that cold is a distinct and independent power in nature. With the Stoic philosophers, he regards air as by its constitution cold and dark; and hence water drawn from a well freezes on being exposed to the atmosphere, while rivers overshadowed by high banks seldom freeze, and even where their surface congeals, the heat is not exhaled, but only driven down nearer the bottom.

It is contrary to sound physics to admit more principles than are indispensably required, and this argument alone may be sufficient for the rejection of cold as a distinct power in nature. What we term cold, in reference to our feelings, is merely the diminution of heat; but the existence and materiality of heat rest on a very different foundation. The introduction of heat into a body is accompanied by the infusion of a certain extrinsic repulsive force, and its passage through the mass is connected with a series of depending internal motions, which imply the regular expenditure of time and velocity. The contraction which follows on the diminution of heat is due to the mutual attractive powers of the particles of the substratum, now exerted with less opposition. That expansion, again, which some fluids manifest in the act of congelation, proceeds from the operation of the principle of crystallization, with the recondite nature of which we are still unacquainted.

The notion that cold has a separate and independent existence, appears, however, to receive some countenance from the elegant experiment of collecting and concentrating the frigorific impressions in the focus of a metallic reflector. This curious fact, which is one of the oldest in physical science, has lately been revived, and combined with circumstances of peculiar interest. The experiment was first mentioned about the year 1590, by Baptista Porta, in the enlarged edition of his *Magia Naturalis*, when the four books of which it originally consisted were augmented to twenty, at the very time that his ingenious countryman Sanctorio had invented and applied to medical purposes the air thermometer. Porta relates, that, if a shut eye be held in the focus of a speculum, before which is placed a ball of snow, intense cold will be felt on the eye-lid.* Cavalieri, the celebrated discoverer of infinitesimals, in his work on the conic sections, printed in 1632, and entitled *Lo Specchio Ustorio*, extended the experiment to all impressions which

* Si quis candelam in loco, ubi spectabilis res locari debet, apposuerit, accedet candela per aerem usque ad oculos, ut illos calore, et lumine offendet; hoc autem mirabilius erit, ut calor, ita frigus reflectitur, si eo loco nix objiciatur, si oculum retigerit, quia sensibilis, etiam frigus percipiet.

he conceived to be propagated in straight lines, not only to those of heat and cold, but to those of sound, and even smell.* It was afterwards frequently repeated at Florence, by the Academy del Cimento, with the important addition of the thermometer, which that learned body had the merit of introducing into practice. Similar experiments were next performed by Mariotte in France.

* Cavalieri mentions that, with a spherical speculum made of lead, and indifferently polished, he was able to inflame dry substances by the reflexion of a charcoal fire ; and that, with a deep truncated parabolic speculum, badly polished, he produced the same effect in the open focus, from a small fire of wood at the distance of five feet. Esperienza di questo hò fatto io, che con vno Specchio sferico di piōbo ancor mal polito, hò acceso il fuoco nella materia arida al fuoco di carboni ; e di piu l'hò fatta con la superficie parabolica, cioè con vn cánone parabolico, che hauea il suo vicino alla cima, essendo esso specchio parabolico trōcato pur nella cima, qual' era di stagno, e mal polito, tal che opponendolo al fuoco, ò alla fiamma di ben poca legna, nella distanza di tre braccia, ponendo la mano lì, dou'era la parte trōcato, et il foco della parabola, non vi si potea sostenere, anzi vi s'accese fuoco ; la qual cosa potria alcuno applicare al riscaldamēto delle stanze, ò alle distillationi. Pp. 85, 86.—In general, says this ingenious mathematician and philosopher, the same form of speculum which concentrates light and heat must likewise collect cold, which spreads from its source, from a mass of snow for instance, in straight lines. The hyperbola is, therefore, the figure which he thinks the best adapted for the purpose ; and he proposes this for condensing the smells radiated from an odoriferous substance. Hora dunque basterà quello, che si è detto di sopra intorno al lume, e calore, *potendo noi nell'istesso tempo intendere le medesime cose anco per il freddo, che dilatádosi dal corpo freddo ad ognii positione per linea retta, e perciò nell' infinite linee, che si partono dal corpo freddo, come dalla neue,* essendoui dentro le parallele, che sono vnite dallo specchio parabolico, e le diuergenti, che sono vnite dall' elittico, e le conuergenti vnite dall' iperbolico, con opporre alcun di questi specchi ad vna massa di neue, ò di ghiacco, sentiremo nel loro foco essere il freddo fatto molto gagliardo, ma per questo effetto sarà più atto l'iper-

Specula and burning glasses appear in the sequel to have been allowed to fall into great neglect. We find scarcely any mention of their application to physical researches, till, after the lapse of more than half a century, Kraft repeated, at St Petersburg, during the severe winter of 1740, the frigorific experiment of the Italian philosophers, with a reflector belonging to the cabinet of the Imperial Academy. Ambitious to operate on a grand scale, he selected three huge blocks of clear ice, nearly of a cubical form, each side being two, four, and five feet; but, to save the trouble of transporting them, he carried the speculum out of doors. No sensible effect, however, was then perceived by him, though he used the air thermometer on account of its extreme delicacy. In 1744, this academician again resumed the observation, and with scarcely better success, having obtained only a doubtful cold of three degrees. The cause of the failure was evidently his performing the experiment out of doors, and not in a warm room. The blocks of ice had, by long standing, acquired almost the same temperature as their ambient medium. Had the air happened to become suddenly colder, they might, from their relative condition, have excited impressions even of heat, and thus have perplexed philosophy for many years afterwards.

Such unsatisfactory results, from the action of a mass of ice of above a ton weight, seem for a long time to have shaken the belief in former experiments; and the subject was almost forgotten, when Pictet of Geneva, in 1781, repeated the original observation, on a small scale, indeed,

---

bolico di tutti, come quello, che raccoglierà maggior quantità di linee fredde; e questo basti ancora circa il freddo, potendosi forsi in vn certo modo creder, *che alle effetto accadesse anco intorno à gli odori, prouando noi dilatarsi pur quelli dalli corpi odoriferi verso ogni banda.*—Id. p. 128.

but with entire success. Since that time a pair of brass reflectors, with a wire case for holding charcoal or snow, has been deemed an essential apparatus in every physical cabinet. The concentration of cold in the focus of a speculum always excites surprise; and the experiment is often exhibited with a sort of mysterious air, as if it established, or at least rendered probable, the distinct and material existence of cold; but, in fact, it is not more difficult to conceive the impressions of cold to be collected than those of heat. Both those impressions are only relative to the temperature of the atmosphere, which serves as the medium of their transmission. The one process terminates with the deposition of a portion of heat, the other with its abstraction.

The diminution of heat, or the increase of cold, is produced in nature under four different circumstances: 1. By the obliquity or absence of the sun; 2. by the tenuity of the higher atmosphere; 3. by the evaporation which takes place in dry air; and, 4. by the chilling impression shot downwards from a clear and serene sky.

1. In our temperate climates the thermometer in winter very seldom descends fifteen degrees on Fahrenheit's scale, below the point of congelation; but in the higher latitudes the intensity of the cold is often far greater. In the northern parts of Sweden and Russia, the rivers and ordinary lakes are frozen to the depth of several feet; wine, and even ardent spirits, become converted into a spongy mass of ice, and, as the cold still augments, it penetrates the living forests, and congeals the very sap of the trees, which occasionally burst from this internal expansion with tremendous noise. The Baltic Sea has been repeatedly covered with a solid floor of ice, capable of transporting whole armies, with all their stores and engines of war. Those waters, indeed, are only brackish; but the more northern ocean itself has often been frozen to a very considerable thickness. In

Siberia and Hudson's Bay, and even in the northern provinces of Sweden, mercury has been at some times observed contracted by exposure into a solid semi-metal; and, consequently, the cold which then prevailed must have exceeded seventy-one degrees, or thirty-nine below the commencement of Fahrenheit's scale.

2. In elevated tracts the increase of cold is very striking. Even at an altitude of three miles and a half, the air is generally sixty-eight degrees colder than at the level of the sea. On the summit of the Andes, therefore, a thermometer would often sink perhaps under the beginning of Fahrenheit's scale; and it seems probable that mercury would naturally freeze in winter on the top of Mont Blanc. Mountains are hence regarded as the grand stores or depositaries of cold in the milder climates. In every country, therefore, the air of subterraneous caves, and the water of deep springs or wells, are during the summer months comparatively cold. Hence the obvious advantage of cellars, in addition to their preserving an uniform temperature, which is so favourable to the ripening of the liquors deposited in them. But the air at the bottom of an open and very deep pit must be colder than the mean state of the ground; for in all the changes which take place at the surface, the cold air will descend, and the warm air still float over the mouth of the pit. The wealthier classes of antiquity were accustomed, accordingly, to cool the wine for their tables by suspending it for some time in a bucket let down near the surface of profound wells.

3. Evaporation is a natural process, by which heat is powerfully abstracted from the exhaling moisture, while this assumes a gaseous constitution in the act of combining with dry air. The fact seems to have been known in the warm regions of the East at a very early period of society, suggested probably by the familiar use of a rude unglazed pot-

tery for all culinary purposes. The Egyptians, and other inhabitants of the sultry shores of the Levant, have, from the remotest ages, cooled the water for drinking in their porous jars. Athenæus reports, from a history of Protagorides, that King Antiochus had always a provision for his table prepared in that way. The water having been carefully decanted from its sediment into earthen pitchers (ὑδρειαις κεραμειαις), these were transported to the highest part of his palace, and exposed to the clear and keen atmosphere (εξαιθριαζουσιν), two boys being appointed to watch them the whole night, and kept constantly wetting their sides. This labour of sprinkling the surface of the jars seems to have been afterwards spared, in consequence probably of the adoption of a more porous kind of earthenware. Galen, in his *Commentary on Hippocrates,* relates, that he witnessed the mode of cooling water, which was practised in his time, not only at Alexandria, but over all Egypt. The water, having been previously boiled, was poured at sunset into shallow pans (αγγοις οστρακινοις), which were then carried to the house tops, and there exposed during the whole night to the wind; and, to preserve the cold thus acquired, the pans were removed at day-break, and placed on the shaded ground, surrounded by leaves of trees, prunings of vines, lettuce, or other slow-conducting substances.

The bottles or bags made of goat-skins, in which the wandering Arabs are wont to carry their scanty provision of water, allowing a small portion of the liquid to transude and exhale, render it by consequence comparatively cool, and better fitted to mitigate or allay the intolerable thirst created in traversing their sandy deserts. In Guinea, it is customary to fill gourds or calabashes with water.

The Moors introduced into Spain a sort of unglazed earthen jugs, named *bucaros* or *alcarrazas,* which, being filled with water, present to the atmosphere a surface constantly humid, and furnish by evaporation, during the dry and hot weather, a refreshing beverage. The same practice has been adopted by degrees in various parts of the south of Europe. In India, during certain months, the apartments are kept comparatively cool, by dashing water against the matting of reeds or bamboos which line the doors and outside of the walls. Even the more luxurious mariners, in their voyages between the tropics, are accustomed to cool their wines by lapping the bottle with wet flannel, and suspending it from the yard or under the cabin windows. In all such cases the effect is accelerated, though not augmented, by the swiftness of the current of air. What have been called *Egyptian coolers,* and lately produced by our potters, are less perfect in their operation. Being very thick, they require only to be soaked in water, and the evaporation from their surface cools the adjacent air. On the inside, however, where the bottle is placed, the action, in consequence of the confined humidity, must be enfeebled. In damp weather, these vessels, it is evident, are entirely useless.

The natives of India likewise are enabled, by directing a skilful process of evaporation, to procure for themselves a supply of ice during their short winter. In the upper country, not far however from Calcutta, a large open plain being selected, three or four excavations are made in it, about thirty feet square and two feet deep, and the bottom covered to the thickness of nearly a foot with sugar canes or dried stalks of Indian corn. On this bed are placed rows of small unglazed earthen pans, about an inch and quarter deep, and extremely porous. In the dusk of the evening, during the months of December, January, and February, these are filled with soft water previously boiled and suffered to cool; when the weather is very fine and clear, a great part of the water becomes frozen during the night. The pans are regularly visited at sunrise, and their contents thrown into baskets which retain the ice. These are now carried to a conservatory, made by sinking a pit fourteen or fifteen feet deep, lined with straw under a layer of coarse blanketing. The small sheets of ice are thrown down into the cavity, and rammed into a solid mass. The mouth of the pit is then closed up with straw and blankets, and sheltered by a thatched roof.

4. It was stated in the article CLIMATE, that impressions of cold are constantly showered down from a clear and azure sky. These effects are no doubt more conspicuous in the finer regions of the globe. Accordingly, they did not escape the observations of the ancients, but gave rise to opinions which were embodied in the language of poetry. The term αηρ was applied only to the grosser part of the atmosphere, while the highest portion of it, free from clouds and vapour, and bordering on the pure fields of æther, received the kindred appellation of αιθρια. But this word and its derivatives have always been associated with ideas of *cold* We have seen that the verb εξαιθριαζω is used by Athenæus to signify the cooling of a body by exposure under a serene sky. Homer uses the term αιθρος, in speaking of the reception of his hero, when overcome with *cold* and toil.* The same poet of nature applies the epithet αιθρηγενης, or αιθρηγενετης, or *frigorific*, to Boreas, the north wind.† The chorus in the *Antigone* of Sophocles deprecates the pelting storm, and likewise the cold (αιθρια) of inhospitable frozen

---

* Αιθρῳ και καματῳ δεδμημενον ηγεν ες οικον· *Odyss.* lib xiv. 318.

† Ὡς δ' ὁτε ταρφειαι νιφαδες Διος εκποτεονται,
Ψυχραι ὑπαι ῥιπης αιθρηγενεος Βορεαο. *Iliad.* lib. xix. 357–8.
Και Βορεης αιθρηγενετης, μεγα κυμα κυλινδων. *Odyss.* lib. v. 296.

tracts.* The word αιθριος is employed by Herodotus to signify a chill as well as a dry atmosphere.† Of the same import is the expression in Horace, *sub Jove frigido.*

In the finer climates, especially, transpiercing cold is, therefore, felt at night under the clear and sparkling canopy of heaven. The natives carefully avoid exposing themselves to this supposed celestial influence; yet a thin shed of palm leaves is sufficient at once to screen them from the scorching rays of the sun, and to shelter them against the chilling impressions rained from the higher atmosphere. The captains of the French gallies in the Mediterranean used formerly to cool their wines in summer by hanging the flasks all night from the masts. At daybreak they were taken down and lapped in several folds of flannel, to preserve them in the same state. The frigorific impression of a serene and azure sky must undoubtedly have concurred with the power of evaporation, in augmenting the energy of the process of nocturnal cooling practised anciently in Egypt, and now systematically pursued in the higher grounds of India. As the chilness accumulated in the ground is greatest in clear nights, when the moon shines brightest, it seemed very natural to impute this effect partly to some influence emanating from that feeble luminary. It was long imagined that the lunar beams are essentially cold; and some philosophers, at no remote period, have attempted even to prove the fact by experiment. Mr Boyle, though he rejected judicial astrology, was yet disposed to admit the notion of stellar influences.

The obvious mode of cooling water, or other liquids, by the infusion of ice or snow, was practised in the warmer

---

* δυσαυλων παγων αιθρια
και δυσομβρα φευγειν βελη. *Antigone*, 357.

† Θερμοτερον γαρ δη εστι το ὑδωρ της τε αιθριας και της δροσου. *Euterpe.*

countries from the earliest ages. It is even mentioned in the Proverbs of Solomon, "*As the cold of snow in the time of harvest,* so is a faithful messenger to them that send him." Aristotle, presuming that the finer parts of water are dissipated by congelation, maintained that it is pernicious to drink melted snow. This speculative opinion seems not, however, to have been regarded by the ancients. Theocritus calls snow-water an *ambrosial drink,* ποτος αμβροσιος. Xenophon mentions the practice of cooling wine by the addition of snow. It is related by the historians of Alexander the Great, that in his Indian expedition, when he laid siege to the city of Petra, he commanded thirty pits to be dug and filled with snow, which was covered over with oak branches. The luxurious Romans had excavations regularly formed for keeping snow the whole year, chaff and other light substances being employed to preclude from it the access of heat. But as the snow preserved in this way could not escape being soiled, instead of mixing it directly in the drinking cup, a more refined practice was introduced of surrounding the silver goblet which contained the liquor with a mass of the melting snow. This improvement was ascribed to the profligate emperor Nero. Similar modes of storing up snow have been adopted in all the warm countries. The caves on the sides of Mount Ætna are considered as natural magazines for supplying a material, which is not only carried down to Palermo and Messina, but even shipped to the island of Malta. The Italians formerly cooled their wine by setting the large glass flasks containing it in wide vessels of wood or cork, the intervening space being filled with snow on which water was poured.

Saltpetre or nitre being almost a natural production of the East, its property of rendering water cold by solution was probably known from a very remote period to the oriental nations. This process of cooling is described in the *Institutes* of Akbar as the discovery of that enlightened prince, who governed India with parental mildness from the year 1560 to 1605. One part of nitre is directed to be thrown into a vessel containing two parts of water; and a gugglet of pewter or silver filled with pure water, and having its mouth close stopped, is then stirred quickly in the mixture for the space of a quarter of an hour.

The frigorific property of nitre was probably first communicated from India or Persia to Europe, and seems to have become known to the Italians about the middle of the sixteenth century. As early as the year 1550, all the rich families in Rome cooled the liquors for their tables by dissolving that salt in water. Into a vessel of cold water the nitre was gradually added in the proportion of a fourth or fifth part, while a globular bottle with a long neck, containing the wine or water to be cooled, was whirled rapidly round its axis. The salt, being afterwards recovered by crystallization, would always serve the same purpose again with undiminished effect. In India every family of distinction keeps a domestic, whose sole employment is to cool liquors by this process; but nitre being cheap in that country, it is used in larger proportions, and the water charged with it is allowed to become a perquisite of the operator.

The application of salts to produce cold was extended by Boyle, and afterwards more successfully by Fahrenheit. But within these twenty years Mr Walker of Oxford, and Professor Lowitz of St Petersburgh, have resumed the subject, and produced compound saline powders possessed of intense frigorific power. The solution of salts in water expanding that liquid, augments its capacity for heat, and consequently depresses its temperature. This effect is likewise the greater in proportion to the quantity of saline matter which can be dissolved. But after water is saturated with one species of salt, it can still absorb some portion of another. Hence the frigorific effects of solution are always increased by employing a compound dry powder. Nitre and sal-ammoniac, or the nitrate of potash and the muriate of soda, in equal parts, added in the form of a dry powder to three parts by weight of water, will sink Fahrenheit's thermometer forty degrees. But equal parts of the muriate of ammonia and of the nitrate of potash, with one part and a half of the sulphate of soda, or common Glauber's salt, will cool down three parts of water forty-six degrees. A still greater effect, amounting to fifty-seven degrees, is produced by dissolving equal parts of the nitrate of ammonia and of the carbonate of soda in one part of water. The frigorific action is in general augmented by throwing the desiccated powder into dilute acid instead of water. Thus three parts of the phosphate of soda, and two parts of the nitrate of ammonia, joined to rather more than one part of weak nitric acid, will sink the thermometer seventy-one degrees.

These changes induced on the temperature of the liquid menstruum are no doubt considerable, yet they are still only transient, and the process requires some address and manipulation, not always readily attained. But the principle of evaporation, when rightly understood, leads to a far easier mode of cooling liquids, which may be prolonged at pleasure. A close investigation of that principle, at the very commencement of his philosophical labours, has conducted Professor Leslie through the whole train of his discoveries on the subject of refrigeration. The process of evaporation had not then been examined with attention. The depression of temperature which always accompanies it was hastily supposed to be proportional to the rate with which the moisture is dissipated, and to be therefore augmented by every circumstance that can accelerate this effect. If water, contained in a porous vessel, expose on all sides its surface to a current of air, it will cool down to a certain point, and there its temperature will remain stationary. The rapidity of the current must no doubt hasten the period of equilibrium, but the degree of cold thus induced will be still the same. A little reflection may discover how this happens. Though the humid surface has ceased to grow colder, the dispersion of invisible vapour, and the corresponding abstraction of heat, still continues without intermission. The same medium, therefore, which transports the vapour, must also furnish the portion of heat required for its incessant formation. In fact, after the water has been once cooled down, each portion of the ambient air which comes to touch the evaporating surface must, from its contact with a substance so greatly denser than itself, be likewise cooled to the same standard, and must hence communicate to the liquid its surplus share of heat, or the difference between the prior and the subsequent state of the solvent, which is proportional to the diminution of temperature it has suffered. Every shell of air which encircles in succession the humid mass, while it absorbs, along with the moisture which it dissolves, the measure of heat necessary to convert this into steam, does at the same instant thus deposit an equal measure of its own heat on the chill exhaling surface. The abstraction of heat by vaporization on the one hand, and, on the other, its deposition at the surface of contact, are therefore opposite contemporaneous acts, which soon produce a mutual balance, and thereafter the temperature induced continues without the smallest alteration. A rapid circulation of the evaporating medium may quicken the operation of those causes; but so long as it possesses the same drying quality, it cannot in any degree derange the resulting temperature. The heat deposited by the air on the humid surface becomes thus an accurate measure of the heat spent in vaporizing the portion of moisture required for the saturation of that solvent at its lowered temperature. The dryness of the air is therefore, under all circumstances, precisely indicated by the depression of temperature produced on a humid surface which has been exposed freely to its action.

Guided by these views, Mr Leslie was enabled to construct a correct *hygrometer*, that should indicate the *dryness* of the air, from the diminution of temperature which a small body of water, exposed on all sides, suffers by evaporation. His efforts again to improve this instrument led him next to the invention of the *differential thermometer*, which was converted into an hygrometer by having one of its balls covered with cambric, lint, or tissue paper, capable of being easily wetted. Reduced to such a delicate and commodious form, it detected with the utmost precision, and under all circumstances, the relative condition of the air in regard to dryness.

It appears that absorbent substances, exposing a broad surface, are capable of assimilating to their previous state the air confined over them. Flannel, for instance, which has been intensely dried, will support a remarkable degree of dryness in a close receiver. The trap-rock and compound clays, brayed into a coarse powder, and desiccated before a strong fire, will exert a more powerful and extended action. But dried oat-meal will act with equal energy, and for a longer time. Of the saline substances, the muriate of lime absorbs moisture with the greatest and most protracted force. After it has become drenched with humidity, it may likewise be recovered again, though the process of restoring it unaltered is rather troublesome. But the best and most powerful absorbent is the concentrated sulphuric acid, or the oil of vitriol of commerce, which continues for a long time to attract moisture with almost undiminished force, and possesses besides the valuable property, after it has become charged with humidity, of being easily restored again by the application of heat to its original strength.

To cool water in any climate or state of the atmosphere,

we have only therefore to put it into a small porous vessel, presenting on all sides a humid surface, and to suspend this within a close wide cistern, of which the bottom is covered with a layer of sulphuric acid. The broad surface of the acid absorbing the moisture as fast as it diffuses itself through the confined air, keeps that medium constantly at a point of extreme dryness, and thus enables it to support, with undiminished vigour, the process of evaporation.

In practice, the cistern or refrigeratory, having a broad cylindrical form, from twelve to sixteen inches in diameter, and composed of dense well-glazed earthenware, (see Plate III. Fig. 7.) is placed in a cellar, or other cool place, and charged with sulphuric acid to the height of about half an inch from the bottom. One of the porous earthen pots, being filled up to the lip with water fresh drawn from the well, is set upon a low porcelain stand in the middle of the cistern, to which the lid or cover is then carefully adapted. In the space of from three to perhaps five hours, the cooling is nearly completed, and the pot should now be removed; for though the water will be kept at the same degree of coldness, as long as it remains shut up within the refrigeratory, the acid would be unnecessarily weakened by the incessant absorption of moisture.

The production of cold is greater when the cistern is large, or when a small pot is used, insomuch that the effect will be diminished one half, if the humid surface should equal that of the acid, the opposite actions of such surfaces inducing an exactly intermediate state with respect to dryness and moisture in the condition of the aerial medium. The power of evaporation is also diminished in the low temperatures. Thus, if the atmosphere were at 95° by Fahrenheit, the water within the refrigeratory might be cooled down 36°, or brought to 59°; but if the thermometer be at 50°, the water can be cooled only 18°, which brings it to the freezing point. This seems to be a very convenient property, since the power of the refrigeratory is always the greatest at the season when its application is most wanted.

It is easy, therefore, by such means to cool water in our climate at all times to near the freezing point, and, even under the torrid zone, to reduce it to the temperature of 60, which in those regions is sufficient, perhaps, for essential comfort.

By supplying a succession of porous earthen pots, the acid will continue to act with scarcely diminished force, till it has absorbed half its weight of moisture; during which time it will have assisted in cooling about fifty times that quantity of the water exposed to evaporation. At this stage the dilute acid should be drawn off, and a charge of concentrated acid again introduced into the refrigeratory.

This method of procuring cold, it will be readily perceived, could be employed with advantage for various domestic purposes. For instance, butter may in summer be kept cool for the table, by putting it, after being washed with water, into a wet porous pot, and shutting this up for a couple of hours in the refrigeratory. To cool wine sufficiently, one bottle only is used at a time in the smallest refrigeratory. A sheath of stocking or flannel, previously soaked in water, being drawn over the body of the bottle, it is laid in a reclined position on one of the porcelain sliders, near the surface of the acid, and allowed to remain shut up during the space of three or four hours.

The refrigerating combination here employed produces its effect by a sort of invisible distillation, carried on by the play and circulation of the included air. The minute portions of moisture which successively combine with the contiguous medium, must abstract from the mass of water as much heat as would support them in the state of vapour, or would in ordinary cases convert them into steam. This vapour again being conveyed through the air, is attracted by the sulphuric acid, and, recovering its liquid constitution, deposits the heat which it had borne away. The acid is therefore warmed at the expense of the water subjected to evaporation, and the whole performance of the apparatus consists in a mere transfer and interchange of condition.

### CONGELATION

Is the passage of any substance from the liquid to the solid form, in consequence of the abstraction of heat. The conversion of water into ice could not fail to draw the notice of men in all ages. The minute and divided fragments of the same production, which descends from the clouds in the shape of snow or hail, displayed the various powers of nature. The ancients imagined that water which has lain for ages in a frozen state, acquires at last a permanent consolidation. They extended accordingly the name of ice (*κρυσταλλος* or *crystal*) to the pure and pellucid kind of quartz which often occurs on the sides of lofty mountains, near the boundary of perpetual congelation.

It was early remarked, that melted ice has the lightness and quality of boiled water. In fact, the portion of air combined with ordinary water is discharged in the act of freezing, as well as in that of boiling. Water thus deprived of its air is, therefore, prepared for a readier congelation. The ancients accordingly, we have seen, always boiled the water which they designed afterwards to cool. Aristotle relates, in his *Meteorology*, that the fishermen who cast their nets in the Pontine Lake, used to carry in close vessels boiled water, for the purpose of sprinkling the reeds, that these might quickly freeze together, and cease to disturb the fish by their rustling noise. The expulsion of air from water, during the progress of congelation, was afterwards fully proved by Mariotte, one of the earliest members of the

French Academy of Sciences. If two wine glasses, filled, the one with water from the well, and the other with water recently boiled, be exposed to the frost, the ice of the latter will seem almost uniformly pellucid, while the ice of the former will appear charged with small air-bubbles, crowding towards the centre of the mass, to which they are driven by the advance of the congelation.

That congelation shoots at angles of 120 degrees, was first observed in the beginning of the seventeenth century by the great Kepler; and this ardent and inventive genius, in an elaborate dissertation, which he printed as a new year's present, investigated the various forms and modifications of the icy crystals. The subject was next discussed by Des Cartes and Bartholinus, and about a century afterwards resumed by Mairan, and may be considered as a step towards the general theory of crystallography, which has been since reared by the patience and ingenuity of Haüy.

Other liquids, such as vinegar, dilute mineral acids, weak spirits, and saline solutions, are likewise capable of being frozen; but they yield an ice distinctly different from that of pure water, resembling an aggregation rather than an uniform solid, and wanting consistency, strength, and clearness. The frost appears to seize on the water only, and to fill the compound liquid with close spicular shoots, entangling the stronger acid or brine in their interstices. It was a mistake, therefore, to assert that the ice of sea-water is really fresh. In the process of melting, some portion of the brine may probably flow off, but the residue still is always brackish. This fact is even positively stated by the missionary Crantz, in his accurate account of Greenland. The very intelligent and enterprising navigator, Mr Scoresby, reckons the specific gravity of the spongy salt water ice to be .873, while that of fresh water ice amounted to .937.

The ancients were altogether unacquainted with artificial

congelation, and with any cold, indeed, below that of freezing. The application of nitre to the cooling of water seems, before the close of the sixteenth century, to have suggested to the Italians the experiment of mixing it with snow. A very intense degree of cold was thus generated, capable of converting speedily into solid ice a body of water contained in a smaller vessel immersed in the dissolving mixture. Sanctorio, who may be regarded as the father of modern physics, mentions, in his *Commentary on Avicenna*, that he produced the same effect by employing common salt instead of nitre, in the proportion of the third part of the snow, and had repeatedly performed the experiment in the presence of a numerous auditory.

From Italy this discovery was gradually communicated over the rest of Europe. In the course of the seventeenth century, iced creams, fruits, and various confitures, were first produced on the tables of the luxurious. The famous coffee-house, *Procope*, was founded at Paris in 1660, by a Florentine of that name, a vender of lemonade, who was very successful in the art of preparing rich ices. Thirty years afterwards, the use of such artificial delicacies in that city had become quite common.

The cold resulting from the addition of saline powders to snow or pounded ice, depends on the powerful attraction of those salts, which restores the frozen mass to its liquid form, and therefore augments its capacity for heat. Fahrenheit fixed the commencement of his thermometrical scale at the temperature of the compound of salt and snow, conceiving it to be the lowest possible ; but much lower degrees of cold are now produced. One part of the muriate of soda, or purified common salt, being added to two parts of dry snow or pounded ice, will sink the thermometer five degrees below zero. One part of sal-ammoniac, and two of common salt, joined to five parts of snow, will bring it seven degrees lower; but equal parts of the nitrate of ammonia and common salt, joined to two parts and a half of snow, will depress the thermometer twenty-five degrees below the freezing point.

Still more intense cold might be produced, if the ingredients were, before their mixture, cooled down to congelation. Thus, five parts of the muriate of lime, added to four parts of snow, will sink the thermometer to forty degrees below the beginning of the scale, or the limit of freezing mercury; and if the muriate of lime were crystallized, the effect would be ten degrees more. The same extreme energy is exerted on adding four parts of dry caustic potash to three parts of snow.

The mineral acids, likewise, in a diluted state, produce similar effects. Two parts of weak sulphuric acid, joined to three of snow, will sink the thermometer to twenty-three degrees below zero. The muriatic and nitric acids, in nearly the same proportions, will depress it from four to seven degrees more. By repeating the applications, therefore, a most intense cold may be created; yet, to succeed completely, a skilful manipulation is required. The saline matters should be reduced to a fine powder, and the freezing mixtures should be made in very thin vessels, not larger than will barely hold them. In this way, by successive stages of cooling, Mr Walker once obtained the enormous cold of ninety-one degrees below the commencement of Fahrenheit's scale.

The mere evaporation of some very volatile liquids is sufficient to produce excessive cold. Thus, if a thermometer, having its bulb covered with lint, be dipped in the common or sulphuric ether, it will, on exposure to the air, sink perhaps thirty or forty degrees. This effect is augmented under the receiver of an air-pump. If a narrow thin tube of glass, filled with water, and cased on the outside with lint soaked in ether, be suspended above the pump, and the exhaustion quickly made, a cylinder of ice will be formed.

The same property is manifested in a higher degree by a singular liquid, discovered by Lampadius in 1796, by distilling a mixture of pyrites and charcoal. It was called at first the *alcohol of sulphur*, but now more appropriately the *sulphuret of carbon*. According to Dr Marcet, who has completed the investigation of its properties, a thermometer having its bulb covered with lint wetted by this liquid, and held in the open air, will sink not fewer than sixty degrees; but if the same experiment be performed within the exhausted receiver of an air-pump, the alcohol of the barometer will even descend to eighty-two degrees below zero. It must be observed, however, that these effects produced by the evaporation of ether, and of the sulphuret of carbon, are quite evanescent, and that the receiver becomes soon charged with their fumes, which then prevent any further action. Those fumes likewise corrode the valves of the pump, and soon render it quite useless. Neither ether, therefore, nor the sulphuret of carbon, could be applied in practice with any sort of advantage to the production of ice, even on the smallest scale.

We have now to relate a discovery which will enable human skill to command the refrigerating powers of nature, and, by the help of an adequate machinery, to create cold, and produce ice, on a large scale, at all seasons, and in the hottest climates of the globe; but to explain this interesting subject with greater clearness and accuracy, it is requisite to trace the successive advances which conducted to the result. Where a conclusion appears simple, the careless observer is apt to suppose it easily attained; yet, though sound philosophy tends always to simplification, the rare quality of simplicity is scarcely ever the flash of intuition, but the slow fruit of close and patient investigation. In pursuing the researches with his hygrometer, Professor Leslie was early induced to inquire into the condition of the higher atmosphere, and its relations to humidity. He thus detected a fact of great importance in meteorology, and pointing at various ulterior views.

As rarefaction enlarges the capacity of air for heat, so it likewise augments the disposition to hold moisture; at the same time that the removal of the ordinary pressure facilitates the expansion of the liquid matter, and its conversion into a gaseous form. Accordingly, if the hygrometer be suspended within a large receiver, from which a certain portion of air is quickly abstracted, it will sink with rapidity. In summer, the additional dryness thus produced, amounts to about fifty hygrometric degrees each time the air has its rarefaction doubled; so that, supposing the operation of exhausting to be performed with expedition, and the residuum reduced to a sixty-fourth part, the hygrometer would mark a descent of 300°. But this effect is only momentary, for the thin air very soon becomes charged with moisture, and, consequently, ceases to act on the wet ball of the hygrometer. The cold, however, excited on the surface of that ball, by such intense evaporation, will have previously frozen the coating.

The increased power of aqueous solution which air acquires, as it grows thinner, being ascertained, and carefully investigated, the object was to combine the action of absorbent with the transient dryness produced within a receiver by rarefaction. The sentient ball of the hygrometer being covered with dry salt of tartar, the instrument first indicated increasing dryness, and afterwards, as the rarefaction proceeded, it changed its course, and marked humidity. The same variation of effect nearly was observed when the hygrometer had been wetted as usual with pure water, and a broad saucer containing the mild vegetable alkali was placed on the plate of the air-pump. It was thus proved, that the action of this imperfect absorbent is soon overpowered by the tendency to vaporization in attenuated air, and that, beyond a certain limit, it surrenders its latent moisture.

Mr Leslie resolved, therefore, to try the effect of sulphuric acid, whose peculiar energy as an absorbent he had, under other circumstances, already ascertained; but various incidents prevented him, for a considerable time, from resuming his philosophical inquiries. At last he began those projected experiments, and was almost immediately rewarded by the disclosure of a property, the application of which blazed on his fancy. In the month of June 1810, having introduced a surface of sulphuric acid under the receiver of an air-pump, he perceived with pleasure that this substance only superadded its peculiar attraction for moisture to the ordinary effects resulting from the progress of exhaustion; and, what was still more important, that it continued to support with undiminished energy the dryness thus created. The attenuated air was not suffered, as before, to grow charged with humidity; but each portion of that medium, as fast as it became saturated by touching the wet ball of the hygrometer, transported its vapour to the acid, and was thence sent back denuded of the load, and fitted again to renew its attack with fresh vigour. By this perpetual circulation, therefore, between the exhaling and the absorbing surface, the diffuse residuum of air is maintained constantly at the same state of dryness. The sentient ball of the hygrometer, which had been covered with several folds of wetted tissue paper, was observed, at an early stage of the operation, suddenly to lose its blue tint, and assume a dull white, while the coloured liquor sprung upwards in the stem, where it continued for the space of a minute stationary, and again slowly subsided. The act of congelation had, therefore, at this moment taken place, and the paper remained frozen several minutes, till its congealed moisture was entirely dispersed. Pursuing this decisive intimation, the hygrometer was removed, and a watch-glass, filled with water, substituted in its place. By a few strokes of the pump, the whole was converted into a solid cake of ice, which being left in the rare medium, continued to evaporate, and, after the interval of perhaps an hour, totally disappeared. A small cup for holding the water was next adopted, and the whole apparatus gradually enlarged.

The powers both of vaporization and of absorption being greatly augmented in the higher temperatures, the same limit of cold nearly is, in all cases, attained by a certain measure of exhaustion. When the air has been rarefied 250 times, the utmost that, under such circumstances, can perhaps be effected, the surface of evaporation is cooled down 120 degrees of Fahrenheit in winter, and would probably sink near 200 in summer. Nay, a far less tenuity of the medium, when combined with the action of sulphuric acid, is capable of producing and supporting a very intense cold. If the air be rarefied only 50 times, a depression of temperature will be produced, amounting to 80 or even 100 degrees of Fahrenheit's scale.

We are thus enabled, in the hottest weather, to freeze a mass of water, and to keep it frozen, till it gradually wastes away by a continued but invisible process of evaporation. The only thing required is, that the surface of the acid should approach tolerably near to that of the water, and should have a greater extent; for otherwise, the moisture would exhale more copiously than it could be transferred and absorbed; and, consequently, the dryness of the rarefied medium would become reduced, and its evaporating energy essentially impaired. The acid should be poured to the depth perhaps of half an inch, in a broad flat dish, which is covered by a receiver of a form nearly hemispherical; the water exposed to congelation may be contained in a shallow cup, about half the width of the dish, and having its rim supported by a narrow porcelain ring, upheld above the surface of the acid by three slender feet. (See fig. 1, and 2, Plate III.) It is of consequence that the water should be insulated as much as possible, or should present only a humid surface to the contact of the surrounding medium; for the dry sides of the cup might receive, from communication with the external air, such accessions of heat as greatly to diminish, if not to counteract, the refrigerating effects of evaporation. This inconvenience, however, is in a great measure obviated, by investing the cup with an outer case at the interval of about half an inch. If both the cup and its case consist of glass, the process of congelation is viewed most completely; yet when they are formed of a bright metal, the effect appears on the whole more striking. But the preferable mode, and that which prevents any waste of the powers of refrigeration, is to expose the water in a pan of porous earthenware. If common water be used, it will evolve air-bubbles very copiously, as the exhaustion proceeds; in a few minutes, and long before the limit of rarefaction has been attained, the icy spiculæ will shoot beautifully through the liquid mass, and entwine it with a reticulated contexture. As the process of congelation goes forward, a new discharge of air from the substance of the water takes place, and marks the regular advances of consolidation; but after the water has all become solid ice, which, unless it exceed the depth of an inch, may generally be effected in less than half an hour, the circle of evaporation, and subsequent absorption, is still maintained. A minute film of ice, abstracting from the internal mass a redoubled share of heat, passes, by invisible transitions, successively into the state of water and of steam, which, dissolving in the thin ambient air, is conveyed to the acid, where it again assumes the liquid form, and, in the act of combination, likewise surrenders its heat.

In performing this experiment, the object is generally to seek at first to push the rarefaction as far as the circumstances will admit. But the disposition of the water to fill the receiver with vapour, being only in part subdued by the action of the sulphuric acid, a limit is soon opposed to the progress of exhaustion; and the included air can seldom be rarefied above a hundred times, or till its elasticity can support no more than a column of mercury, about three tenths of an inch in height. A smaller rarefaction, perhaps from ten times to twenty times, will be found sufficient to support congelation after it has once taken place. The ice then becomes rounded by degrees at the edges, and wastes away insensibly, its surface being incessantly corroded by the play of the ambient air, and the minute exhalations conveyed by an invisible process to the sulphuric acid, which, from its absorbing the vapour, is all the time maintained above the temperature of the apartment. The ice, kept in this way, suffers a very slow consumption; for a lump of it, about a pound in weight, and two inches thick, is sometimes not entirely gone in the space of eight or ten days. During the whole progress of its wasting, the ice still commonly retains an uniform transparent consistence; but, in a more advanced stage, it occasionally betrays a sort of honey-combed appearance, owing to the minute cavities formed by globules of air, set loose in the act of freezing, yet entangled in the mass, and which are afterwards enlarged by the erosion of the solvent medium.

But almost every practical object is attained through far inferior powers of refrigeration. Water is the most easily frozen, by leaving it, perhaps for the space of an hour, to the slow action of air that has been rarefied only in a very moderate degree. This process meets with less impediment; and the ice formed by it appears likewise more compact, when the water has been already purged of the greater part of its combined air, either by distillation, or by long continued boiling. The water which has undergone such operation should be introduced as quickly as possible into a decanter, and filled up close to the stopper, else it will attract air most greedily, and return nearly to its former state in the course of a few hours.

The most elegant and instructive mode of effecting artificial congelation, is to perform the process under the transferrer of an air-pump. A thick but clear glass cup being selected, of about two or three inches in diameter, has its lips ground flat, and covered occasionally, though not absolutely shut, with a broad circular lid of plate glass, which is suspended horizontally from a rod passing through a collar of leather. (Fig. 6. Plate III.) This cup is nearly filled with fresh distilled water, and supported by a slender metallic ring, with glass feet, about an inch above the surface of a body of sulphuric acid, perhaps three quarters of an inch in thickness, and occupying the bottom of a deep glass basin that has a diameter of nearly seven inches. In this state the receiver being adapted, and the lid pressed down to cover the mouth of the cup, the transferrer is screwed to the air-pump, and the rarefaction, under those circumstances, pushed so far as to leave only about the hundred and fiftieth part of a residuum; and the cock being turned to secure that exhaustion, the compound apparatus is then detached from the pump and removed to some convenient apartment. As long as the cup is covered, the water will remain quite unaltered; but on drawing up the rod half an inch or more, to admit the play of a rare medium, a bundle of spicular ice will, after the lapse perhaps of five minutes, dart suddenly through the whole of the liquid mass; and the consolidation will afterwards descend regularly, thickening the horizontal stratum by insensible gradations, and forming in its progress a beautiful transparent cake. On letting down the cover again, the process of evaporation being now checked or almost entirely stopped, the ice returns slowly into its former liquid condition. In this way the same portion of water may, even at distant intervals of time, be repeatedly congealed and thawed successively twenty or thirty times. During the first operations of freezing, some air is liberated; but this extrication diminishes at each subsequent act; and the ice, free from the smallest specks, resembles a piece of the purest crystal.

This artificial freezing of water in a cup of glass or metal affords the best opportunity of examining the progress of crystallization. The appearance presented, however, is extremely various. When the frigorific action is most intense, the congelation sweeps at once over the whole surface of the water, obscuring it like a cloud; but, in general, the process advances more slowly; bundles of spiculæ, from different points, sometimes from the centre, though commonly from the sides of the cup, stretching out and spreading by degrees with a sort of feathered texture. (Plate III. fig. 4.) By this combined operation, the surface of the water soon becomes an uniform sheet of ice. Yet the effect is at times singularly varied; the spicular shoots, advancing in different directions, come to inclose, near the middle of the cup, a rectilineal space, which, by unequal though continued encroachment, is reduced to a triangle; and the mass below, being partly frozen, and therefore expanded, the water is gradually squeezed up through the orifice, and forms by congelation a regular pyramid, rising by successive steps; or, if the projecting force be greater, and the whole more contracted, it will dart off like a pillar. The radiating or feathered lines which at first mark the frozen surface, are only the edges of very thin plates of ice, implanted at determinate angles, but each parcel composed of parallel planes. This internal formation appears very conspicuous in the congealed mass which has been removed from a metallic cup, before it is entirely consolidated. Sea-water will freeze with almost equal ease, but it forms an incompact ice, like congealed syrup, or what is commonly called *water-ice.*

When cups of glass or metal are used, the cold excited at the open surface of the liquid extends its influence gradually downwards. But if the water be exposed in a porous vessel, the process of evaporation, then taking effect on all sides, proceeds with a nearly regular consolidation towards the centre of the mass, thickening rather faster at the bottom, from its proximity to the action of the absorbent, and leaving sometimes a reticulated space near the middle of the upper surface, through which the air, disengaged by the progress of congelation, makes its escape.

When very feeble powers of refrigeration are employed, a most singular and beautiful appearance is, in course of time, slowly produced. If a pan of porous earthenware, from four to six inches wide, be filled to the utmost with common water till it rise above the lips, and planted above a dish of ten or twelve inches diameter, containing a body of sulphuric acid, and then a broad round receiver placed over it; on reducing the included air to some limit between the twentieth and the fifth part of its usual density, according to the coldness of the apartment, the liquid mass will, in the space of an hour or two, become entwined with icy shoots, which gradually enlarge and acquire more solidity, but always leave the fabric loose and unfrozen below. The icy crust which covers the rim, now receiving continual accessions from beneath, rises perpendicularly by insensible degrees. From each point on the rough surface of the vessel, filaments of ice, like bundles of spun glass, are protruded, fed by the humidity conveyed through its substance, and forming in their aggregation a fine silvery surface, analogous to that of fibrous gypsum or satin spar. At the same time, another similar growth, though of less extent, takes place on the under side of the pan, so that continuous icy threads might appear vertically to transpierce the ware. The whole of the bottom becomes likewise covered over with elegant icy foliations. (Plate III. fig. 5.) Twenty or thirty hours may be required to produce those singular effects; but the upper body of ice continues to rise for the space of several days, till it forms a circular wall of near three inches in height, leaving an interior grotto lined with fantastic groups of icicles. In the meanwhile, the exfoliations have disappeared from the under side, and the outer incrustation is reduced, by the absorbing process, to a narrow ring. The icy wall now suffers a regular waste from external erosion, and its fibrous structure becomes rounded and less apparent. Of its altitude, however, it loses but little for some time; and even a deposition of congealed films along its coping or upper edge seems to take place at a certain stage of the process. This curious effect is owing to a circumstance which, as it serves to explain some of the grand productions of nature, particularly the *icebergs* of the arctic circle, merits particular attention. The circular margin of the ice, being nearer the action of the sulphuric acid than its inner cavity, must suffer, by direct evaporation, a greater loss of heat; and, consequently, each portion of thin air that rises from the low cavity, being chilled in passing over the colder ledge, must deposit a minute corresponding share of its moisture, which instantly attaches itself and incrusts the ring. Whatever inequalities existed at first in the surface of the ice, will hence continually increase.

Artificial congelation is always most commodiously performed on a large scale. Since the extreme of rarefaction is not wanted, the air-pump employed in the process admits of being considerably simplified, and rendered vastly more expeditious in its operation. Two or three minutes at most will be sufficient for procuring the degree of exhaustion required, and the combined powers of evaporation and absorption will afterwards gradually produce their capital effect. In general, plates of about a foot in diameter should be preferred, which can be connected at pleasure with the main body of the pump. The dish holding the sulphuric acid is nearly as wide as the flat receiver; and a set of evaporating pans belongs to it, of different sizes, from seven to three inches in diameter, which are severally to be used according to circumstances. The largest pan is employed in the cold season, and the smaller ones may be successively taken as the season becomes sultry. On the whole, it is better not to overstrain the operation, and rat er to divide the water under different receivers, if unusual powers of refrigeration should be required. As soon as the air is partly extracted from one receiver, the communication is immediately stopped with the barrel of the pump, and the process of exhaustion is repeated on another. In this way, any number of receivers, it is evident, may be connected with the same machine. If we suppose but six of these to be used, the labour of a quarter of an hour will set as many evaporating pans in full action, and may, therefore, in less than an hour afterwards, produce nearly six pounds of solid ice. The waste which the water sustains during this conversion is extremely small, seldom indeed amounting to the fiftieth part of the whole. Nor, t after multiplied repetitions, is the action of the sulphuric acid considerably enfeebled by its aqueous absorption. At first that diminution is hardly perceptible, not being the hundredth part when the acid has acquired as much as the tenth of its weight of water. But such influence gains rapidly, and rises with accelerated progression. When the quantity of moisture absorbed amounts to the fourth part by weight of the acid, the power of supporting cold is diminished by the twentieth; and, after the weights of both these come to be equal, the refrigerating energy is reduced to less than the half. Sulphuric acid is hence capable of effecting the congelation of more than twenty times its weight of water, before it has imbibed near an equal bulk of the liquid, or has lost about the eighth part of its refrigerating power. The acid should then be removed, and concentrated anew by slow distillation.

When the exhaling and absorbing surfaces are rightly disposed and apportioned, the moderate rarefaction, from twenty to forty times, which is adequate to the freezing of water, may be readily procured by the condensation of steam. In all manufactures where the steam-engine is employed, ice may therefore, at all times, be formed in any quantity, and with very little additional expence. It is only required to bring a narrow pipe from the condensing vessel, and to direct it along a range of receivers, under each of which the water and the acid are severally placed. These receivers, with which the pipe communicates by distinct apertures, may, for the sake of economy, be constructed of cast iron, and adapted with hinges to the rim of a broad shallow dish of the same metal, but lined with lead to hold the acid.

The combined powers of rarefaction and absorption are capable of generating much greater effects than the mere freezing of water. Such frigorific energy, however, is at all times sufficient for effecting the congelation of mercury. Accordingly, if mercury, contained in a hollow pear-shaped piece of ice, be suspended by cross threads near a broad surface of sulphuric acid under a receiver; on urging the rarefaction, it will become frozen, and may remain in that solid state for the space of several hours. But this very striking experiment is easily performed without any foreign aid.

Having introduced mercury into the large bulb of a thermometer, and attached the tube to the rod of a transferrer, let this be placed over the wide dish containing sulphuric acid, in the midst of which is planted a very small tumbler nearly filled with water. After the included air has been rarefied about fifty times, let the bulb be dipped repeatedly into the very cold but unfrozen water, and again drawn up about an inch; in this way it will become incrusted with successive coats of ice, to the thickness perhaps of the twentieth part of an inch. The water being now removed, the pendant icicle cut away from the bulb, and its surface smoothed by the touch of a warm finger, the transferrer is again replaced, the bulb let down within half an inch of the acid, and the exhaustion pushed to the utmost. When the syphon-gage has come to stand under the tenth of an inch, the icy crust starts into divided fissures, and the mercury, having gradually descended in the tube till it reach its point of congelation, or 39 degrees below zero, sinks by a sudden contraction almost into the cavity of the bulb; and the apparatus being then removed and the ball broken, the metal appears a solid shining mass, that will bear the stroke of a hammer.

But a still greater degree of cold may be created, by applying the same process likewise to cool the atmosphere which encircles the apparatus itself. A glass matrass was blown nearly of a hemispherical s ape, its bottom quite flat, and about three inches in diameter, an its neck about half an inch wide and cut square over. The whole was covered with a coat of patent lint, which takes up water very copiously. A portion of sulphuric acid was next introduced, forming a layer of perhaps a quarter of an inch thick; and a spirit of wine thermometer, having its bulb also cased with wetted lint, was then inserted within the matrass, a brass ring attached to the tube securing it in the right position. Things being thus arranged, the matrass or flat bottle, with its thermometer, was placed on a slender stool with glass feet, about an inch above the sulphuric acid in the broad basin, and the large receiver luted over it. The air was then partly extracted, till the gage came below one inch. In a few minutes the lint was frozen entirely, and looked white. After an interval of a quarter of an hour, to allow time for the evaporation of that icy coat to cool down the interior apparatus, the pump was again urged, and the exhaustion pushed to about three tenths of an inch. In a short while the inclosed thermometer sank not fewer than 180 degrees, and remained stationary till the ice had wasted away.

It is obvious, therefore, that the refrigerating powers could be pushed still further by a judicious combination of the apparatus. An idea of the mode of proceeding may be formed from the inspection of Plate III. fig. 8. It would be easy to show that the maximum effect is obtained, when the dimensions of the successive cases rise in a geometrical progression. The action, however, is not doubled for each additional case, but increased rather more than one half.

These plans are difficult in the execution, and, though they enlarge our conceptions of the extent of the descending scale of heat, yet they furnish merely speculative results. A very important practical improvement has been lately made in the process of artificial congelation. Sulphuric acid is certainly a cheap and most powerful agent of absorption; but the danger in using such a corrosive liquid, especially by unskilful persons, formed always a serious obstacle to its general adoption. Mr Leslie had early noticed the remarkable absorbent quality of our mouldering whinstone or porphyritic trap; and in April 1817 he substituted that material, grossly pounded and dried before a kitchen fire, instead of sulphuric acid, and actually froze a body of water with great facility. This earth will attract the fiftieth part of its weight of moisture before its absorbent power is reduced to the one half, and is hence capable of freezing the sixth part of its weight of water. It may be repeatedly dried, and will, after each operation, act with the same energy as at first.

But an absorbent still more convenient and powerful afterwards occurred to Mr Leslie; namely, *parched oatmeal.* With a body of oatmeal of a foot in diameter, and rather more than an inch deep, he froze a pound and a quarter of water, contained in a hemispherical porous cup. The meal is easily dried and restored again to its action. In a hot climate the exposure to the sun alone might prove sufficient. By the help of this simple material, therefore, ice may easily and safely be produced in any climate, and even at sea.

Other substances could no doubt be employed as absorbents. But, except the muriate of lime, or what is called the *oil of salt* desiccated, none hold out any prospect of success. Dried common salt will barely effect congelation; and stucco, or the sulphate of lime, deprived of its water of crystallization, which might seem to promise a powerful absorption, has scarcely any efficacy whatever.

## Explanation of Plate III.

Fig. 1, represents a large air-pump intended for the purpose of freezing water, consisting of six receivers, each of them having a broad glass saucer for holding sulphuric acid, and a small porous earthen cup containing the water.

Fig. 2, a section of the above, showing the communication between the receivers and the body of the pump.

Fig. 3, the lever key for opening and shutting the cocks.

Fig. 4, the more ordinary appearance of the surface of the water in the porous cup, at the moment when the act of congelation begins.

Fig. 5, the very singular kind of ice, striated, columnar, and cavernous, which, under a slight rarefaction, but in cold weather, rises slowly and changes its form by degrees, while part of the remaining water is drawn through the substance of the cup, and covers the outside with a thick icy collar above those irregular foliations.

Fig. 6, represents an elegant mode of almost instantaneously freezing within a transferrer. Above the saucer of the sulphuric acid is placed a glass cup holding water, and the air having been previously exhausted, and the instrument detached from the pump, on pulling up the rod, the water, now left exposed to the most powerful evaporation, quickly runs into spicular ice, which gradually increases and consolidates into a pure transparent mass. The lid being let down again upon the glass cup, the action ceases, and the ice returns slowly to the state of water.

Fig. 7, a refrigeratory for cooling water at all times to a moderate degree, without the operation of an air-pump ; a body of sulphuric acid lying at the bottom of the pan, while a porous vessel containing water is set in the centre of the refrigeratory, and the air is confined about it by replacing the lid.

Fig. 8, exhibits a system of vessels for producing spontaneously great cold. It consists of a series of leaden or pewter vessels, placed one within the other, and whose surfaces form a descending geometrical progression, being covered externally with a soft wet lint, and holding each of them a portion of sulphuric acid. The powerful evaporation maintained by this arrangement causes the interior vessels to become successively colder, and thus augments by a repeated multiplication the final effect. Placed under the receiver of the air-pump, this system of evaporating vessels, with no very high degree of exhaustion, and at all seasons, excites ultimately the most intense cold yet produced, far exceeding what is required for the congelation of mercury.

# DEW.

---

Dew is the humidity which the air, under certain circumstances, deposits, in the form of minute globules, on the surfaces of the bodies in contact with it. The Greek term *δροσος* was evidently derived from *ὑδωρ*, *aqua*, implying simply *watering* or *humification.* The Latin name *ros* is of the same descent. Our English word is obviously borrowed from the German *thau,* akin to the verb which signifies to *melt,* and conveying the idea, therefore, in the Shaksperian phraseology, of air "*melting, thawing, and resolving itself into a dew.*" The Swedish term *dag* is no doubt of the same origin, though it likewise denotes low *mist* or floating vapour. It is remarkable that the French language, though certainly not remarkable for its copiousness, has two distinct terms for dew: *serein* for the humidity which collects in the evening; and *rosée,* for what appears accumulated in the morning; the latter being derived from the Latin word *ros,* and the former intimating that *clearness* and *serenity* of the sky which is most conducive to the formation of dew.

When the atmosphere has a temperature below the point of congelation, the *dew* which might adhere to the substances exposed to it passes into the form of *hoar-frost.*

This was called by the Greeks πάχνη, from its *hard* or *consolidated* nature. The French term is exactly the same compound as our own, *white frost* or *gelée blanche.* But the German language has a simple and primitive word to denote it, *reif;* which, in the Swedish, has been slightly modified into *rim,* a word likewise adopted by the older English writers, and still retained in the Scottish dialect, or dilated into *rime-frost,* and thence probably corrupted into *raw-frost.*

As dew appears to collect only during fine clear nights, when the heavens glow with sparkling constellations, the ancients, in the infancy of science, imagined it to be actually shed from the stars, and, therefore, to partake of a pure and celestial essence. Hence the vulgar notion that *dew falls,* which has prevailed through all ages, and continues to tincture every language.* The mythologists described dew as the daughter of Jove and of the Moon;† and Plutarch asserts it to be most abundant in the time of full moon. The lunar beams themselves were supposed to contribute some influence, being of a cold nature, and therefore possessed of a humifying quality.‡ The moon, it was imagined, performed merely the office of an imperfect mirror reflecting the softened lustre of the sun without any portion of his heat.

The dew of heaven has always been regarded as a fluid of the purest and most translucid nature. Hence it was celebrated for that abstergent property which, according to

---

* Gesner, Facciolatus, and other lexicographers, blend this idea in their definition of dew. Ros, *Humor cælo defluens noctu, cum sudum est, et omnis aura quiescit; qui si gelu concrescat, est* PRUINA.

† Διὸς θυγάτηρ ἔρσα τρέφει καὶ σελάνας δίας.

‡ Statius has the expression *rorifera luna.*

the vulgar persuasion, enables it to remove all spots and stains, and to impart to the skin the bloom and freshness of virgin beauty. Like the elixir of later times, it was conceived to possess the power of extending the duration of human life; and Ammianus Marcellinus ascribes the longevity and robust health of mountaineers, in comparison with the inhabitants of the plains, chiefly to the frequent aspersion of dew on their gelid bodies. Dew was also employed as a most powerful agent, in all their operations, by the alchemists; some of whom pretended that it possessed such a subtle and penetrating efficacy as to be capable of dissolving gold itself. Following out the same idea, the people of remote antiquity fancied that the external application of dew had some virtue in correcting any disposition to corpulence. The ladies of those days, anxious to preserve their fine forms, procured this celestial wash, by exposing clothes or fleeces of wool to the humifaction of the night. It was likewise imagined that grasshoppers feed wholly on dew, and owe their lean features perhaps to such spare diet.*

The philosophers of Greece, after genuine knowledge had illumined that interesting region, entertained far juster notions concerning the nature and formation of dew. Aristotle, whose universal genius ranged over both the physical and the intellectual world, studied facts closely, and sought to reason accurately from the phenomena actually observed. In his book *De Mundo*, he defines "dew to be humidity detached in minute particles from the clear chill atmosphere."† In his treatise of *Meteorology*, he states that "dew is only formed beneath a calm and cloudless sky, but never in windy

---

* *Rore aluntur cicadæ.* PLIN. xi. 26. *Dumque thymo pascentur apes, dum rore cicadæ.* VIRG. Eclog.

† Δροσος εστιν ὑγρον εξ αιθριας κατα συστασιν λεπτον.

weather."* He further subjoins, that it collects in low places, and not on the summits of mountains. Vapour, which, according to him, is only [illegible] combined with water, rises in the atmosphere during the day ; but when the cold begins to prevail at night, it again discharges its humidity. This vapour, however, he thinks can never ascend high above the surface of the earth, both because it must soon lose its buoyant heat, and because in lofty situations it would be scattered and dissolved by the agitation of the air. Dew is hence most copious in fine weather, and in low damp situations. A north wind checks its production, but a gentle southern gale, charged with humidity, will occasion a copious deposit. When a more intense cold prevails in the atmosphere, the vapour precipitates its humidity in a congealed form, and the dew passes immediately into hoar-frost. Cold occasions this consolidation. Dew has hence the same relation to hoar-frost that rain bears to snow, the frozen mass of clouds constituting the one, and attenuated low vapour, seized by frost, forming the other. The heat of the sun's rays thus first raises the vapour from below ; but, in all its subsequent changes and modifications, the moon and stars, contrary to the earlier and more popular notions, exert no sort of influence.

The Aristotelian opinions seem to have given place among the Romans to the ruder notions which prevailed in remote antiquity respecting the mode of its formation. Pliny invariably speaks of a dew as *falling* from the heavens ; *cum ros cecidisset.* We might expect, therefore, that the poets would continue in their verses to perpetuate the same idea.

---

* The same observation is repeated by Pliny, *Nat. Hist.* xviii. 29.—*Neque in nube neque in flatu cadunt rores ;* and in another place, still more explicitly, thus: *Rores neque gelu, neque ardoribus, neque ventis, nec nisi serenâ nocte.* Id. ii. 70.

Sparsaque cœlesti rore madebit humus. OVID, Fast. i 312.

Vitreoque madentia rore,
Tempora noctis eunt. Id. Fast. iii. 880.

———— qua prata jacent,
Quæ rorifera mulcens aura,
Zephyrus vernas evocat herbas. SENECA Trag.

Hinc ubi roriferis terram nox obruit umbris. LUCRET. vi. 4. 64.

Virgil marks the cold which always accompanies the formation of dew, and which, when it becomes more intense, converts the lucid globules into spicular shoots of hoar-frost.

Cum primum gelidos rores aurora remittit. Eclog. viii. 15.

The opinion, that dew falls from the sky, maintained its credit during the course of the middle ages. The alchemists even carried this idea so far as to fancy, that since the dew gradually evanishes in the progress of the day, under the action of the solar rays, it then merely seeks by sympathy to regain its native seat in the highest heavens. Nay, some of those ingenious enthusiasts have not scrupled, in confirmation of their wild hypothesis, broadly to assert, that a few drops of morning dew, being inclosed in an empty egg-shell, which is placed at the foot of a ladder, resting against the roof of a house, the shell will become buoyant while the sun shines, and will mount along the ladder till it reaches the very top. The famous Van Helmont, who refined on the notions of the alchemists, considered the lights of heaven as of two distinct kinds; the one which flows from the sun, and rules the day, being intrinsically warm, and possessing masculine virtue; the other, which rules the night, and emanates from the moon and stars, being of a feminine nature, and having a cool or refrigerating influence. This cold light, he imagined, produces the purest essence of water, which is stored in the moon, to recruit the waste of the nether world; and he supposed the allegory of Jacob's ladder might represent that perpetual ascent and descent of aqueous matter, by which the revolution of the system is constantly maintained. That the moon's beams are naturally cold, he thought sufficiently established by the prevailing belief of the common people, who carefully avoid sleeping in the open air, without some screen to protect them from the chilling impressions which are shot down upon the ground.

Baptista Porta asserted that air is actually converted into water, from the accession of cold; and thought this transmutation proved by the fact, that, on the approach of severe weather, the windows of an apartment have the inside of the panes of glass covered with moisture. Gaspar Schott, as late nearly as the middle of the seventeenth century, was so much persuaded of the coldness of the moon's rays, that he stoutly appealed to the effects of their concentration in the focus of a reflector. This experiment, however, was made sixty years before, and with an opposite result, by the famous Sanctorio, the founder of scientific medicine, and the inventor of the thermometer. He actually employed the air thermometer for the first time; but this very ingenious inquirer must have been deceived by some extraneous circumstances, when he saw the liquor sink so considerably as he asserts, under the calorific action of the moon's light. The philosophical ideas of Sanctorio were perhaps, in some instances, too refined for his age, and the vulgar notion concerning the production of dew continued afterwards, for more than an hundred years, to be still generally retained. But the progress of horticulture, near the latter part of the seventeenth century, brought out some unexpected facts, which seemed at variance with the popular belief. It was remarked that a bell-glass, being placed in the evening over a plant, was in the morning profusely covered with dew on the inside, though scarcely any moisture appeared to adhere to the external surface. The humidity which formed the minute globules must, therefore, have risen from the plant or the ground, and adhered against the glass.

Such, however, was the very slow advance of sound philosophy, that Perlicius, who proposed what he calls a *Drososcope*, consisting of an oblique balance, playing in a soft rack, for indicating the quantity of dew accumulated in the absence of the observer, concludes the discourse which, under the direction of Professor Weidler, he delivered in scholastic form, on taking out his degree at Wurtemberg in 1727, with the general inference, that dew descends to us from the atmosphere of Jupiter, Venus, the Moon, Mars, and Saturn; but that, though it falls from the air, it by no means originates in this fluid. He had found, in the month of August, that 250 grains of dew formed on the surface of a square foot in the country, while only ninety-four, and at other times seventy-six, or even sixty-four grains, were deposited in the town.

The first person, however, who appears to have openly rejected the inveterate opinion of the descent of dew, was Gersten, another German professor, who made several experiments on the subject, and printed at Giessen, in 1733, an academical thesis, in which he advanced the opposite hypothesis. He found that all plants exhale, in various proportions, the moisture which forms the aqueous deposit; and remarked, that plates of copper exposed during the night have only their under surface bedewed. This dissertation led the celebrated Professor Muschenbroeck to repeat the same observations at Utrecht. Having obtained similar results, he communicated the main facts to his Parisian correspondent, M. du Fay, who planned immediately a series of experiments on a large scale. This ingenious philosopher caused two tall ladders to be set up, reclining against each other, in a vacant space, remote from all trees and lofty buildings; and on the 25th October 1736, at four o'clock in the evening, the weather being clear and calm, he laid panes of glass on the steps at the different heights of six, thirteen, seventeen, twenty-five, and thirty-one feet above the ground. These he visited at certain intervals during the progress of the night. By five o'clock, a pane close to the ground had its under side completely wet, while its upper side was only slightly dewed. At six o'clock, the pane, six feet above the surface, was covered with dew; and at seven o'clock, the effect had reached the highest pane. During the whole night, the dew continued to form; but it appeared always more copious on the lower panes.

These facts might be deemed sufficiently conclusive; but M. du Fay sought likewise to ascertain the relative quantities of moisture deposited at different altitudes. He procured several rectangular bits of green cloth, cut six inches long and four inches broad, and adjusted all to the same weight. These he suspended in horizontal positions, at four o'clock in the evening, one of them only half a foot above the ground, and the rest at the heights of six, thirteen, and twenty-five feet. On weighing them next morning, he found that they had respectively imbibed fifty-three, sixty-six, fifty-six, and fifty-four grains of dew. The subsequent night having proved windy, they gained only seven, nine, ten, and six grains. It was evident, therefore, that dew is formed not only sooner, but more copiously, near the surface of the ground, than at greater elevations.

To determine still more precisely the several quantities of moisture imbibed at different heights, M. du Fay took three linen towels, each three and a half feet long and two and a half broad, and having dried them thoroughly in the sun, he stretched them horizontally at one foot, and seventeen, and twenty-eight feet above the ground. After exposure during the whole night, the air being quite clear and calm, the lowest one was found to have gained 3060 grains, the next only 2346, and the highest 2742. There occurs some slight anomaly in this result; but on reducing the greatest effect to English measures, it corresponds almost exactly to a cubic inch of dew for each square foot of the surface. This might appear to be rather a low estimate for the climate of Paris in the autumnal season, since, at the same rate, it would give only a deposit of about two and a half inches during the whole year.

In the mean time, Muschenbroeck made observations on the humifaction of substances placed above the leaden platform of his observatory at Utrecht. The dew formed in such a situation, he thought, could not have risen from the ground, but must have fallen from the atmosphere. But, pushing these experiments further, he was conducted to a most curious and interesting discovery. He found that dew forms in very different proportions on different bodies, and that it will scarcely adhere to a surface of polished metal, while it streams profusely over glass or porcelain. Even the colour of the substance appeared, in some instances, to alter the effect. Thus, a piece of red morocco leather acquired, by exposure through the night, twice as much dew as another piece of the same size, whether black or blue. A closer examination, however, convinced him that this modification was not caused by the mere colour itself, but was occasioned by the addition or infusion of the matter employed to produce it.

M. du Fay, on his part, repeated those experiments with the same success. He likewise performed others of a similar description. Electricity was at this period in high vogue, and he had distinguished himself in that department. It therefore struck M. du Fay that the disposition of certain bodies to attract or to repel dew was somehow connected with the distinction into electrics and conductors. In the prosecution of this idea, he sought to compare the humifying action of vitreous with that of resinous substances. He took a tin basin, of the same shape and dimensions as one of glass, and having coated it on both sides with forty or fifty applications of a solution of shell-lac in alcohol, he exposed the two vessels during a fine clear night; but the surface of the glass was found to gather twice as much dew as the resinous coating. In the course of his researches, he likewise noticed another curious fact, the explication of which must be referred to a recent discovery. Having selected two large and equal watch-glasses, he set their convex faces horizontally, the one on a porcelain, and the other on a silver saucer, and exposed them both, in this position, to the influence of the night air; when the former was always observed to collect five or six times more dew than the latter. The metallic surface must have, therefore, in some way or other, contributed by its presence to check the precipitation of humidity, though its direct action is confined to a very narrow limit, since, under a resinous coating of only the hundredth part of an inch in thickness, the effect is precisely the same as with a cake of wax. In confirmation of this remark, having composed a square plate, by cementing at the edge a polished rectangle of brass, six inches long and three inches broad, to a similar piece of glass, he found, on exposing it to the atmosphere at night, the vitreous surface covered as usual with dew, while the metallic one was scarcely at all affected; but, on laying another slip of glass across these plates, the end which rested on the brass remained quite dry, while the other end soon became profusely wetted.

It seemed to follow from the experiments of Muschenbroeck and Du Fay, that, strictly speaking, dew neither falls nor rises, but, according to the doctrine of Aristotle, only separates, under a certain change of circumstances, from the air, and attaches itself to some substances in preference to others. The theory of vapour, proposed afterwards by Le Roy of Montpellier, threw further light on the subject. Moisture is suspended in the atmosphere by a real chemical solution, in the same manner as nitre and other salts are dissolved in water. The solvent energy is in both cases augmented by the addition of heat. A rise of temperature enables the air to support a larger portion of humidity, while the decrease of it enfeebles the attractive power, and occasions a precipitation in the shape of mist or dew. This perspicuous explication, as we have seen, had been already anticipated, though but vaguely stated, by Aristotle.

A deposition entirely similar to dew or hoar-frost, is hence formed, whenever the air becomes suddenly chilled, by touching any surface much colder than itself, and not consisting of polished metal. Thus, the walls of long passages, vaults, or massive buildings, generally drip with wet during the early part of the summer, before the external heat has sufficiently penetrated. In like manner, it is observed, that when a severe and long-continued frost is succeeded by a thaw, the backs of houses are quickly incrusted with shoots of hoary icicles.

The formation of dew is hence easily produced artificially in a close room, without waiting for exposure under a clear nocturnal sky. If a caraffe, filled with water from a spring or well, be carried into a warm apartment, the outside of the glass will become soon covered over with an aqueous deposit, which must increase till the body of water has acquired nearly the temperature of the encircling air, and will afterwards gradually disappear. But if a piece of tin-foil be applied to the bottle, it will remain dry, while the rest of the surface appears humified with dew.

It is a curious fact, that air always begins to deposit its moisture on glass, even before it has reached the point of saturation, or become absolutely damp. This property, and the circumstances connected with it, Professor Leslie discovered, in the year 1798, by help of his hygrometer, which he had already brought nearly to a state of perfection. On exposing wine-glasses at the approach of evening, they were soon covered with dew, while this instrument still indicated several degrees of dryness. The difference was yet greater in summer than in winter. The hygrometer being inclosed within a small glass receiver, placed on a wetted plate likewise of glass, stood, after the whole internal surface had become lined with dew, at 5° in a very cold room, but at 15° when the apartment was kept warm.

The general observation is explained by Professor Leslie's researches into the propagation of heat. No adjoining substances can ever come into absolute contact; but air approaches much nearer to the boundary of glass, porcelain, or paper, than to a surface of polished metal. By an extension, therefore, of the principle of capillary action, the suspended aqueous particles, which have a strong adhesion to glass, to which they are brought so close, readily detach themselves from their union with the air. But the same particles being held back from the proximity of a metallic surface, to which they have little attraction, are never deposited on it unless the air is actually overloaded with them. The hygrometer will accordingly reach the absolute zero when it is shut up within a case of polished tin.

Professor Leslie afterwards made several occasional observations relative to the production of dew, particularly during a series of clear warm days in the month of May 1801, at Easbury, in Dorsetshire, where he placed his instruments both on the lawn and on the balustrade of a tower sixty feet in height. Without entering into details, it may be sufficient here to mention the principal results. The periodical variations, both of the hygrometer and thermometer, were much greater near the surface than at some elevation. On the approach of sunset, the thermometer on the ground sunk rapidly; the hygrometer relapsed to about five degrees, and the dew began to form on the blades of grass. During the night, the thermometer descended still lower, the hygrometer indicated absolute humidity, and the lawn was covered with a profusion of dew. But a little after sunrise, the thermometer again mounted, the hygrometer began to act, and the sheet of moisture gradually exhaled. In the progress of the day, the heat and dryness increased; and, about two o'clock, the thermometer and hygrometer, both of them screened from the direct action of the sun, stood generally, the former at 75°, and the latter at 85°. But, on the top of the tower, all those changes were less violent. The thermometer, which at that altitude seldom rose in the course of the day to 70°, or the hygrometer to 65°, indicated, as night again closed, a depression, though very moderate in comparison with what was shown at the surface of the ground. The thermometer stood several degrees higher than below, the hygrometer remained at a dryness of 15° or 20°, and no dew was deposited on the balustrade. But similar differences of effect, although on a smaller scale, were exhibited at very moderate heights. On lifting those instruments in the evening only a foot from the ground, the thermometer would rise a degree or two, and the hygrometer mount from the verge of moisture to perhaps 10°. At an elevation of four feet, those changes were nearly doubled. The dew thus always began, as in Du Fay's experiments, to form at the surface of the earth, and continued to mount upwards with the progress of the night.

It is hence easy to explain the general phenomena of dew. "In fine calm weather, after the rays of the declining sun have ceased to warm the surface of the ground, the descent of the higher mass of air gradually chills the undermost stratum, and disposes it to dampness, till their continued intermixture produces a fog, or low cloud. Such fogs are, towards the evening, often observed gathering in narrow vales, or along the course of sluggish rivers, and generally hovering within a few inches of the surface. But in all situations, these watery deposits, either to a greater or a less degree, occur in the same disposition of the atmosphere. The minute suspended globules, attaching themselves to the projecting points of the herbage, form dew in mild weather, or shoot into hoar-frost when cold predominates. They collect most readily on glass, but seem to be repelled by a bright surface of metal."* The unequal heating of the surface during the day thus occasions, on statical principles, a perpetual interchange between the higher and the lower atmosphere, which is prolonged through the night, the warm portions of air still continuing to ascend, and leaving their place to be occupied by the descent of similar cold portions of that fluid. This vertical play is a provision of nature for the attempering of the diurnal vicissitudes of climate. "In clear and calm weather, the air is always drier near the surface during the day than at a certain height above the ground, but it becomes damper on the approach of evening, while, at some elevation, it retains a moderate degree of dryness through the whole of the night. If the sky be clouded, less alteration is betrayed in the state of the air, both during the progress of the day and at different distances from the ground; and if wind prevail, the lower strata of the atmosphere, thus agitated and intermingled, will be reduced to a still nearer equality of condition."†

* Leslie *On the Relations of Air to Heat and Moisture*, p. 132.
† *Ibid.* p. 92.

The descent of chill air caused by superior density, explains the formation of dew in low situations, and its progressive elevation as the cold accumulates. But some further explication was wanted to reconcile the concluding observations of M. du Fay. The subject was in consequence resumed by M. Benedict Prevost, who performed a curious set of experiments, described in a Memoir read before the Philosophical Society of Montauban in 1803. The results are certainly perplexing, and would almost seem anomalous. 1. Tin or copper foil, and gold or silver leaf, being applied to plates of glass, and exposed to dew, were observed, as before, to remain generally dry, while the vitreous surface became bathed with moisture. 2. After exposure to the night air, not only a dry border appeared, extending a little way beyond the film of metal, but the side opposite to that coating continued still dry, though all the rest of the glass was profusely wetted. 3. A piece of glass, being laid above the metallic leaf, destroyed its effect. 4. A rectangular piece of tin-foil being pasted on the inside, at the top of a pane of glass, in a window having a northern exposure, and a similar piece applied at the bottom on the outside; when the dewing began first against the inside, the interior coating appeared wetter than the naked surface, and the portion of this immediately behind the exterior coating seemed always drier than the rest. The facts were exactly reversed when the dewing commenced on the outside of the window. 5. Opposite to the middle of a rectangular leaf of metal, a similar but smaller piece being applied on the outside of the pane, when the dew began to form within the apartment, the space behind the exterior coating still remained dry. 6. In all cases, whether on the inside or the outside of the pane, on covering the metallic leaf with a piece of glass of the same dimensions, the effect was exactly the same as if no metal had been interposed. 7. Similar appearances were produced by combining gilt paper or quicksilvered glass, the results depending wholly on the nature of the extreme surfaces, according as they consisted of metal, or of glass or paper.

On reviewing these curious facts, M. Prevost was struck with their apparent analogy to the phenomena of electricity. He thought they might all be comprised under a single proposition: *That glass which separates two masses of air of unequal temperatures, attracts or repels humidity according as it is armed with metal on the hot or on the cold side.* To account for these very singular yet interesting facts, he proposed a random and strained hypothesis, grounded on some loose notions of chemical affinities. But we need not stop to examine it.

Dr Thomas Young, in his *Lectures on Natural Philosophy*, published in 1806, concludes a short abstract of the experiments of Prevost with suggesting, that they would derive their explication from Professor Leslie's *Discoveries on Heat.* The anticipation was perfectly just, though the discoveries themselves required then a little further extension to embrace the whole phenomena. Professor Leslie had carefully investigated the laws which modify the propagation of hot or cold pulses through an aerial medium from a solid or a liquid boundary. But he did not contemplate the pulsation excited at the conterminous surface of two strata of air having different temperatures. It was indeed impossible to devise an experiment in which the opposite layers of fluid could be kept distinct, for the warmer portions of air would seek always to rise, while its colder and denser portions would endeavour to sink downwards, and thus form, by insensible shades, a vertical gradation of temperature. But though the pulsatory action excited at each successive horizontal stratum might singly escape observation, it seemed probable that the accumulated impressions transmitted from numerous boundaries would become very sensible. Accordingly, in a close heated room, the *pyroscope*, or differential thermometer, having one of its balls gilt, which is susceptible of such pulses only, marks, near the floor, perhaps four or five degrees of calorific impression, yet, when lifted higher, it indicates an effect always diminishing in proportion to the proximity of the ceiling. The entire action exerted, or the amount of the intermediate energies, was, therefore, as the excess of the temperature of the stratum of air next the ceiling above that of the stratum in which the instrument happened to be placed. Carried out of doors in clear and calm weather, after the sun had withdrawn his beams, it betrayed a much stronger tendency the contrary way, and marked a copious frigorific impression, evidently produced by the coldness which must pervade the upper regions of the atmosphere. But to fit the pyroscope for making observations during the day, it was converted into the *Æthrioscope*, in which the influence of light is neutralized; a combination of great delicacy, and, therefore, a valuable acquisition to meteorological science. (See the article CLIMATE.) The application of this new instrument has not only ascertained the existence, but measured the intensity, of the cold pulses which are at all times darted downwards from the successive strata of air, though often partially intercepted by clouds, or more completely obstructed by low fogs. But since the spheroidal cup, which concentrates the various oblique impressions on the upper ball of the æthrioscope, can do little more than double the direct action against a horizontal surface, it may hence be computed that, in fine bright evenings, those cold pulses rained from the sky are sufficient alone to depress the temperature of the ground, according to the seasons, sometimes eight degrees, but generally about three degrees, by Fahrenheit's scale. The blades of grass, thus chilled from exposure, cool in their turn the damp air which touches them, and cause it to drop its moisture. For the same reason, the naked ball of the æthrioscope, as it is still more cooled, appears much sooner affected, being commonly covered with profuse liquid globules long before the dew has begun to form on the surface of the ground.

All the difficulties and seeming anomalies of the observations of Du Fay and Prevost now vanish away. The various phenomena proceed chiefly from the cold induced by exposure under a clear sky; but other causes will often essentially modify the results. 1. The impression received on a plate of polished metal scarcely amounts to the tenth part of what is communicated to a surface of glass, wood, cloth, paper, earth, or grass. 2. When the action continues the same, the corresponding depression of temperature yet depends on the slowness with which the cold is subsequently dispersed. In calm weather a plate of glass, or a sheet of paper, if covered on both sides with a leaf of metal, will gain or lose heat twice as slow as before, and if coated only on one side, its progress will be a half slower. But high winds greatly assist the dispersion of heat, and often reduce the effects of external impressions to the third or fourth part of their ordinary measure.

Hence the reason why scarcely any dew is formed in windy weather, though the sky be clear; for the frigorific pulses must then have little efficacy, not cooling the ground perhaps more than one or two degrees. In the last observation of Du Fay, the slip of glass laid across the rectangle, composed of alternate bars of glass and of brass, being greatly chilled by exposure, had by contact communicated its coldness to the matter under it, and thus enabled the metal to assist in the deposition of dew. In Prevost's second experiment, the metallic leaf, being scarcely affected by the frigorific impressions, checked by its presence the progress of cold along the vitreous surface, and therefore maintained a dry border all around it. Hence, in his third experiment, a piece of glass covering the metal received the entire impressions, and restored the former effect. The application of a metallic coating against the inside of a pane, must, in the fourth experiment, have augmented by one half the efficacy of the external pulses of cold, and thus made the dew to attach more profusely. For a like reason, while a leaf of metal on the outside of a pane became, in the fifth experiment, slightly dewed, the addition of a smaller metallic leaf against the inside increased the effect, by promoting the accumulation of cold.

The remarkable experiments which the late Dr Patrick Wilson, professor of practical astronomy in the university of Glasgow, performed during the severe frost of January 1780, are easily explained on the same principles. In the declivity of a garden, a thermometer laid, in a clear star light, on the surface of the snow, stood from eight to ten degrees lower than when suspended at the height of a few inches. This excessive cold was evidently not occasioned by evaporation; for, on blowing with bellows against the bulb when it lay on the snow, so far from sinking more, the mercury actually rose two degrees higher than its station in the free air. The intensity was no doubt in part owing to the low position of the snow; for a thermometer suspended at a pole projecting from a window twenty-four feet above the surface indicated four degrees less cold than below. But, besides the accumulating action of the descent of cold air, the snow must have been also chilled extremely by the frigorific pulses darted from an azure sky. This inference, though not perceived at the time, or, indeed, likely to have been admitted then as philosophical, is distinctly supported by an experiment of Dr Wilson. Having screened a spot of the garden by a sort of sharp roof formed with two inclined sheets of brown paper, and laid a thermometer under it on the surface of the snow, the instrument soon marked six degrees of less cold than before, or than another exposed at only a short distance. But this open screen, since it could not impede the mere descent and influx of cold air, must have intercepted a more powerful frigorific influence.

Dr Wilson afterwards performed other similar experiments, which are detailed in his paper on hoar-frost, drawn up in 1788, and inserted in the first volume of the *Transactions of the Royal Society of Edinburgh*. He made the important remark, that during a fog there was no difference of temperature between the surface of snow and the incumbent air. But he neglected to pursue the consequences, and was disposed, from the various facts which he had observed, to conclude vaguely that hoar-frost is always accompanied by a production of cold.

About the same time Mr Six of Canterbury, the inventor of the self-registering thermometer, employed that very useful instrument in making similar but more extensive observations. He found, in a clear summer evening, his thermometer, when laid on the grass, to sink five degrees lower than when suspended freely near the surface. But he had occasion afterwards to remark still greater differences. On a clear and still night in winter, the thermometer which had been supported in the air, fell no fewer than thirteen and a half degrees when placed flat on a meadow. He likewise noticed, as Dr Wilson had done, that thick fogs always impede, and often wholly prevent, the peculiar cooling of the ground.

It seemed therefore ascertained that, in the absence of the sun, the surface of the earth, and especially its projecting herbage, acquire, in calm weather, from the mere aspect of a bright and unclouded sky, a very notable degree of cold. This cold appears likewise connected evidently with the formation of dew. But what is the nature of that relation? Is the coolness contracted by substances on exposure to the nocturnal air, to be considered as the effect or as the cause of their dewing? The former opinion, we have seen, was espoused by Dr Wilson, though sound theory should make us expect that the deposition of dew, or the conversion of humidity from a gaseous to the liquid state, must, on the contrary, occasion a small extrication of heat. But constant experience shows that cold bodies, not sheathed with metallic lustre, become always sprinkled with minute aqueous globules, from the contact of damp air. The simplest truths, however, are very seldom the soonest perceived, and the late ingenious and learned Dr Wells has the merit of being the first who distinctly attributed the formation of dew to the previous cold induced on the ground from the aspect of the sky. He had early conceived an opposite idea, but a closer examination of the subject led him to adopt juster views. Being once engaged in the research, he prosecuted his observations with assiduity and ardour for upwards of two years, at a friend's villa on the skirts of London, in spite of his professional avocations, and at the evident risk of his precarious health. The numerous facts thus collected are detailed in his *Essay on Dew*, which appeared in 1814, and immediately attracted a very considerable share of public notice. This little work, however, does not add much to our stock of accurate information, but it is rendered interesting by the variety of collateral objects which it embraces. The experiments themselves rarely display address or delicacy; and Dr Wells, without ever employing the hygrometer or the pyroscope, instruments which he could have then easily procured, generally contents himself with stating merely rude approximations. Fortunately, such coarse results were sufficient to support the main principle, for otherwise they would have required much correction. But we must still regret that the worthy author should have frequently trusted to conjectural reasoning, instead of appealing to direct experiment.

The chief observations collected by Dr Wells may be reduced to a narrow compass. The coldness of the objects exposed was always found to *precede* the formation of dew, which continued, in favourable circumstances, to accumulate somewhat progressively during the whole night, so that, from midnight to sunrise, the deposition was even greater than from sunset to midnight. Dew was more abundant in spring and autumn than at other seasons, and it was always very copious when the atmosphere inclined to humidity; for instance, in clear nights succeeding to misty mornings, or in clear mornings succeeding to misty nights.

The coldness which bodies contract from exposure must be augmented by every circumstance which retards the communication of heat. Hence loose and spongy materials are mostly affected. Thus, in a clear night the grass was twelve degrees colder than the garden mould, and sixteen and a half degrees colder than a hard gravel walk. In another bright evening, the surface of snow being nine degrees colder than the air, a piece of swandown laid on it became still four degrees colder. Again, a lock of wool placed on a small table in the garden became nine and a half degrees colder than the air, while swandown, in the same situation, acquired a coldness of eleven and a half degrees.

The quantities of dew which attach to different substances appear to follow the proportions of their relative coldness. Parcels of wool, each weighing ten grains, being teased out into flattened balls of two and a half inches diameter, and laid on a grass plot, on a gravel walk, and on fresh garden mould, acquired, during a clear calm night, respectively, sixteen, nine, and eight grains of humidity. In another favourable night ten grains of wool laid on the table attracted sixteen grains of dew, while another similar parcel, suspended at the same height in the free air, acquired only ten grains; but the former must have also been much colder than the latter, since its confined situation, unlike the open exposition, would check the dissipation of the frigorific impressions. Hence dew is always denser on grass than on the leaves of shrubs.

But the cooling of substances from exposure, though one great source of dew, is not the only cause of its formation. In low fogs, while the ground is scarcely colder than the incumbent bed of air, the humidity yet settles profusely on all bodies, even on the polished surface of metals. From Six's experiments, it appears that, from the height of 200 feet, the temperature of the atmosphere in fine evenings decreases regularly about ten degrees, the colder and therefore denser portions being always thrown down to the surface. Hence the reason of the ancient remark, that dew is more copious in low vales than on the tops of hills. But the observations of Dr Wells serve to confirm the general statement. A lock of wool exposed on a table, imbibed, during a clear night, sixteen grains of dew; but a similar parcel, placed immediately under the table, and consequently screened from the aspect of the sky, attracted four grains. In the latter case, the mere accumulation of cold air below must have occasioned the aqueous deposition.

It might perhaps have been judged sufficient if Dr Wells had contented himself with assuming the coldness induced on the ground as merely an experimental fact. At any rate, we cannot help regretting that he should have sought the explication of this primary phenomenon from the very loose, cumbrous, and visionary hypothesis of M. Prevost of Geneva, concerning what is gratuitously called *radiant heat*. We are at a loss, indeed, to conceive how a speculation so repugnant to all the principles of sound philosophy should at this time have procured any favour, unless it proceeds from the blind admiration which the multitude are prone to entertain for whatever lulls the reasoning faculty, and appears cloudy and mysterious. (See METEOROLOGY.)

# METEOROLOGY.

FROM μετεωρος, *sublimis*, is that science which describes and explains the various phenomena which occur in the region of our atmosphere. It has deeply engaged the attention of men in every state of society, from the roving savage to the refined votary of wealth and pleasure. The moment we cross our doors, we commit ourselves to the influence of the weather. But the hardier classes of the community, the shepherd, the ploughman, and the mariner, whose labour creates or procures the staple articles of life, are always exposed, by their occupation, to the mercy of the elements. They were hence led, by the strongest motives, to examine closely the varying appearance of the sky, and to distinguish certain minute alterations which commonly precede the more important changes. No doubt they would often mistake the indication of those aspects, and would infer conclusions from mere casual circumstances. Those tokens which portend the more violent convulsions of the atmosphere, the pelting storm or the careering tempest, are generally of a decided character; but the symptoms which go before the ordinary fluctuations of the weather, can only be dimly conjectured by long experience and sagacious observation. This shadowy knowledge, this dubious and very limited anticipation of the changes which arrive in the medium we breathe, is merely the fruit of personal assiduity and application. Like the conclusions which men of discernment form to themselves on the subject of physiognomy, it is scarcely communicable; it receives therefore no accessions in the progress of ages, but perishes with the individual.

The vague rules which experience had formed on the subject of atmospheric changes were adopted by the philosophers of antiquity, and incorporated into their cosmological systems. But in attempting to explain the phenomena, they sought merely to satisfy the imagination; and the supposition of certain elements, each endued with peculiar properties, appeared, on a superficial view, to connect harmoniously the general facts.

In the infancy of science, however, it was very difficult to distinguish between such appearances as are only incidental, and those constant presages which invariably anticipate their effects. The different fluctuations that happen in the lower atmosphere seemed dependent on the influence of the heavenly bodies, which are continually altering their places. Not only the sun and moon appeared to rule the seasons, but the planets, and even the fixed stars, were conceived to retard or accelerate, by their feeble action, the revolution of the grand cycle of changes. The application of the term *aspect* to the positions of those remote bodies, implied a principle of intelligence, and the *stellar influence* shed by them might therefore be deemed capable of producing all the variety of effects. Hence the origin of *Physical Astrology,* which has, during so many ages, maintained ascendency in the world, and still colours the verses of our poets, and pervades the creed of the vulgar.

The heat evolved by the luminous particles transmitted from the sun, as they lose themselves among the lower strata of our atmosphere, or under the surface of the land

and the sea, varying in its intensity, according to the obliquity of their incidence, and the extent of medium which they traverse, produces the most important effects in the great economy of the globe. But the correct knowledge we have at length acquired concerning the mutual action of bodies, contracts the limits of the supposed celestial influence almost to a point. The light of the moon does not amount to the 100,000th part of that of the sun, and the heat which this very feeble illumination excites or communicates has never been detected by the most delicate instruments, or the best-devised experiments. But all the rays shot from the planets, and from the whole constellations of fixed stars, are quite lost in the contrast with those lunar beams. Their combined impressions, during the lapse of countless ages, on the atmospheric temperature, would hence elude the utmost powers of calculation.

It is only by their attractive force that the heavenly bodies, except the warming energy of the sun, can ever affect the constitution of this globe. As, by their action unequally exerted over its surface, they raise tides in the ocean, so they must likewise agitate the atmosphere with a corresponding reciprocation. This effect, however, depends much more on the proximity of the disturbing body, than on its magnitude or density. Thus the sun, with twenty-three million times the mass of the moon, yet being about four hundred times more distant, has only about the third part of her influence in causing the tides. But, according to the computations of Laplace, the joint action of the sun and moon is only capable of producing a tropical wind, flowing westward at the rate of about four miles a day; a quantity evidently too small to be ever subjected to observation, and certainly insufficient to occasion any immediate and practical results.

It may be calculated that Jupiter and Venus, in those

parts of their orbits which approach the nearest to the earth, affect the aerial tides by a force 76,000 and 82,000 times less than the sun. Those planets could therefore excite a most minute shifting in the atmosphere, limited to a foot in the space of fourteen or fifteen days, or about a single mile in twenty years. As the mass of Mars is unknown, its disturbing force cannot be estimated, though it is less than that of Jupiter. But the combined influence of all the rest of the solar system is, from their great distance, incomparably smaller.

Nothing can be so utterly groundless, therefore, as the disposition to refer the ordinary changes of the weather to the influence of the moon. But, compared with this, the fancied efficacy of the stellar aspects, which was for ages firmly believed by the learned world, vanishes into the shadow of a vision. Nor has the most elaborate examination of numerous registers of the weather, disclosed any precise and constant connection between the phases of the moon or the positions of the stars, and the succession of atmospherical phenomena. The inferences which seem to indicate a different conclusion may be fairly disregarded, as drawn only from a loose and imperfect inspection of facts.

It cannot be disputed, however, that all the changes which happen in the mass of our atmosphere, involved, capricious, and irregular as they may appear, are yet the necessary results of principles as fixed, and perhaps as simple, as those which direct the revolutions of the solar system. Could we unravel the intricate maze, we might trace the action of each distinct cause, and hence deduce the ultimate effects arising from their combined operation. With the possession of such data, we might safely predict the state of the weather at any future period, as we now calculate an eclipse of the sun or moon, or foretell a conjunction of the planets. After the protracted chain of combination has been completed, the same series of events must again be repeated through the boundless lapse of ages. As all the anomalies of the planets are periodical, so likewise the successive varieties follow some cycle of vast extent. In the remotest annals of atmospherical phenomena, we might descry a gleam of futurity, and read the changes which still lie hid in the womb of time.

But, besides this great cycle, there probably are much shorter interior and subordinate periods, in which the weather nearly, though not absolutely, returns after the same order. Whatever principles concur to modify its succession, the revolutions of the sun and moon are certainly the primary and predominant causes. The character of the weather is accordingly strongly marked by the vicissitudes of day and night, and by the annual repetition of summer and winter. The menstrual action of the moon escapes correct observation; but, accumulated during a sufficient time, it may possibly produce a decided influence. Every nineteen years the motions of the moon come to coincide with those of the sun, and her nodes perform their revolutions in nearly the same time, or about eighteen years. This period, or the half of it, has therefore not improbably some slight connection with the usual series of the changes of the weather.

"If the sages of antiquity be justly blamed for adopting implicitly the notions and prejudices of the vulgar, those of modern times may be charged with too eager a disposition to reject whatever savours of popular opinion. A collection of the numerous rules and remarks formed in the course of ages among different orders of men, deeply interested, by their occupations, in watching the changes of the sky, would undoubtedly contain some important truths, which the diligence and sagacity of the philosopher might discriminate, and employ for the basis of beneficial speculations. The most sanguine can hardly uphold the prospect, that mankind will ever arrive at such a pitch of knowledge as to be capable of prognosticating the future modifications of the atmosphere, with the same precision with which they can foretell the successive revolutions of the heavenly bodies; yet the motions, however irregular in appearance, which prevail in the element that we breathe, are, equally with those performed in the regions of space, the result of certain fixed laws. The variable aspect of the sky proceeds partly from the direct action of the sun-beams, but principally from the winds which they excite and maintain. The unequal gravitation of the different portions of our atmosphere to the sun, and to the moon especially, must occasion some small effect in producing or altering the aerial currents; and even the disturbing forces of the planets have a remote share, how minute soever, in the formation of meteors. Nor can we hesitate to conclude, with the late ingenious and elegant M. Bailly, whose fate demands the tribute of a tear, that the notions, so widely spread among men, of the aspects and influences of the celestial bodies, are only the corrupted remains of astronomical science, already advanced to high perfection in some distant age of the world. If motions were to rise and cease instantaneously with the operation of their causes, the same succession of seasons would exactly attend on each revolution of the sun; but the currents of air acquire velocity by degrees, and thenceforth continue to flow till their force is spent. The varied face of our terraqueous globe will, therefore, modify the direction, the power, and the duration of the winds, raised by the action of the solar rays. Hence an extreme complication of causes, which will produce an immense series of fluctuating events. That profound geometer, M. de Lagrange, has established by demonstration, that all the changes arising from the disturbing forces in the planetary system, are subjected to certain vast cycles, on the return of which the same motions are perpetually renewed. Similar periods, but of an extent that affrights the imagination, probably regulate the modifications of the atmosphere; for, whenever a coincidence of circumstances prevails, the series of appearances must inevitably recur. The aggregate labours, indeed, of men, continually transform the face of our globe, and consequently alter the operations of natural causes; but, if the agency of animals be stimulated and determined solely by the influence of external objects, it follows that the re-actions of living beings are comprehended in the same necessary system, and that all the events within the immeasurable circuit of the universe are the successive evolution of an extended series, which, at the returns of some vast period, repeats its eternal round during the endless flux of time. Besides the grand cycle, there must evidently be many intermediate smaller periods, at the lapse of which our atmosphere will present nearly, though not exactly, the same fleeting aspects. Whether these bear any decided relation to the lunar revolutions, cannot with certainty be affirmed. A copious collection of registers, formed in the course of ages, will probably, at some future time, lead to the discovery of certain remarkable periods, which will enable men to conjecture, with tolerable precision, the succeeding changes of the weather. It would be most advisable, perhaps, to begin the inquiry with the tropical countries, in which the seasons are more uniform, and to advance by degrees into the temperate climates. In the mean time, our prognostications may be greatly assisted by observing and studying the concatenation of *phenomena*. Certain coincidences of aspects mark the near recurrence of some small portions of the general series, and afford scope for the doctrine of chances." (*Monthly Review for September* 1795, Art. I. p. 14.)

Such views of the cyclical returns of the varied seasons may expand the imagination, but will be considered rather as curious speculations, than as definite conclusions which can ever be reduced to real and actual application. To improve meteorology, we must submit to tread a humbler path; we must study closely, and by the aid of delicate instruments, the constitution and modifications of the atmosphere, and pursue, under various circumstances, and in different parts of the globe, a lengthened train of careful and precise observations. The ordinary registers of the weather are of a very modern date, and, besides being for the most part unskilfully kept, they seldom record more than the state of the barometer and thermometer, which afford not any complete indication of the disposition of the atmosphere. No wonder than meteorology, though cultivated from the earliest times, has advanced the most slowly to the perfection of science.

To prepare the way for establishing meteorology on a solid basis, we shall *first* inquire into the extent and constitution of the medium which we breathe; and shall *next* review the different philosophical instruments which assist external observation, and indicate at all times the exact condition and qualities of that mutable fluid.

The ancients imagined that our atmosphere, the seat of care, mortality, and corruption, reached as far as the moon, beyond which was a boundless expanse of resplendent ether, the abode of celestial beings, exempt by their nature from all anxiety, and absorbed in the enjoyment of eternal bliss. But the discovery of the weight and pressure of the air destroyed at once this magnificent vision. Comparing the length of the mercurial column with the density of the aerial medium, it followed, that if the atmosphere had been an uniform fluid, it could not exceed the elevation of five miles. But the air being very dilateable, the higher portions of the mass which covers our globe, sustaining a diminished pressure, must swell upwards, and occupy a proportionally wider space. This property hence removes the boundary of the atmosphere to a much greater elevation. By means of an excellent pneumatic machine, air can be rarefied about a thousand times; but this degree of rarefaction would not occur below the height of forty-two miles in the atmosphere. Such is nearly the limit deduced from another consideration, first proposed by the famous Kepler. This most original and inventive philosopher conceived, that the depression of the sun below the horizon, when twilight closes around us, might furnish the data for discovering the altitude of the portion of sky which reflects his latest parting rays. Let C (Plate VI. fig. 2.) be the position of a spectator on the surface of the earth, and A represent the point where the sun sets; the tangent AB will mark the track of his lowest ray, which illuminates the upper part of the atmosphere at B, whence the emission of a secondary ray BC will barely reach the ground at A. But assuming the mean estimate of astronomers, that twilight expires when the solar depression or the arc CA amounts to eighteen degrees. In the right-angled triangle COB, the base OC, or the radius of the globe, being 3956, and the acute angle COB nine degrees, or the half of AOC, the hypotenuse OB is easily found, and hence about 49 miles is the excess BD of CB above CD, or the elevation of the boundary of the atmosphere, illumined by the latest beams of the sun. A correction might be required for the deviation of the rays from their rectilineal path, in consequence of the unequal refraction of the different strata of the atmosphere; but in a question of this kind, resting on rather loose or doubtful observations, it seems superfluous to affect any delicacy of calculation.

The inference drawn from the limits of twilight is not so conclusive as might at first appear. The feeble slanting rays, shed from the higher regions of the atmosphere, may not have been received directly from the sun, but derived, after repeated reflexions, from the more distant parts of the sky. The very ingenious Lambert attempted to distinguish, besides the primary twilight, a secondary, and even a ternary *crepusculum*. It is easy to see, that the solar ray SA, (Plate VI. fig. 3.), after grazing along the surface, will illuminate the upper atmosphere at B, from which some light will be darted in the direction of the tangent BCD, to tinge another elevated portion of the sky at D, which may cast a few expiring rays to the spectator at E, or shoot onwards to the opposite sky at F, and thence reach, in a state of almost extinction, the ground at G. Whilst the first *crepusculum*, therefore, sets in the west, the second will travel like a bow over the heavens, followed at a regular distance by the dusky veil of the third, or the vanishing train of the fourth. But it may be computed, from the vast length of the tracks which the light would have to traverse, that those shades would in succession be ten thousand times darker. The clearest sky, however, on the close of evening, never appears marked by such contrasted boundaries. The vault of heaven seems to darken by insensible transitions, from the western to the eastern horizon. It is most probable, therefore, that the diminution of light, after the close of evening, is owing to the prodigious rarefaction of much higher portions of the atmosphere, which still catch some faint rays of the sun, without being able, from excessive attenuation, to reflect them efficiently to the earth. But since, unless the sky be overcast, there is total darkness in no climate, even at midnight, we may therefore infer, that the body of air extends to such an altitude, as to receive the most dilute glimmer, after the sun has attained his utmost obliquity, and sunk ninety degrees below the horizon. It would thence follow, that the elevation of the atmosphere must be equal at least to 1638 miles, or the excess of the hypotenuse of an isosceles right-angled triangle, having 3956 miles, or the radius of the earth, for its base.

This very great extension of a rare expansive atmosphere appears conformable to the general phenomena. But the thin investure of our globe, at least near the equator, may stretch out much farther; and yet its elevation can never exceed a certain absolute limit. The highest portions of the atmosphere, which is carried round in 23 hours and 56 minutes, by the rotation of the earth about its axis, would be projected into space, if their centrifugal force at that distance were not less than their gravitation towards the centre. But the centrifugal force is directly as the distance, while the power of gravity is as its square. Consequently, when the centrifugal force at the distance of 6.6 radii of the earth is augmented as many times, the corresponding gravitation is diminished by its square or 43.7 times, their relative proportion being thus changed to 289. Now, the centrifugal force being only the 289th part of gravity at the surface of the equator, it will therefore just balance this power at the distance of 6.6 radii from the centre, or at the elevation of 22,200 miles. On this hypothesis, Plate VI. fig. 7, will represent our globe encompassed by its atmosphere, of which the equatorial diameter extends from A and B.

Such is the extreme boundary of atmospheric expansion. Though it surpasses all our ordinary conceptions of the space occupied by the dilateable fluid, it yet scarcely exceeds the twentieth part of the distance of the moon, which was held by the ancients to communicate with our atmosphere. If it really spreads out to the limit now assigned, it must, in its remote verge, attain a degree of tenuity which would utterly baffle imagination to conceive. Perhaps the fluid itself may change in those lofty regions, and pass into a sort of ethereal essence, more analogous to diffuse light than to a mass of air.

The constitution of the atmosphere forms the next object of inquiry. The analysis of that rare medium is one of the finest discoveries of modern chemistry. It appears to consist of two distinct expansible fluids combined in different proportions, a single portion of oxygen gas being united to three parts by weight, or four parts by bulk, of azote. There is also a very slight admixture of carbonic acid gas, amounting to perhaps the thousandth part of the whole. It may be doubted, however, whether this analysis be complete. The combination of those gases obtained artificially, generates a fluid in which we can hardly recognise the ordinary qualities of atmospheric air. Some fugacious elements must therefore escape during the process of decomposition. Indeed, the air may be considered as an universal solvent. It is the medium of all smells, and must therefore dissolve the different odorous effluvia, and transmit them to the olfactory nerves. The presence of moisture may perhaps assist the solution, but the mass of air is still the great receptacle of those diffusive emanations. We can readily distinguish several earths and stones by their difference of smell. Nay, the metals themselves, especially when rubbed, emit peculiar odours. What can be more variously contrasted, for instance, than the smells of iron, of tin, and of copper? The air must hence actually dissolve some traces of those metals, highly attenuated indeed, and almost evanescent. If a lump of *assafœtida* loses but a grain by exposure for several weeks, a bright surface of copper may, in similar circumstances, suffer the waste of only the thousandth part of a grain. The metal, if not encrusted by oxidation, would consequently experience a certain diminution, however small, in the course of ages.

The atmosphere is thus charged with emanations of all the various substances which it sweeps. To detect such dilute corpuscles, wholly transcends the powers of chemical analysis.

It seems probable, that the air holds some matters in more copious solution than others; and the phenomena of the aerolites lead us to suspect that it attracts iron and nickel with greater force than the rest of the metals. The quantity of those adventitious particles contained in a given mass of air may be exceedingly small, and yet the aggregate weight diffused through the whole atmosphere would form a considerable amount. It appears from numerous endiometrical observations, which agree tolerably well when performed in a similar way, that the constitution of our atmosphere is the same in all places on the surface of the earth, and at every elevation which has been yet explored. Such experiments have been made at very distant points, repeated on the summit of the loftiest mountains, and applied to portions of air brought down by balloons from the altitude of five miles. The result is what we should expect from the perpetual agitation and commixing of the lower strata of the atmosphere.

But a variety of circumstances render it extremely probable, that an expanse, far above the region of the clouds, is filled by some peculiar fluid, very different from the grosser element spread below. The shooting stars which are seen every clear night, the *bolades* or fire-balls, and which must traverse the sky at the height of several hundred miles, all seem to indicate the existence of a very ignitible medium. Nor is it difficult to conceive how such a collection of highly inflammable fluids could be formed. Not to mention the multiplied processes of art which emit those products, the great laboratory of nature is incessantly at work in generating and pouring forth hydrogen gas, and its various compounds. The volcanic mountains cover a considerable portion of the surface of the globe; and their innumerable spiracles, with scarcely any interruption, continue to discharge immense streams of inflammable aerial fluids, a great part of which escape conflagration. But, as hydrogen gas has little attraction to common air, it rises upwards by its buoyancy, without suffering much loss in the passage through that fluid. The largeness of their volume, and the celerity of their projection, conspire to favour the ascent of those inflammable gases to the loftiest regions of the atmosphere.

A comparison of the several quantities of astronomical refraction, at different altitudes, points to a similar conclusion. The refraction which the rays of light suffer in slanting across the higher regions of the air, is greater than that calculation assigns to the corresponding density of the medium. But the discrepancy would entirely disappear, if we suppose those strata to consist of hydrogen gas, which is known to possess, in a remarkable degree, the power of refracting.

It seems very probable, that the higher range of atmosphere has a sort of phosphorescent quality, or shines with a certain feeble light, for some time, after it has been heated or excited by the incident rays of the sun. Such may be the principal cause, and not any reflex illumination, of that lucid glow which, even at midnight, is diffused over the clear canopy of heaven.

The phenomenon of what is called the *zodiacal light* might, perhaps, be traced to the same source. That remarkable appearance, which is more conspicuous in the finer climates, and near the vernal equinox, on the approach of evening, has often been ascribed to the extension of a supposed luminous atmosphere about the sun. But Laplace has shown, that such an atmosphere, far from stretching to our earth, would not even reach the orbit of Mercury. The zodiacal light must therefore have only a terrestrial origin. Supposing the uppermost regions of our atmosphere to consist of diffuse inflammable matter, we might infer from analogy, that, like the calcined sulphate of barytes and other incinerated substances, they are endued with a phosphorescent quality, and capable of scattering a lambent gleam on being excited by the beams of the sun. But this luminary darting perpendicular rays, will evidently affect most powerfully the range of atmosphere which occurs in his diurnal track. The expansion of the gaseous investure of our globe above the equator will hence, from its vast elevation, be descried in places even beyond the tropics, glowing with a gentle flame. The luminous cone which converges from the sun to the distance of perhaps 30 or 40 degrees in the circle of the equator, gradually contracts and grows fainter in proportion as that body sinks farther below the horizon.

It thus appears, that the opinion entertained by the ancient philosophers concerning the existence of a vast shining æthereal expanse beyond our atmosphere, is, with some modifications, consonant to the principles of sound philosophy. This is not the first occasion in which we have to admire, through the vale of poetical imagery, the sagacity and penetration of those early sages. It would be weakness to expect nice conclusions in the infancy of science; but it is arrogant presumption to regard all the efforts of unaided genius with disdain. Seldom has a discovery been made without some distant ray of anticipation.

Having ventured to state that the highest region of the atmosphere is probably occupied by some very diffuse phosphorescent gas, we shall hazard a conjecture which will appear bolder, and even paradoxical,—that perhaps air, in its most concentrated state, occupies the bottom of the ocean, and forms a vast bed, over which the incumbent waters roll. Air has actually been condensed above a hundred times, and during this process it betrayed no deviation from the fundamental law, that its elasticity is directly proportional to its density. There seems no reason, therefore, to doubt, that if an adequate compressive force could be exerted, air might be reduced to the thousandth part of its ordinary volume. But this elastic fluid would then be denser than water, and, consequently, instead of rising, would fall through the liquid. Suppose, for instance, a bladder filled with air, and having a small bullet attached to it, were thrown into the sea; in continuing to sink, it would reach a depth where the enormous weight of the column of water would compress it to the same density with the surrounding mass; and if the bullet were now disengaged, the bladder would remain suspended in that stratum, or if carried a little lower, it would precipitate itself to the bottom.

To form some estimate of this singular event, a simple calculation will be required. Air of the ordinary temperature is 840 times lighter than distilled water, and is therefore 865 times lighter than sea water, assuming the density of this to be 1.03. But the mean pressure of the atmosphere being equal to that of a column of thirty-four feet of distilled water, is hence equal to the weight of a column of $32\frac{3}{4}$ feet of sea-water. Wherefore $32\frac{3}{4} \times 864$, or 28,296 feet, is the depth of the ocean where the necessary compression would obtain. But a small correction must be applied, on account of the augmented density of the sea water itself under the load of such a column. The logarithm of this density is found very nearly by multiplying the height of the column by six, and striking off seven decimal places; whence the *modulus* of the compression of water may be reckoned 723,824 feet. Of this large number, the former depth is about the twenty-fifth part; consequently, to procure an equilibrium between the condensed air and the corresponding stratum of sea water, it is requisite that the air should be contracted one twenty-fifth part more, or reduced to 901 times less than its ordinary volume. But $32\frac{3}{4} \times 900 = 29,475$ feet, from which deduct the fiftieth part for the mean condensation of the column, and there remains 28,885 feet, the correct depth.

This computation is to be considered as only a near approximation, yet sufficiently accurate for the object in view. Nor shall we fatigue our readers by the investigation of a strict formula, including exponentials. It is enough to mark the conclusion, that any portion of air carried five miles and a half below the surface of the sea will never ascend again. Now, this limit is only half the depth which the theory of tides assigns to the waters of the ocean. There is more difficulty in conceiving by what process air can be conveyed to its abyss. Increase of pressure, however, enables water to hold a larger share of air; and the effect is hence the same as an augmented attraction. The minute globules of air may therefore be gradually drawn downwards from stratum to stratum, till they are at last detached from the body of water by their own superior density. The precipitation and accumulation of concentrated air under the ocean would thus be the results of some unceasing operation. Such a process may perhaps constitute a part of the great economy of nature. It seems probable that the existence of a subaqueous bed of air is necessary to feed the numerous fires which continually rage in the bowels of the earth, and occasionally burst forth on the surface in volcanic spiracles.

The variable disposition of the atmosphere is the main cause of all the meteorological phenomena. It is of importance, therefore, to examine the theory and application of the different instruments which can be employed to explore the state of that medium. Some of these have been long in common use; but others, of a more delicate and refined construction, are only beginning to be known, and promise, when generally adopted, to expand our views, by opening a store of new and ulterior prospects. The ordinary observations are confined to the weight and temperature of the air. There are other data still wanted to determine at any time the actual condition of that medium. The dryness or humidity of the atmosphere, its brightness or degree of illumination, the different depth of the cerulean hue of the sky, and the variable disposition to chill the surface of the earth by impressions of cold transmitted from the higher regions; these objects of inquiry should be conjoined with others of a more practical tendency, depending immediately on the mutable state of the weather. Such are the attempts to measure the daily evaporation from the ground, to register the quantity of rain which falls, and to mark the direction and indicate the force or velocity of wind. A complete apparatus of meteorological instruments will, therefore, include primarily the *Barometer, Thermometer, Hygrometer, Photometer, Æthrioscope, Cyanometer;* and comprehend likewise the *Atmometer, Rain-Gage, Drosometer,* and *Anemometer*. We shall review this series in the order of enumeration.

I. BAROMETER.—The capital discovery of the weight and pressure of the atmosphere, achieved by Torricelli in 1643, was the first step in the progress of meteorology to the rank of a science. Prior to that memorable period, it rested on principles altogether loose and conjectural. But the construction of the barometer, as an accurate instrument of observation, soon disclosed what was passing in the more elevated and distant regions of the atmosphere. The column of mercury, it was perceived, seldom remained long stationary, but rose and fell through a sensible space, according to the different state of the heavens. In fine calm weather, it generally stood high, but commonly subsided a short while before rain or wind. The barometer came, therefore, to be regarded as a *weather-glass*, announcing the proximate changes of the sky, and owed its general reception to this belief. Its indications, however, were found by farther experience to be liable to much uncertainty. Though the extreme depression of the mercurial column appears invariably to foretell the ravage of a hurricane, yet its partial variations often pass away without being followed either by rain or wind. The barometer evidently can mark only the pressure of the atmosphere, and intimate, by consequence, the state of the weather, in as far as it depends on that cause.

Philosophers have eagerly sought to explain the fluctuations of the mercurial column. They have tried every principle that might appear to exert any influence in modifying the local weight of the atmosphere; but their very numerous attempts, it must be confessed, have hitherto proved singularly unsuccessful. It was requisite to show that such causes would not only give results of the kind expected, but were, besides, fully adequate to the production of the phenomena. In most instances, however, either none of those effects could have followed, or they would occur in a very inferior degree and disproportionate extent.

All the proposed explications of the changes of the barometer may, perhaps, be referred to three distinct sources:— 1. The action of heat on the atmosphere; 2. the influence of moisture on that fluid; and 3. the impression made by its rapid motion in wind. Now, the heating or cooling of the air above any given place, could in no way affect its pressure. The only change occasioned would be in the density of the fluid; and, by the influx of the portion below, or the efflux of a similar portion above, the pressure of the atmospheric column would soon become adjusted to an equilibrium with the surrounding mass. The diurnal variations of the barometer within the tropics, seldom exceeding the twentieth part of an inch, as they follow a different course from the progress of heat, are obviously not derived immediately from this cause. According to the accurate observations of Humboldt in South America, at all seasons the barometer falls from nine o'clock in the morning till four in the afternoon; it then rises till eleven at night, and from this time it again descends till half-past four in the morning, and next mounts till nine o'clock.

The transition of the air from a state of dryness to humidity, seemed to furnish the most plausible explication of the changes of the mercurial column. But the indication of the barometer was, in this case, distinctly at variance with the ordinary feelings of men. Those who suffer under a delicate or enfeebled constitution, are accustomed, in damp weather, to complain of the air as heavier and less elastic. This language is, no doubt, metaphorical only, and descriptive of the disordered state of the nervous system; but it shows the utter fallacy of trusting, in philosophical matters, to the loose results of vulgar observation and experience. Moisture, so far from loading the air by its weight, communicates, like heat, increased expansion and elasticity. But even supposing a column of air to become suddenly charged with humidity, before its subsequent dilatation had, by diffusing it, produced an equilibrium, still the additional pressure would have been extremely small, not exceeding, at a moderate computation, the fifteenth part of a mercurial inch. Any transition of that medium from dryness to humidity, would be quite inadequate, therefore, to the production of the effects actually observed.

Some philosophers imagine that moisture, in separating from the air, ceases to press that fluid by its gravitation, and would hence explain the fall of the barometer on the formation of clouds and the precipitation of rain. But when the aqueous particles are disunited in their solvent, whether dispersed in minute globules, or collected in large drops, they must evidently descend till they acquire the celerity sufficient to maintain a resistance in the medium equal to their weight. The pressure of the whole atmospheric mass upon the surface of the earth must therefore continue exactly the same under all the changes of the constitution of the medium.

The combined action of winds seems at first to promise the most satisfactory explication of the variations of the barometer. It is evident that opposite currents rushing to the same quarter will occasion an accumulation of the air; and, on the other hand, different streams flowing from any point must reduce the vertical column. But such conclusions are quite vague, without being subjected to the ordeal of calculation. Now, it is easy to compute that a concourse of winds, blowing at the rate of fifty miles an hour, and hence approaching to the violence of a hurricane, would be required to raise the barometer only the tenth part of an inch. The utmost power of a tempest could not, therefore, affect the mercurial column the twentieth part of what is frequently remarked in such circumstances. But trifling as this influence appears, it would be still at variance with actual observation. So far from rising in strong winds, the barometer almost invariably sinks; and instead of continuing depressed in the place where those currents originate, and where a calm must prevail, it generally stands high.

To explain the descent of the barometer during wind, a very ingenious idea had been proposed, which, being apparently confirmed by experiment, has obtained general reception. It is conceived, that a current of air, in sweeping over the surface of the earth, must cease to exert any vertical pressure. But this assumption can hardly be reconciled with any strict principle in science, for the particles of air will not for a moment cease to gravitate, nor will any horizontal motion of them produce the slightest derangement in a perpendicular direction. A remarkable experiment, however, was made by the ingenious Mr Hauksbee, about the beginning of the eighteenth century, "showing," as he says, "the cause of the descent of the mercury in the barometer in a storm." Having connected the square box cisterns of two barometers by a horizontal brass pipe of three feet in length, he inserted in the side of one of the boxes another pipe opening outwards, and connected the opposite side with a pipe attached to the neck of a large globular receiver, into which three or four charges of atmosphere had been compressed. On turning the stop-cock, the imprisoned air rushed with vehemence over the surface of the mercury in the cistern, and effected its escape, while both columns fell simultaneously about two inches, and gradually rose again as the force of the blast diminished. From this experiment, Mr Hauksbee formally derives four distinct corollaries:— 1. "That we have hence a clear and natural account of the descent and vibrations of the mercury in violent storms and hurricanes;" 2. "That not only the different forces, but also the different directions, of the winds, are capable of producing a difference of the subsidence of the mercury;" 3. "That strong winds may affect the animal economy upon this very account, of their altering the pressure of the atmosphere;" and 4. "That the weight of the atmosphere being diminished in one place, it is also as much diminished at the same time in another place which holds a communication with the former."

This experiment has a specious appearance, and might even seem to warrant the conclusion drawn from it. But a closer examination dispels the illusion. Since the air had been condensed four times, it must issue from the receiver with the immense velocity of 2700 feet in a second, or the double of that with which air of the ordinary density would rush into a vacuum. This is a rapidity, however, twenty times greater than the most tremendous hurricane. The very small change of the four hundredth part of an atmosphere would hence have been sufficient to produce the strongest wind ever observed, and therefore its influence in passing over the mercurial surface must have been quite insignificant.

But the experiment itself was absolutely fallacious. The peculiar result proceeded from a mere casual circumstance, the exit-pipe from the mercurial cistern being wider than the pipe which introduced the current of air. This incidental arrangement is not mentioned in the description of the apparatus, and we have, therefore, caused the original figure to be engraved on a reduced scale, not only to prove the fact which we have just stated, but also to warn experimenters of the necessity of noting scrupulously every accessary circumstance blended in their operations. (Plate IV. fig. 2.) It is easy to perceive that the tube G, which discharges the air from the box F, is much wider than the tube E, which conveyed it from the receiver A. This air, previously condensed, and still restrained in its passage through E, on entering the cavity of the box, immediately expands beyond the limit of equilibrium, and finding an easy escape through G, allows that state of dilatation over the mercury during the time of the horizontal flow. But the air contained in the other cistern K must, from its communication by the slender pipe I, suffer a like expansion, and, consequently, the columns L and H will, in the same time, subside equally.

Such is unquestionably the true explication of this curious fact. Were any confirmation needed, it could easily be derived from a very simple experiment. Let A (Plate IV. fig. 3.) be any cylinder of tin, suppose three inches long and two inches in diameter, having an open pipe inserted at B, a quarter of an inch wide, and perhaps two inches long, and another opposite pipe inserted at C, about three eighths or half an inch wide, and one inch long; at right angles to these, a recurved glass tube or syphon, G H E, of the tenth of an inch bore, is cemented below, descending twelve inches, and rising again six inches to the open swell at E, which contains coloured water terminating at F. Holding the cylinder in a horizontal position, and applying the mouth at B, let a sudden blast be injected into the cavity; the water will rise instantly to G, thus showing the diminished pressure, and, consequently, the rarefaction of the air above it. But if a cap D, with a narrow projection of perhaps only the eighth part of an inch, be adapted to the exit-pipe C, on repeating the experiment, an opposite effect will take place, and the column of water, so far from mounting to G, will now sink to H. It is evidently the difficulty of the escape through D that occasions the accumulation of the air within the cylinder, and the consequent depression of the water in the syphon. These different results are perfectly analogous to the local fall or rise of the surface of a river, occasioned by the widening or contracting of its channel.

After the complete failure of so many theories for explaining the variations of the barometer, which it would be tiresome to enumerate, we may be charged with presumption in attempting the solution of a problem that has racked and exhausted so much ingenuity. Yet in all these various efforts, a principle seems to have been overlooked, of extensive influence, accordant with the general phenomena, and sufficient, we think, on a close investigation, to produce the measure of effect which is required.

It is obvious that a horizontal current of air must, from the globular form of the earth, continually deflect from its rectilineal course. But such a deflection, being precisely of the same nature as a centrifugal force, must hence diminish the weight or pressure of the fluid. The only question is to determine the amount of that disturbing influence. Though it should appear quite inconsiderable in the interval of a short space, it may yet accumulate to a very notable quantity through the wide extent over which the same wind is known to travel. Suppose a current to begin to flow from

A (Plate VI. fig. 9.) in the direction of a tangent, it will successively bend from a rectilineal track at the points B, C, D, E, F, G, &c. on the surface of the earth. The particles of the fluid are, therefore, drawn incessantly from their course by the action of gravity. Their vertical pressure is consequently diminished by the force spent in producing this deflection. Wherefore, during the prevalence of the wind, the atmospheric column will press with inferior weight at B than at A, at C than at B, at D than at C; thus gradually decreasing through the whole chain. Suppose the intervals AB, BC, CD, DE, &c. to be each of them a mile, and that the current reaches the points B, C, D, E, &c. in successive minutes, a celerity which frequently happens; the deflection at B, owing to the curvature of the earth, would be eight inches, or two-thirds of a foot; but the space through which a body would descend in a minute, by the action of gravity, is $60 \times 60 \times 16 = 57{,}600$ feet, or 86,400 times greater than the deviation from the tangent. Wherefore the atmospheric pressure would, on that hypothesis, be diminished by the 86,400th part for each interval of a mile from A to G. In the space of 288 miles this diminution would consequently be the 300th part of the incumbent weight; and over an extent of 2880 miles it would amount to the 30th part. If we assume the very probable estimate, that storms involve the whole region of the clouds, or attain an elevation of near three miles, the diminution of pressure, occasioned by a long series of deflections in the stream, would affect one half of the atmosphere. Wherefore, a wind which has blown over a tract of 2880 miles, at the rate of 60 miles an hour, might cause the mercurial column to subside half an inch. If the velocity of the wind were doubled, which is probably the limit of the most tremendous hurricane, the fall of the barometer would be four times greater, and amount to two inches.

That the same powerful wind can sweep over an immense track of surface, is well ascertained. The effects of a hurricane originating in the West Indies, at the distance of 5000 miles, have often been felt on our shores. But a wind arising at A must evidently be followed, at succeeding intervals, by a current from K, L, M, &c. the range of influence being thus extended over a larger track. During the flow of the air, the depression of the barometer at G will be maintained, or rather augmented.

These conclusions are perfectly accordant with the facts observed. It appears, from comparing the most accurate registers, that the pressure of the atmosphere is subject to very nearly the same variations through a vast extent. A barometer may be considered as in unison with another placed at the distance of perhaps 200 or 300 miles, and the mercurial columns in both of them rise and fall by an almost simultaneous movement. But, in stormy weather, the separated barometers, however they may approximate in their indications, cannot absolutely correspond. The gradation of atmospheric pressure from G to F, from F to E, &c. may be very slow; and yet the minute differences may accumulate to a very sensible amount in the long range from G to A, or to M. Thus, though a barometer may stand only the twentieth part of an inch lower than another separated 200 miles, it would be half an inch below a barometer at the distance of 400 miles.

The theory now stated seems to furnish the most satisfactory explication of a great variety of phenomena. Thus the minute diurnal alternations to which the barometer is subject within the tropics, are caused by the influence of the land and sea breezes. During the heat of the day those gentle airs blow from the ocean, and, therefore, near the shore, and in the interior of the islands, the mercurial column subsides. But after the vigour of the sun's beams has declined, they flow in an opposite direction, and consequently the mercury rises again. The fluctuations are only partial, however, because such breezes hold a very short sway, and do not perhaps extend beyond 100 or 200 miles.

The common remark, that a north-east wind, so far from depressing the mercurial column, generally causes it to rise, might appear at variance with the principles which we have advanced. But that wind which really comes from the north has probably a short course, and may be dependent on a more extensive current which maintains a flow in the higher regions towards the poles. The existence of opposite streams, though incompatible with the supposition of a wide-spreading hurricane, may yet be admitted in local and partial derangements of the atmosphere.

Hitherto we have contemplated the winds as describing only arcs of great circles about the earth; but they may be constrained to bend in smaller circles, and perhaps trace the parallels of latitude. Their flexure from a rectilineal course being in this case augmented, they would in the same extent exert an influence proportionally greater on the barometer. Such, near the arctic regions, may be the effect of a westerly wind of no very distant origin.

The same principles will explain the phenomena of eddies, whirlwinds, and tornados. Suppose (Plate VI. fig. 4.) a horizontal stream of air, rushing from A to B, meets a contrary current blowing from C to D, and that some obstacle E occurs in their line of separation. The flow will evidently diverge on both sides of E (Plate VI. figs. 5. and 6.) till it swells by degrees into a vertiginous revolution. Such is the origin of the famous *malstroem*, or whirlpool, on the coast of Norway, occasioned by the meeting of opposite tides. The aerial vortices are evidently produced by a similar cause. If the opposite streams have equal force, the circulation will be maintained in the same spot; but if the one current flows with greater rapidity than the other, it will transport the vertiginous motion by the excess of its celerity. A motion of progression and revolution is thus upheld at the same time. Such appearances are frequently witnessed in summer, especially in the hotter climates. A whirlwind arises on a sudden, and runs over the surface of the ground, drawing into its vortex all the light substances which occur in the track. Immense havoc is often thus produced in the fields of rice and plantations of sugar-cane. It is easy to show, that if $r$ denote in feet the radius of the extreme circle described by the whirlwind, and $t$ the time of circumvolution in seconds, the elasticity, or pressure of the column at the verge, will suffer a diminution corresponding to the fraction $\frac{5r}{4t^2}$. The amount of this diminution over the whole base would be reduced to three-fourths; and consequently $h$, expressing the height of the revolving column of air, $\frac{15rh}{16t^2}$ would represent the mean effect of centrifugal action. Suppose the whirlwind to have an elevation of 200 feet and a radius of fifty, and to circulate in three seconds, the diminished pressure would be equal to the weight of a column of $\frac{15.50.200}{16.9}$, or 1040 feet. This example, assuming a celerity of sixty-five miles an hour, might be reckoned an extreme case, but it would occasion the mercury to sink in the barometer more than an inch, or 1.12.

The formula for the depression of the barometer caused by a rapid whirl or tornado in the lower atmosphere, may be changed into an expression that shall embrace the velocity instead of the time of revolution. This velocity in miles each hour being denoted by $v$, the diminished pressure of the vertiginous column will be $\frac{3v^2h}{50r}$. It thus appears that the effect is directly as the square of the celerity, and inversely as the radius of circumvolution. And since the rapidity of whirl can never exceed certain limits, the action of a tornado must diminish in proportion to the extent of surface which it occupies. Suppose the height of the cylinder of air to be 300 feet, the radius of the sweep 500 feet, and the celerity of its extreme circulation eighty miles an hour: Then $\frac{3.6400.300}{50.500} = 230$ feet, which corresponds to a descent of 1.247, or about a quarter of an inch in the mercurial column.

II. The THERMOMETER.—The invention of this instrument has not only dispelled the illusions which formerly prevailed on the subject of heat, but has mightily contributed to extend our acquaintance with the actual condition of the atmosphere. A lump of iron brought into the house, from its exposure during a frosty night, feels intensely cold, yet becomes gradually warmer; and if it be put into the fire, it will soon grow extremely hot, till it acquires the faculty of burning. On removing it from the fire, and laying it again out of doors, it will, through all the steps of a contrary progress, relapse imperceptibly into its former state. This obvious fact, and many others of a similar kind, might have been sufficient to show that *hot* and *cold* are nearly relative conditions, which every substance is capable of assuming. The schools upheld very different notions, which, after a long series of descents, have at last found shelter among the vulgar. It was maintained that, of the four elements, air and fire are hot, and water and earth essentially cold; and that the compound bodies, from their constitution, partake of those qualities in different degrees. While the senses were the sole arbiters of heat and cold, substances became classed according to the sensations which they excited. If a body, such as lead, rapidly abstracted heat from the touch, it was reckoned cold. But the same quality was likewise bestowed on other substances of a very different kind, for instance, on vinegar and hemlock, because they affect the stomach with a sort of chilling sensation. Soft wood and feathers might be deemed warm, since they draw off heat feebly from the touch; but pepper was also ranged in the same class, because it stimulates the organs of taste. Such a confusion of ideas disfigured physical science, and perverted the practice of medicine.

The construction of the air-thermometer, by the famous Sanctorio of Padua, in 1590, was the commencement of a salutary revolution. As it was at first intended for exploring disease, it has ultimately rendered signal service to the medical practitioners. About twenty years afterwards, those instruments were manufactured by Drebbel, who carried them from Holland into England. They were very rude, however, and adapted to no constant scale, but regarded merely as *weather-glasses*. The discovery of the barometer opened new views, and showed that the former instruments marked the conjoint effects of heat and pressure. To distinguish the separate influence of heat, it became hence requisite to employ some different fluid from air. Alcohol was preferred, as being very expansible; and, from the year 1655, thermometers consisting of rather wider tubes, terminating in balls or large bulbs filled with that liquid, were manufactured by Italian artists, who imitated an arbitrary standard adopted by the *Academia del Cimento.* Such thermometers were clumsy, and susceptible of only a low range. Römer, who made the fine discovery of the progressive motion of light, proposed mercury, as a fluid sufficiently expansible, and capable of bearing an intense degree of heat. The origin of the scale was then fixed at the melting point of ice or snow, and the scale itself divided into degrees corresponding to the ten-thousandth part of the capacity of the bulb. Yet this mode of construction seemed tedious, and liable to some inaccuracy. The capital improvement was made in 1724, by Fahrenheit, who took another standard point from the boiling of water under the mean pressure of the atmosphere. For many years that ingenious artist manufactured thermometers in Amsterdam on correct principles, very neat and small, adapted especially to medical purposes. The multitudes of young physicians who at that period studied in Holland, quickly dispersed them to every part of the globe. The observations thus obtained gave juster ideas of the comparative temperatures of different climates, and in many cases reduced the exaggerations of travellers to moderate bounds. It thence appeared that the heat of the torrid zone was not so excessive, nor the cold of the arctic regions so intense, as had been commonly represented. The tropical plants could, therefore, enjoy in our hot-houses all the warmth of their native climate. The thermometer was first applied to direct the operations of horticulture, and afterwards extended to regulate the process of brewing and other arts more immediately depending on practical chemistry.

Quicksilver was deemed preferable to alcohol in the filling of the thermometers, not only because of the wide range which it embraces, but on account of the remarkable property which it was afterwards found to possess, of expanding equably with equal accessions of heat. But it failed in the lower part of the scale. The reported sinking of the mercurial thermometer in Siberia appeared to indicate an intensity of cold beyond all conception. But the discovery made by Professor Braun, in 1759, of the actual congelation of mercury itself, reduced the extent of refrigeration to moderate limits. This fluid metal suffers a large contraction in passing into the state of solidity; and, therefore, though it freezes about thirty-nine degrees below the zero of Fahrenheit, it yet shrinks through a space of more than a hundred degrees before it becomes fixed. As alcohol has never been congealed, though brought, in some experiments we could mention, to 150° below zero, thermometers filled with it are now employed to explore, if not to measure, intense cold.

Metallic thermometers are likewise well adapted for examining the state of the atmosphere. They are commonly constructed on the principle of the compensation balance of a chronometer, a spiral or circular spring being composed of two soldered opposite plates of distinct metals, for instance of brass and steel, or of zinc and platinum, which expand very differently under the action of heat, and therefore continually change their incurvation. Some instruments of this kind, made by Breguet at Paris, are remarkably elegant; and those more lately manufactured by that ingenious artist, with a very slender form, surpass all other thermometers in the exquisite sensibility of their indications. Fig. 3, Plate V. represents the first form, reduced to about the fourth part of its natural size. The main piece of mechanism is a circular spring, fixed at one end, and composed of steel and brass soldered together. The other end carries a clip, that acts on a short train of wheels, which turn an index on the dial-plate, the extent of the scale, including the range of atmospheric temperature. The dial with its index are shown by A; and B exhibits generally the interior mechanism. Fig. 4, Plate IV. delineates the latest improvement. To prevent the cracking or dislocation which a large motion of the spring sometimes occasions at the junction of the two metals, a thin plate of a third metal is interposed. Three plates, consisting of platinum, gold, and silver, are united by the rolling press into a single riband, of the thickness of only the 1200th part of an inch. The instrument is formed with about twenty-seven spires or circumvolutions of this very slender spring, bearing an index which travels in its circle over fifty centesimal degrees, or ninety degrees by Fahrenheit's scale. The metallic film is, from its extreme tenuity, almost instantly penetrated by the impressions of heat or cold, and the sensibility of this thermometer accordingly surpasses all conception.

It is of much importance, in keeping meteorological journals, to have thermometers that shall indicate the extreme changes which occur during the absence of the observer, such as the greatest heat of the day, and the lowest cold of the night. For this object the metallic thermometers are easily adapted, since their index may push forward or draw back any moveable mark, and thus indicate the limits of its variation. Large mercurial thermometers, also, if mounted like the wheel barometers, to turn an index, will answer a similar purpose. But smaller instruments, though of a more complex construction, have been generally preferred. A sufficient account of the self-registering thermometer proposed by Lord Charles Cavendish in 1757, and of the more complete instrument described by Mr Six in 1782, may be found in any work which treats of the subject. The latter has now come into pretty general use, though we are sorry to remark that it seems to have fallen into the hands of very inferior artists, the scale consisting merely of boxwood, rudely and inaccurately subdivided.

Both these instruments have been employed to ascertain the coldness of the ocean at great depths. It becomes requisite, however, to make some allowance for the contraction which the glass bulbs must suffer under the enormous compression of the superincumbent columns of water. This can easily be computed, from the effect of perhaps ten atmospheres in a condensing engine; and such corrections were actually applied to the observations made in Captain Phipps' voyage in 1773 towards the north pole. It is a matter of equal surprise and regret, that all such niceties were overlooked in the late expeditions of Captain Ross and Captain Parry to the arctic regions.

The simplest, and by far the best, self-registering thermometers, are those invented and constructed by the late Dr John Rutherford, of Middle Balilish, and first described in the Transactions of the Royal Society of Edinburgh for the year 1794. The one, which marks the *minimum*, is filled with alcohol, and the other, which indicates the *maximum*, is filled with quicksilver; and they are both attached to the same frame, or, what is still better, affixed to separate frames, placed nearly horizontal, or rather elevated about five degrees, to prevent the separation of the thread of liquid. The tubes have bores from the twenty-fifth to the fifteenth part of an inch wide, and include a minute tapered or conical piece of ivory, or of white or blue enamel, about half an inch long. This mark having in either thermometer its base turned towards the bulb, is drawn to the lowest point by the alcohol, which again freely passes it; but it is always pushed forward to the highest limit by the mercury, which afterwards leaves it. These marks, however, are now made cylindrical, a little thickened at the ends, and about three eighths of an inch in length. (See Plate V. figs. 1. and 2.) Fig. 1. exhibits the alcohol or *minimum* thermometer; and fig. 2. the one which indicates the *maximum*, and is filled with quicksilver.

This instrument was first employed by Dr Rutherford, in some interesting experiments, to ascertain the temperature of germination, and determine the depth to which frost penetrates into the ground in different soils and situations, at the suggestion of his neighbour Dr Coventry, the very able and intelligent professor of agriculture in the University of Edinburgh, both of them proprietors in the small county of Kinross. The few instruments of this kind, for some time circulated, were made by the hands of the inventor; but artists have since learned to imitate and improve the original construction, and nothing is wanted now but to promote their diffusion. We regret that this very useful register-thermometer seems to be not yet so well known in London as it deserves.

The minutest changes which take place in the constitution of bodies are, for the most part, attended by corresponding alterations of temperature. To explore those abstruser operations of nature, which often betray the influence of more extensive principles, it was now requisite to improve the delicacy of the thermometer. To this object the writer of the present article had early turned his attention. At first he enlarged the capacity of the bulb, and thus procured degrees of such a size on the stem as to be capable of a very distinct subdivision. These instruments, however, received their impressions very slowly; and therefore tubes of extremely fine bores being selected, he had small bulbs blown, and filled with quicksilver, not in the ordinary way, but by the aid of a compressing force. With such exquisite thermometers, it was easy to procure much nicer observations, and to detect even the finer modifications of corpuscular action. But, to include the usual range of temperature, it was necessary to draw the stem to an immoderate length. The attempt to remedy this inconvenience led to the construction of the first kind of *differential thermometer*. The main object was evidently attained, if the mercury should always be made to start from some given point. The tube was therefore left open, and a cap adapted to the top, containing a surplus portion of the fluid. When the thermometer was kept inverted, this mercury closed round the orifice, and joined the thread in the stem, as in Plate V. fig. 11. But when the instrument was reversed, the excluded fluid instantly separated, and sunk into the cavity under the top of the tube, as in fig. 10. In this situation, the descent of the mercurial thread in the stem marked the depression of temperature.

A differential thermometer of this construction was used for three or four years, in a variety of experiments, without ever failing. But an open thermometer must be liable to some uncertainty, and unavoidably subject to a continual deterioration. The loose mercury will undergo a slow oxidation, while the bore of the tube is apt to be soiled and tarnished by the insensible introduction of moisture. Such an instrument, however, is well adapted for many researches.

In the beginning of the year 1795, another form of the differential thermometer was devised, which, from its simplicity, its delicacy, and extensive application, has contributed essentially to the progress of physical and chemical science. This instrument is now so generally known, that a few remarks on its general construction will be here judged sufficient. The differential thermometer is a modification of the air-thermometer, but susceptible of the impressions of heat only, and exempt altogether from the influence of the variations of atmospheric pressure. The tube to which the scale is applied has a bore of equal calibre, from the fiftieth to the fifteenth part of an inch wide, the other branch being commonly wider; the terminating balls are not less than four tenths of an inch in diameter, and seldom exceeding an inch and a half. A glass tube, terminated by a ball containing air, is joined hermetically, or by the flame of a lamp, to another longer tube, terminated by a similar ball containing air also, but including a small portion of some coloured liquid. The tubes are then bent, generally into a recurved or double stem, like the letter U, and the liquid is adjusted to the proper height by making bubbles of air pass from the one ball to the other, from the little enlargement of the bore left at the junction of the tubes. If both balls have the same temperature, the liquid must evidently remain stationary; but if the ball of the shorter tube be warmed, the air expanding and exerting more elasticity, will depress the liquid in the stem; or if this ball be cooled, the air, by its contraction, allows the liquid to ascend, from the superior elasticity of the air contained in the opposite ball. The fall or rise of the liquid will, therefore, mark the excess of heat or cold in the adjacent ball, and the space through which it moves will measure the precise difference of temperature.*

---

* It is not difficult to compute, in general, the size of the scale of the differential thermometer. Let the diameters of the two balls be expressed in inches by $a$ and $b$, the diameter of the bore of the tube being denoted by $d$, and the measure of a centesimal degree by $x$. The capacities of those balls, $\pi$ representing the ratio of the circumference to the diameter of a circle, w uld hence be $\frac{\pi a^3}{6}$ and $\frac{\pi b^3}{6}$, and the portion of liquid raised through the space $x$ would be $\frac{d^2 v}{4}$. Consequently the opposite elasticities exerted are $\frac{3d^2x}{2a^3}$ and $\frac{3d^2x}{2b^3}$, or, expressed by the altitudes of a column of sulphuric acid, of which 220 inches may be reckoned equivalent to the weight of the atmosphere, they become $\frac{660d^2x}{2a^3} + \frac{660d^2x}{2b^3}$, or $\frac{330d^2x}{a^3} + \frac{330d^2x}{b^3}$; wherefore, since air expands one 250th part for each centesimal degree, $\frac{330d^2a^3x + 330d^2b^3x}{a^3b^3} + x = \frac{220}{250}$, and, by reduction, $x = \frac{22a^3b^3}{8250\,(d^2a^3 + d^2b^3) + 25a^3b^3}$. This corresponds to ten of the millesimal degrees adopted for the differential thermometer, but the length of an hundred of those degrees, which may be preferred as a larger basis for the scale, will be $\frac{44a^3b^3}{1650\,(d^2a^3 + d^2b^3) + 5a^3b^3}$.

Alcohol and other volatile fluids were avoided in filling the tube, lest their vapour should affect unequally the elasticity of the air contained in the balls, and thus disturb the accuracy of the indication. Linseed oil tinged with alkanet root was first used; but it was found to be sluggish in its movements, leaving along the inside of the tube a sort of trail, which sometimes collected into globules. But the great objection to this and other fixed oils was, that they did not remain at the same point of the scale, but slowly shifted their place, owing, apparently, to a partial absorption of air in one of the balls, while their orange tint was found to fade away by exposure to the light. Deliquiate potash, coloured with archil, was next employed, and with tolerable success. By degrees, however, it deposited the colouring matter, and became almost limpid. Hydrogen gas, instead of common air, was then adopted for filling the balls. This mode of construction prevented the deposition of the colouring matter; but it was experienced to be troublesome, and attended with

---

Suppose $a = \frac{3}{4}$, $b = \frac{4}{3}$, and $d = \frac{1}{35}$, of an inch; then

$$\frac{44 \times \frac{27}{64} \times \frac{64}{27}}{1650\left(\frac{1}{1225}\right)\left(\frac{27}{64} \times \frac{64}{27}\right) + 5 \times \frac{27}{64} \times \frac{64}{27}} = \frac{44}{\frac{1650}{1225} \times \frac{4825}{1728} + 5} =$$

$\frac{44}{3.76+5} = 5$ inches. If the balls be of equal diameters, this expression for the length of an hundred degrees in inches will become simply $\frac{44a^3}{3300d^2 + 5a^3}$. Thus, suppose $a = \frac{4}{5}$ and $d = \frac{1}{30}$ of an inch; then $\frac{44 \times .512}{\frac{3300}{900} + 5 \times .512} = 3.62$ inches. If the density of the sulphuric acid were reduced from 1.85 to 1.62, the expression for the length of ten degrees would pass into this very simple form, $\frac{a^3}{756d^2 + a^3}$.

other impediments. After numerous discouraging trials, it was at last discovered that strong sulphuric acid tinged with carmine fulfils every condition, remaining permanently stationary in contact with confined air, and never losing in the slightest degree its colour from the action of the strongest light. Since the year 1801, this liquid has been constantly used in the construction of the differential thermometer. When the exhibition of striking effects rather than scrupulous accuracy is sought for, tinged alcohol, indeed, has in a very few cases been preferred, on account of the great facility and amplitude of its motions. But in every instance, a column of liquid, terminating in the cavity of one of the balls, or in a small cylindrical reservoir under it, was preferred; and not, as proposed by some experimenters, a single drop of the liquid, which forces its way through the bore by successive starts, and is therefore liable to much uncertainty and derangement.

The differential thermometer is capable of some diversity of form. It may consist either of a single branch, pendant or horizontal, or it may be bent into two perpendicular branches, whether contiguous or placed at a short distance. Since the motion of the column in the tube is occasioned by the difference of the elasticity of the air contained in the two balls, it is not essential that those balls should be of equal dimensions; for an equipoise must obtain whenever the augmented elasticity of the air of the hotter ball is balanced by its expansion on the one hand, and the corresponding contraction in the opposite ball joined to the pressure of the ascending column of the sulphuric acid. In general the balls are blown to a certain degree unequal, either to suit the particular instrument, or to please the eye; but in making observations, it must be kept invariably in the position for which it was designed, whether vertical or horizontal, since the pressure of the balancing column would be affected by an obliquity,

The differential thermometer, in its pendant form, and extended to a suitable length, from one foot, perhaps, to three or four feet, may be employed with great advantage by comparing the different temperatures of adjacent strata of air near the surface of the earth during the progress and decline of the day; and to detect the variations in a cloudy or a clear sky, and those occasioned by winds, as modified by the quality of the ground, whether naked or clothed with vegetation. But the main use of this instrument in meteorological researches arises from the various modifications of which it is susceptible. The minute changes of temperature which it marks, discover the existence and intensity of other disturbing causes. It is thus found that fresh ploughed ground is more affected by the solar rays than a green sward, on which also the breeze has little influence.

III. The HYGROMETER.*—The mutable condition of the atmosphere, as it inclines to dryness or humidity, is the main source of all the variety of meteorological phenomena. The changes which it undergoes with respect to moisture, have a marked influence on a very numerous class of substances, and even on the animal frame. But unfortunately those indications are always vague, and often fallacious. To ascertain the portion of humidity which a given quantity of air holds, or is capable of sustaining, is a problem of the first importance; but our advances to the constructing of an instrument fit to measure with accuracy that disposition have been extremely slow.

Most substances of a loose and spongy texture, or possessed of an absorbent quality, are affected, though in very dif-

* From ὑγρὸς, *wet* or *humid*, and μετρον, *a measure*. *Hygrometer* is distinguished from *hygroscope*; the former *measuring* the humidity of the air, the latter only vaguely marking its process.

ferent degrees, by the presence of humidity. Accordingly, the variations, both in weight and bulk, which absorbent bodies undergo, have been employed to indicate the disposition of the air with respect to moisture. For this reason such substances are likewise termed *hygroscopic*, since they are always affected by the state of the ambient medium, though they may not precisely measure its degrees of humidity or dryness. But neither heat nor moisture is passively diffused, or yet shared among different bodies in equal proportions. From some experiments made by Sir John Leslie, it appears that, under a like change of circumstances, a hundred grains of ivory will attract from the atmosphere seven grains of humidity; the same weight of boxwood, fourteen grains; of eider-down, sixteen; of wool, eighteen; and of beech, twenty-eight. Other substances, in their respective measures of absorption, exhibit still wider differences.

The dry or humid state of the air is therefore discovered from the variable weight of certain bodies exposed to its influence. Rock salt has been applied to that purpose; but potash, the muriate of lime, sulphuric acid, and most of the deliquescent substances, whether in a solid or liquid form, act the most powerfully. Other materials of a firm or adhesive consistence manifest the same properties, though in a lower degree. Plates of slate-clay or of unglazed earthenware, the shavings of box or horn, paper or parchment, wool or down, all act as hygroscopes. But these substances, especially the harder kinds of them, unless they be extremely thin, receive their impressions very slowly, and hence they cannot mark with any precision the fleeting and momentary state of the ambient medium. Nor is the weight which they gain by exposure proportioned to the real dampness of the air; for the measures of their successive absorption are found to increase in a most rapid progression as they approach to the point of absolute humidity.

But to weigh the substances with the accuracy befitting such experiments, is a very delicate and troublesome operation. Those thin bodies are liable, besides, to become in time covered with dust, which, while it must evidently augment their weight, cannot be detached from them without injuring their slender texture. The increase of bulk which they acquire from the portion of moisture attracted into their substance, furnishes therefore a more certain and convenient indication of the state of the atmosphere. The solid vegetable and animal fibres are connected by a fine soft netting, in which the power of absorption appears chiefly to reside. Hence the presence of moisture always enlarges the breadth of such substances, without affecting in any sensible degree their length. This effect is visible in the swelling of a door by external dampness, and in the shrinking of a pannel from the opposite cause. But the substances, such as paper or parchment, which have a diffused or interlaced texture, are extended by the absorption of humidity almost equally in every direction. On the contrary, twisted cord and gut, being swelled by moisture, suffer a corresponding longitudinal contraction, accompanied likewise, if not confined, by some uncoiling of their fibres.

All these properties have severally been employed in the construction of hygroscopes. The expansion of the thin cross sections of box or other hard wood, the elongation of the human hair or of a slice of whale-bone, and the untwisting of the wild oat, of cat-gut, of a cord or linen thread, and of a species of grass brought from India, have at different times been used with various success. But the instruments so formed are either extremely dull in their motions, or, if they acquire greater sensibility from the attenuation of their substance, they are likewise rendered the more subject to accidental injury and derangement; and all of them in time appear to lose insensibly their tone and proper action.

An attempt was made by Sir John Leslie to revive the method of measuring the expansion of absorbent cohesive substances by their enlargement of capacity when disposed into a thin shell; and, by successive steps, he carried the hygroscope thus formed to as high a state of improvement as perhaps such an imperfect instrument will admit. A piece of fine-grained ivory, about an inch and a quarter in length, was turned into an elongated spheroid, as thin as possible, weighing only eight or ten grains, but capable of containing, at its greatest expansion, about 300 grains of mercury; and the upper end, which was adapted to the body by means of a delicate screw, had a slender tube inserted, six or eight inches long, and with a bore of nearly the fifteenth part of an inch in diameter. (Plate VII. fig. 10.) The instrument being now fitted together, its elliptical shell was dipped into distilled water, or lapped round with a wet bit of linen, and, after a considerable interval of time, filled with mercury to some convenient point near the bottom of the tube, where is fixed the beginning of the scale. The divisions themselves were ascertained by distinguishing the tube into spaces which correspond each of them to the thousandth part of the entire cavity, and equal to the measure of about three tenths of a grain of mercury. The ordinary range of the scale included about seventy of those divisions. To the upper end of the tube was adapted a small ivory cap, which allowed the penetration of air, but prevented the escape of the mercury, and thereby rendered the instrument tolerably portable.

This hygroscope was largely, though rather slowly, affected by any change in the humidity of the ambient medium. As the air became drier, it attracted a portion of moisture from the shell or bulb of ivory, which, suffering in consequence a contraction, squeezed its contained mercury so much higher in the tube. But if, on the contrary, the air inclined more to dampness, the thin bulb imbibed moisture and swelled proportionally, allowing the quicksilver to subside towards its enlarged cavity. These variations, however, were very far from corresponding with the real measures of atmospheric dryness or humidity. Near the point of extreme dampness, the alterations of the hygroscope were much augmented; but they diminished rapidly as the mercury approached the upper part of the scale. The contraction of the ivory answering to an equal rise in the dryness of the air, was found to be six times greater at the beginning of the scale than at the seventieth hygroscopic division, and seemed in general to be inversely as the number of hygrometric degrees, reckoning from twenty below. Sir John Leslie therefore placed another scale along the opposite side of the tube, the space between the zero and the seventieth division of the hygroscope being distinguished into 100 degrees, and corresponding to the unequal portions from the number twenty to 120 on a logarithmic line. This very singular property will be more easily conceived from the inspection of the figure. The scale might probably be extended farther by continuing the logarithmic divisions. Thus, 320 degrees by the hygrometer would answer to 108 of the hygroscope, or to a contraction of 108 parts in a thousand in the capacity of the bulb. But at the dryness of 300, the contraction of the ivory seemed never to exceed 105.

Boxwood was likewise formed into a hygroscope of the same shape and dimensions; but this absorbent material swells twice as much with moisture as ivory does, and therefore requires its inserted tube to be proportionally longer or wider. Its contractions are still more unequal than those of ivory; for, near the point of extreme humidity, those alterations in the capacity of the bulb appeared to be more than twenty times greater than, under like changes in the condition of the atmosphere, take place towards the upper part of the scale. The space included between the commencement and the hundred and fortieth millesimal division of the scale, might hence be marked with a hundred hygrometric degrees, corresponding to the decreasing portions of a logarithmic line from five to a hundred and five.

In noticing the rapidly declining contractions which ivory and boxwood undergo, Sir John Leslie did not mean, however, to state the quantities with rigorous precision. Much time had been consumed in attempting to trace out the law of those gradations; and such experiments are rendered the more tedious, from the protracted action of the hygroscope, which often continues travelling slowly for the space of a quarter, or even half an hour. This tardiness is indeed the great defect of all instruments of that nature, and utterly disqualifies them for every sort of delicate observation.

The very large expansion which the hygroscope shows on its approach to extreme humidity, explains in a satisfactory manner the injury which furniture and pieces of cabinet work sustain from the prevalence of dampness. On the other hand, the slight alteration which the instrument undergoes in a medium of highly dry atmosphere, seems to have led most philosophers to believe that there is an absolute term of dryness, on the distance of which from the point of extreme moisture they have generally founded the graduations of the different hygroscopes proposed by them. This opinion, however, is far from being correct, and might give occasion to most erroneous conclusions. No bounds can be set to the actual dryness of the air, or the quantity of moisture which it is capable of holding, which, by the joint application of heat and rarefaction, may be pushed to an almost indefinite extent.

The ivory hygroscope, after being for several hours immersed in air remarkably dry, was apt of a sudden to split longitudinally. But if the bulb endured such a range of contraction, it appeared in some instances to take at least *another set*, or to accommodate its constitution, by imperceptible changes, to the state of the surrounding medium.

But though the bulbous hygroscope is, in extreme cases, liable to much uncertainty and some risk, it may yet be used with visible advantage in certain peculiar situations. The very sluggishness of the instrument, when the value of its divisions has been once ascertained, fits it so much the better for indicating the mean results. After being long exposed in situations hardly accessible, it may be conveniently transported for inspection, before it can suffer any sensible change. The hygroscope could be therefore employed with success to discover the degree of humidity which prevails at considerable elevations in the atmosphere. It might be likewise used for ascertaining readily the precise condition of various goods and commodities. Thus, if the bulb were introduced, for the space perhaps of half an hour, into a bag of wool, a sack of corn, or a bale of paper, it would, on being withdrawn from their contact, mark the dryness or humidity of those very absorbent substances.

Other hygroscopic substances have at different times been proposed, which, though possessed of greater sensibility, are yet liable to the same general objections. Thus, quills, reeds, gold-beaters' skin or pellicle, the skins of frogs, or the bladders of rats, were made to act like the bulbs of the thermometers, and to cause, by their contraction or dilatation, as they inclined to dryness or humidity, the included quicksilver to rise or fall in rather a wide tube. These instruments, however, being subject to injury or derangement from the smallest accident, can scarcely be applied to any practical use.

The hygroscopes which depend on the elongation of the fibres, are perhaps on the whole preferable. The slice of whalebone proposed by M. Deluc, and the human hair afterwards employed by M. de Saussure, are both of them sufficiently sensible to external impressions; but the difficulty is to determine the precise relation subsisting between those impressions and the state of the atmosphere. Humidity is not distributed in equal shares through the air, and among the several absorbent substances exposed to its penetration; nor are the degrees of expansion which it communicates, either uniform or proportional to its quantity. The graduation of such instruments, being thus in a great measure arbitrary, can furnish no correct data of the hygrometric state of the atmosphere. The assumption of two fixed extreme points, as the basis of the scale, is evidently erroneous. Air contained within a glass receiver may be rendered as damp as possible, by the copious aspersion of water on the sides; but it can never be absolutely deprived of its moisture, which adheres the more powerfully in proportion as it becomes diminished. Caustic alkalis, concentrated acids, and some of the deliquescent salts, aided by the action of heat, all render the air drier, but without being able to complete the desiccation. By the combined application of other agents, and even by mechanical pressure, the driest air can always be made to deposit some farther portion of moisture.

M. de Saussure directed all the resources of his ingenuity to correct the anomalies of the instrument which he proposed, and at last succeeded, by multiplied precautions, in rendering it as perfect, perhaps, as its nature and composition will admit. The hair-hygroscope (for it is not entitled to the name of hygrometer) certainly shows mobility; but the degrees which it marks can afford no steady or tolerably correct estimate of the dryness of the atmosphere.*

To arrive at an accurate measure of the dryness of the air, it is necessary to pursue a different route. Steam, in whatever way it be formed, whether by the application of heat or the diminution of atmospheric pressure, has nearly the double of the elasticity of common air, or it would, under the same compressing force, occupy about twice as much space. In uniting with that fluid, and forming vapour, it must hence communicate an expansion exactly proportional to the quantity dissolved, or to the share of moisture required for the complete saturation of the air. This principle suggested to Sir John Leslie the means of constructing an accurate hygrometer, to which his researches had been early directed. Inverting a barrel-tumbler, he ground the mouth perfectly flat, and having drilled a hole through the bottom, he cemented into it a syphon-gage, or slender recurved tube, with a narrow bore, and an intermediate swell or cavity, passing through a perforated cap of lead, and holding a portion of nut-oil, tinged with alkanet root. (Plate IV. fig. 6, where it is represented half the natural size.) To form the scale, he divided 4.2 inches into an hundred equal parts, so that each degree corresponded to the ten thousandth part of the air's elasticity, the whole atmospheric pressure being equal to that of a column of the oil 420 inches in height. Having now spread a few drops of water over a surface of plate glass, and slipped the tumbler upon it, the included

* From the few observations which we have made with an instrument of this kind brought from Geneva, we are sorry to say that it falls much below our expectations. Its motions are slow and irregular; and it seems to be little affected by very considerable alterations of the state of the encircling medium.

air quickly dissolved as much moisture as was sufficient for its saturation, and marked the expansion thence acquired, by forcing the oil to rise proportionally. The quantity of effect varied much, but was often very considerable, amounting, in fine weather, to 110 or 120 degrees. This little apparatus appeared to answer the purpose intended; but it was not portable, and it always required some address. It soon gave way, therefore, to other instruments, which promised to be more easily and readily managed.

As an hygrometer of this kind exhibits the actual expansion or increase of elasticity which the air acquires from complete humefaction, it seemed calculated for indicating the variable power of a drier or a moister atmosphere in refracting the rays of light. The barometer and thermometer had long been employed to correct the quantity of refraction; but the application of an accurate hygrometer seemed no less necessary for delicate observations. An instrument of the composition now described was accordingly intrusted, in the course of the year 1794, to the late Dr Maskelyne, and deposited in the Royal Observatory of Greenwich. Other objects, however, interfered, and an investigation, which promised important results, was never prosecuted on a regular and digested plan.

This instrument might be rendered still more accurate, by combining it with the principle of the differential thermometer. Let two similar tumblers, (Plate IV. fig. 5, where it is contracted to about the fourth part of its size,) A and B, inverted and loaded, have their mouths ground to fit a glass plate, the air contained in each of them acting by its elasticity on the column of a recurved tube C, which connects their cavities. Slide both of these tumblers in a dry state upon their bases, dip a hair-pencil in oil, and pass it round the outside of the mouth of A, to make it quite tight; then having removed B, and spread a few drops of water over its place, slip it on the plate again. The air included in B, now absorbing moisture, will continue to expand and to press upon the liquid column, till it has become absolutely saturated. Should any change take place in the temperature of the room during this process, it can have no effect in deranging the indication of the instrument, since it must influence precisely in the same degree the elasticity of the air contained in both balls, and thus produce an exact counterbalance.

This compound instrument is adapted to various delicate physical inquiries. In the union of different substances, a certain alteration of volume, however minute, almost invariably takes place. To ascertain such changes under various circumstances, greatly extends our views of the empire of chemical affinity. For instance, a small bit of fresh charcoal introduced under one of the tumblers will mark its absorption of the air, by the consequent rising of the coloured liquid. But the tumbler being wetted over the inside, and a portion of dry, though not recent, charcoal, with a pared surface, placed within it, after the air has ceased to dilate from humefaction, if the charcoal be brought in contact with water, the liquor will again rise, and continue for some time to indicate a renewed expansion. As the water is imbibed by the charcoal, minute globules of air will appear to escape. But if the water, by its insinuation, had merely dislodged the air from the pores, there would have been no change of volume, and consequently no alteration in the height of the liquid on the scale. Those aerial globules must have, therefore, expanded as they emerged, or they had existed in a state of condensation united to the charcoal.

On the other hand, if a roll of unsized paper, linen rags, slips of wood, or saw-dust, be successively introduced under a tumbler in which a little water is easily shed, the coloured liquor will sink rapidly in proportion as the moisture is absorbed. In this case, there is an obvious diminution of volume, and an union produced between a liquid and a solid, quite analogous to chemical solution; whereas capillary action, such as the ascent of water through sand, is attended by no change whatever of the space occupied by the compound.

Another application of this instrument, was to place under the tumbler separate capsules containing different substances, liquid or solid, which exert a mutual attraction. The included air, in this case, served as the vehicle of transfer, and a sort of distillation was supported, of a peculiar kind. The indications of this invisible process were variable, and often striking. But to enlarge on this subject would be foreign to our purpose. We have only taken the opportunity of noting a few of the results of an inquiry which was not pushed to any extent, but which deserves to be resumed, as likely to open new paths, and to unfold the more abstruse relations subsisting among different bodies.

Philosophers have long entertained very crude notions respecting the union of moisture with air, and the different circumstances which regulate or influence the process of evaporation. Dr Halley supposed that fire, uniting with the particles of water, communicates a vesicular constitution, which enables them to rise and spread through the atmosphere. A similar idea was entertained by Leibnitz, and was extended still farther by Muschenbroeck and Desaguliers. Kratzenstein adopted the vesicular system in 1743, and endeavoured to calculate the attenuation of the fluid which should produce the requisite buoyancy. Hamberger, in 1750, advanced, however, a capital step, by rejecting that hypothesis entirely, and attributing evaporation at once to a real solution of moisture in the air. This simple theory was in the following year explained at large by Le Roi of Montpellier, and fortified by some striking arguments, drawn chiefly from analogy. In pursuance of his theory, he proposed to pour ice-cold water into a deep glass, and to ascertain the dryness of the air, by noting the depressed temperature at which the dewing or deposition of vapour began to form on the outside of the glass, or was again about to disappear. Could this method have been easily and nicely reduced to practice, it might certainly have furnished an accurate estimate of the hygrometer and state of the atmosphere.

Wallerius, in his researches, was drawn aside from the right path, by a fact first noticed by Muschenbroeck, who asserted that evaporation is always more copious from a deep than from a shallow vessel; and this curious and apparently anomalous fact has been confirmed by other subsequent experiments. But he found, from a series of careful observations, that the quantity of water exhaled in the same situation is exactly proportional to the extent of surface exposed. He likewise made some experiments which seemed to furnish an explication of the peculiarity remarked by Muschenbroeck, though their force was not generally perceived at the time. But Richman rather darkened the subject by his strange conclusion, that the measure of evaporation depends on the difference merely between the temperature of the air and of the water, whether in excess or defect.

It was not difficult to perceive that evaporation is promoted by the application of heat, and the agitation of the aerial medium. No attempts, however, were made to determine the relation of that augmented effect to the actual velocity of the wind. But experiments on the influence which an increase of temperature exerts on the quantity of evaporation have been prosecuted with various success by Lambert, Saussure, and Kirwan. The results thus obtained unfortunately differ very widely; and though the researches of the celebrated naturalist of Geneva were those conducted with the most care and address, yet they seem, from the vagueness of their elements, not entitled to much confidence.

If the estimate of the causes which regulate the quantity of evaporation was unsatisfactory, still more perplexing appeared the ordinary account of the source of that depression of temperature which always accompanies the process. It was commonly referred to the operation of different concurring circumstances, among which the agitation of the air was conceived to perform the principal part. The dryness of that medium, on which we shall find the effect alone to depend, was in a great measure overlooked, or was confounded with other occasional agents. An evident confusion of ideas prevailed. The celerity of evaporation was mistaken for its intensity, and the coldness induced on the exhaling surface was viewed as the accumulated effect of a rapid dissipation of moisture. It was not perceived that in a free atmosphere vaporization proceeds with unabated energy, while the corresponding depression of temperature must advance by a relaxing progression, since otherwise the accession of an accelerated movement might push it to any extent. A little reflection, indeed, should have convinced philosophers, that the reduced temperature caused by vaporization must, in given circumstances, have a certain limit beyond which it cannot pass. But simple as this conclusion may now appear, it had escaped the most sagacious inquirers. Even Saussure, a patient and accurate experimenter, persuaded himself that, when water evaporates slowly, the cold produced is scarcely perceptible. To render this sensible, therefore, he thought it necessary to accelerate evaporation. Inserting the ball of a thermometer in a wet sponge, and attaching a cord to its stem, he whirled it briskly round his head, and thus produced a cold of eighteen degrees by Fahrenheit's scale, which he considered as much greater than could be obtained by other processes. To prosecute the inquiry, he had a sort of whirling table constructed, of about five feet diameter, by means of which a thermometer, with its bulb incased in wet sponge, could be made to revolve with the velocity of forty feet in a second. This machine he was at the trouble and expense of having carried up to the summit of the Col du Géant, where, during a residence of several days, he performed a series of interesting and valuable experiments. It did not occur to this philosopher, that by such a contrivance he was only creating to himself a vast deal of unnecessary fatigue, and that his wet thermometer, if left simply at rest for the space of two or three minutes, would have indicated exactly the same results. By all his exertions, he merely shortened the very moderate time required for attaining its extreme limit of depression.

Such were the imperfect notions which still prevailed on the subject of evaporation so late as the year 1796, the date of the publication of the last volume of the *Voyages dans les Alpes*. It is curious to remark, that Dr Black, in his Lectures, never mentions the dryness of the air as any way promoting evaporation, but ascribes the acceleration of the process entirely to the warmth and agitation of that medium. His friend Dr James Hutton, whose acuteness and penetration were conspicuous, had probably studied the phenomenon more closely. "I never had a hygrometer," he says, "but I used to amuse myself in walking in the fields, by observing the temperature of the air with the thermometer, and trying its dryness by the evaporation of water. The method I pursued was this: I had a thermometer included within a glass tube, hermetically sealed; this I held in a proper situation until it acquired the temperature of the atmosphere, and then I dipped it into a little water also cooled to the same temperature. I then exposed my thermometer, with its glass case wetted, to a current of air; and I examined how much the evaporation of the atmosphere, by holding the ball of the thermometer, or the end of the tube in which the ball was enclosed, towards the current of air; and I examined how much the evaporation from that glass tube cooled the ball of the thermometer which was included." He then proceeds to relate some hasty and very inaccurate observations made in this way. The passage now quoted occurs in a quarto volume published in 1792, buried in a repulsive mass of diffuse reasoning and paradoxical speculation, unsupported by any definite experiment. But Dr Hutton had evidently considered it as a conjectural hint, on which he laid little stress; for he speaks immediately afterwards of our possessing accurate hygrometers, which could not be admitted if the assumption he seemed to make had been strictly true.

Whilst such loose and imperfect notions prevailed respecting evaporation, it was expedient to review the process with attention, and analyse the several changes which accompany it. The depression of temperature which it always occasions, had been hastily supposed to be proportional to the rate with which the moisture is dissipated, and to be therefore augmented by every circumstance that can accelerate this effect. But if water contained in a porous vessel expose on all sides its surface to a current of air, it will cool down to a certain point, and there its temperature will remain stationary. The rapidity of the current must no doubt hasten the term of equilibrium, but the degree of cold thus induced will be found still the same. A little reflection may discover how this takes place. It is well known that the conversion of water into steam is, in every case, attended by the absorption of the heat requisite to support a gaseous constitution. Though the humid surface has now ceased to grow colder, the dispersion of invisible vapour, and the corresponding abstraction of heat, still continue without intermission. The same medium, therefore, which transports the vapour, must also furnish the portion of heat required for its incessant formation. In fact, after the water has been once cooled down, each portion of the ambient air which comes to touch the evaporating surface must, from its contact with a substance so greatly denser than itself, be likewise cooled down to the same standard, and must hence communicate to the liquid its surplus heat, or the difference between the prior and the subsequent state of the solvent, which is proportioned to the diminution of temperature it has suffered. Every shell of air that in succession encircles the humid mass, while it absorbs, along with the moisture which it dissolves, the measure of heat to convert this into steam, does at the same instant thus deposite an equal measure of its own heat on the chill exhaling surface. The abstraction of heat by vaporization on the one hand, and its deposition on the other, at the surface of contact, are therefore opposite contemporaneous acts, which soon produce a mutual balance, and thereafter the resulting temperature continues without the smallest alteration. A rapid circulation of the evaporating medium may quicken the operation of those causes; but so long as it possesses the same drying quality, it cannot in any degree derange the resulting temperature. The heat deposited by the air on the humid surface, becomes thus an accurate measure of the heat spent in vaporizing the portion of moisture required for the saturation of that solvent at its lowered temperature. The dryness of the air is, therefore, under all circumstances, precisely indicated by the depression of temperature produced on a humid surface which has been exposed freely to its action.

It may insure perspicuity, however, to recapitulate the great principle on which the formation of the hygrometer depends. When water passes into steam or vapour, it enlarges its capacity, and absorbs a very large share of heat. Any body, therefore, having a wet surface, becomes generally colder, if exposed to the access of air. But this decrease of temperature soon attains a certain limit, where it continues stationary, though the dissipation of moisture still proceeds with undiminished activity. The same medium which transports the vapour, must hence furnish also the portion of heat required for its incessant formation. In fact, after the humid surface has been cooled, each portion of the ambient air which comes to touch it must likewise be cooled down to the level of the dense substratum. The addition of heat at the surface of contact is thus a contemporaneous act with its subtraction by the process of vaporization; and it quickly advances to the same degree of intensity, after which a mutual balance of opposite effects is maintained, and the coolness hence induced continues unaltered. An augmented circulation of the evaporating medium may hasten the process; but while it has the same drying quality, it cannot in the least affect the depression of temperature. As soon as such an equilibrium is attained, the deposition of heat on the humid surface must become just equal to its abstraction. But this deposition is evidently proportional to the diminution of temperature, which is hence a measure of the share of heat abstracted, and therefore of the dryness of the air, or its distance from saturation.

This analysis of the process of evaporation appears so conclusive as to banish all doubt and objection. But it was desirable to confirm the deductions of theory by an appeal to direct observation. Accordingly, on setting the hygrometer upon a table in the middle of a room, and blowing from some distance against the wet ball with a pair of bellows which had acquired the temperature of the apartment, the instrument still indicated the same measure of dryness. The experiment was repeated more accurately on a larger scale, by exposing the hygrometer out of doors to the action of a strong and arid wind, a small screen being interposed and again removed, at short intervals of time, during which alternations no change whatever could be perceived in the quantity of the depression of temperature.

Having, therefore, ascertained the great law of evaporation, and proved that the coldness occasioned by it is not in any degree affected by agitation or other extraneous influence, nothing seemed wanting to construct an *hygrometer* on just principles, but to contrive a thermometer that should mark the smallest alterations of temperature. At first Mr Leslie employed a very delicate thermometer with a short range, open at the top, where a small cap of glass or ivory was fixed, containing a small portion of surplus quicksilver. (Plate V. figs. 16. and 17.) When this thermometer was heated by the hand till the thread of quicksilver filled the whole of the stem, and formed a little globule at the top, it was inverted, as in fig. 17, and all the quicksilver united into one mass; but when it was restored, as in fig. 18, to its first position, the quicksilver fell back from the cap, and lodged about the end of the stem, leaving the bore completely filled. A cup (Plate IV. fig. 1.) made of thin porous earthenware, nearly of the shape of a lady's thimble, but somewhat larger, and filled with water, was exposed to the air, while the thermometer lay beside it in a horizontal position. After a few minutes the thermometer was lifted up and plunged vertically into the cup; and the thread of quicksilver, which had extended through the whole length of the bore, being by this change of position cut off at the top of the tube, immediately contracted, and marked, by the space of its descent, the diminution of temperature in the liquid.

The very severe winter of 1794-5 afforded Sir John Leslie an opportunity of making experiments on the evaporation of ice; in the course of which he was led to the construction of the differential thermometer, now so generally known.

At first he employed it merely as an hygrometer, the one ball being always naked, and the other covered with cambric, wetted as often as occasion required. These balls were about an inch and a half in diameter, and blown to the ends of the same tube, one of them having a projecting point or aperture, which was sealed, after the branches had been bent, and a portion of coloured oil introduced. The graduation of the scale was determined by placing the instrument erect between two boxes, each containing a body of water, which encompassed one of the balls, and by noticing the rise of the oil on changing the relative temperature of the bath. An hygrometer, thus formed, was sent to the late Sir Joseph Banks, in the summer of 1795; and Mr Gilpin, clerk to the Royal Society, deposited a copy of it in their cabinet. But this model is so rude and clumsy, as hardly to recall the original. In the hands of the inventor, however, the instrument was soon improved, and reduced to a convenient and portable size. He had the satisfaction of showing an hygrometer of this construction, in January 1796, to Dr James Hutton, only a few months before the death of that very ingenious philosopher, who was delighted with seeing the application of a principle which he readily comprehended, and which his sagacity had obscurely anticipated.

But the sluggishness of the oil, and its tendency to deposite its colouring matter on exposure to the influence of light, still opposed obstacles to the perfection of this hygrometer. These were entirely removed, however, about the year 1801, by the substitution of concentrated sulphuric acid tinged by carmine. The most powerful and continued action of the solar beams was found to produce no change whatever on that purpurine liquid, when precluded from the access of the external air. Any improvements which have since been effected on this instrument consist chiefly in its mechanical arrangement, in the selection of the tubes, the better proportion of the balls, and the elegance and conciseness of the general shape.

The hygrometer has two distinct forms; the one *portable* and the other *stationary*. The former, (Plate V. fig. 7.) having its balls in the same perpendicular line, is protected by a wooden or metallic case, and fitted for carrying in the pocket; two or three drops of pure water, from the tip of a quill or a hair pencil, being applied to the surface of the covered ball, and the instrument held exactly in a vertical position whenever it is used. The latter form (Plate V. fig. 6.) is susceptible of rather greater accuracy, having its balls bent opposite ways at the same height. In some instances, it is preferable to retain merely the simplest form of the differential thermometer, the vertical stems being more distant, and the balls not reflected. In both these constructions, the two balls, since they occupy the same level, cannot be affected in the smallest degree by the unequal temperatures of the different strata of air in a close heated room. After the covered ball has been wetted, the instrument will continue to perform unimpaired, for the space, perhaps, of two or three hours. The addition of a few drops of water will then restore its action. But the hygrometer may be made to supply itself with moisture. It is only wanted to pass some fibres of floss-silk close over the humid ball, and immerse them at the distance of a few inches in a tall glass decanter, full of water, with a stopper which leaves open a small projecting lip.

The hygrometer has its opposite balls made to exhibit nearly the same colour and opacity, in order to exclude the admixture of photometrical influence, or prevent any derangement which the unequal action of light might otherwise occasion. The naked ball is blown of red, green, or blue glass, and the papered one is externally covered with thin silk, of rather a fainter shade, as it takes a deeper tint when moistened.

The theory which we have given of the hygrometer, corroborated by its accurate performance, might appear complete in every part. But the progress of science commonly detects the existence of some collateral causes which come to mingle their influence with the action of the great dominating principles. In our explication of the hygrometer, we stated that the same air which *abstracts* moisture, and consequently the portion of heat necessary to convert it into the gaseous form, must likewise *communicate* to the wet ball another portion of heat equal to its depression of temperature, which is hence maintained at a constant point. This analysis, however, involves the supposition, that the air conveys heat from bodies merely by its actual transfer. But having discovered that air transmits a certain share, at least, of the heat by a sort of pulsation, or internal tremor, depending on the quality of the surface from which the impression originates, it was requisite to examine anew the process of evaporation. The effect cannot be produced solely by the quickened recession of the contiguous portions of the ambient medium. The conterminous air must communicate heat to the humid surface also by pulsation; and hence the balance of temperature would be liable to incidental variations, if moisture, with its embodied heat, were not likewise abstracted by some corresponding process. And such is the harmonious adaptation of these elements: the discharge of vapour appears to be subject precisely to the same conditions as the emission of heat, and in both cases the proximity of a vitreous or a metallic surface produces effects which are entirely similar. Let two pieces of thin mirror-glass, or what is called Dutch plate, be selected, about four inches and a half square; and having applied a smooth coat of tinfoil, four inches square, to one of these, cover them both with a layer of the thinnest goldbeater's skin, which will adhere closely on being wetted; and after it has again become dry, cut it on each into an exact square of four inches and a quarter: now place the two glass plates horizontally in the opposite scales of a fine balance, and adjust them to an exact counterpoise: then, with a hair pencil, spread two grains of water over the surface of each pellicle: in a few seconds, the plate which is coated with tinfoil will preponderate, and after the former has lost all its moisture, this will be found to retain still three tenths of a grain. The proximity of the subjacent metal to the humid surface, therefore, impedes the process of evaporation, in the ratio of seventeen to twenty; the very same as, in like circumstances, had been ascertained to be the retardation of the efflux of heat. From this and other experiments, we learn, that some constant portion from a humid surface is always abstracted by the pulsation of the aerial medium. The steam exhaled, in uniting with the air, communicates to this elastic fluid a sudden dilatation, which will continue to propagate itself in successive waves.

In further illustration of this matter, cover with a thin pellicle of goldbeater's skin both the balls of the *pyroscope*, or that form of the differential thermometer which has one ball naked and the other enamelled with gold or silver, and wet them equally. The coloured liquid will remain for several minutes stationary at the beginning of the scale, and will then mount slowly, perhaps ten or fifteen degrees. Evaporation had, therefore, produced the same cold or depression of temperature upon the surface of the metal as upon that of the glass; from the glass, however, it was more copious than from the metal, having left the former dry, while the latter still exhaled some portion of moisture. But this action soon ceased, and the liquid fell back to its former level. On applying another pellicle, the liquid continued longer stationary, and rose only about five degrees. With repeated pellicles a difference was perceptible in the time of drying the two balls, till the thickness amounted to the 600th part of an inch.

The method employed for the graduation of the hygrometer is not only very convenient, but susceptible of great accuracy. The instrument, with a temporary scale affixed to it, is introduced into a magazine of dry air, and compared with a standard then put in action. To procure the dryness of the included medium, a flat saucer of thirteen inches in diameter, and holding a body, about half an inch deep, of concentrated sulphuric acid, and set on a ground plate of glass or metal, is covered by a very large inverted receiver, containing more than 1500 cubic inches of air, and having at the top an opening of three inches wide, on which rests a smaller plate, with two or three hooks projecting down from it. The scale of the standard instrument was determined by suspending beside it, under the receiver, two delicate thermometers, one of which had its bulb coated with several folds of wet tissue paper. The descent of the coloured liquid of the hygrometer, corresponding to the difference of ten centesimal degrees of the parallel thermometers, was hence computed, and this length afterwards divided into an hundred equal parts, to form the standard degrees. To graduate any other instrument, it was only requisite to attach a scale of inches, and mark the simultaneous measures when a steady equilibrium had at last obtained. The space of half an hour is generally sufficient to bring this about. A simple proportion, therefore, discovers the length answering to an hundred millesimal degrees, from which the subdivisions of each particular scale are derived.

The condition of the atmosphere with respect to dryness is extremely variable. In our climate, the hygrometer will, during winter, mark from five to twenty-five degrees; but, in the summer months, it will generally range between fifteen and fifty-five degrees, and may even rise, on some particular days, as high as eighty or ninety degrees. On the continent of Europe, it maintains a much greater elevation; and in Upper India it has frequently stood at 160 degrees.

When the indication of the hygrometer does not exceed fifteen degrees, we are directed by our feelings to call the air damp; from thirty to forty degrees, we begin to reckon it dry; and from fifty to sixty degrees, we should account it very dry; and from seventy degrees upwards, we might consider it as intensely dry. A room is not comfortable, or perhaps wholesome, if it has less than thirty degrees of dryness; but the atmosphere of a warm occupied apartment will commonly produce an effect of upwards of fifty degrees.

But this hygrometer will perform its office even if it should be exposed to frost. The moisture spreads over the surface, and, imbibed into the coat of the papered ball, will first cool a few degrees below the freezing point, and then congeal quickly into a solid compound mass. The moment in which congelation begins, a portion of heat liberated in that act brings the ball back to the temperature of freezing; and the coloured liquor, in proportion to the coldness of the external air, starts up in the opposite stem, where it remains at the same height till the process of consolidation is completed. After the icy crust has been formed, evaporation again goes regularly forward; and if new portions of water be applied, the ice will, from the union of those repeated films, acquire a thickness sufficient to last for several days. The temperature of the frozen coat becomes lowered in proportion to the dryness of the atmosphere. The measure of heat deposited on the chill surface by the contact of the ambient air, is then counterbalanced by the two distinct though conjoined measures of heat, abstracted in the successive acts of converting the exterior film of ice into water, and this water into steam; which transformations that minute portion must undergo before it can unite with its gaseous solvent. But the heat required for the melting of ice being about the seventh part of what is consumed in the evaporization of water, it follows that the hygrometer, when the surface of its sentient ball has become frozen, will, in like circumstances, sink more than before, by one degree in seven. This inference is entirely confirmed by observation. Suppose, in frosty weather, the hygrometer, placed on the outside of the window, to stand at twenty-eight degrees, it may continue for some considerable time at that point, until the congelation of its humidity commences; but after this change has been effected, and the equilibrium again restored, the instrument will now mark thirty-two degrees.

The theory of this hygrometer will enable us to determine, not only the relative, but even the absolute dryness of the air, or the quantity of moisture which it can absorb, by comparing the capacity of that solvent with the measure of heat required to convert a given portion of water into steam. To discover the capacity of air is, however, a problem of great difficulty, and it has not perhaps even yet been ascertained with much precision. It was formerly estimated, we are convinced, by far too high. Thus, Dr Crawford made it to be 1.85 times, or nearly double that of water. But, from several concurring observations, we should reckon the capacity of air to be only three tenth parts of that of water. But 600 centesimal degrees, or 6000 on the millesimal scale, being consumed in the vaporization of heat, this measure of heat would prove sufficient to raise an equal mass of air 20,000 millesimal degrees, or those 6000 degrees augmented in the ratio of ten to three. Now, at the state of equilibrium, the quantity of heat that each portion of the aerial medium deposites in touching the chill exhaling surface, or what answers to the depression of temperature which it suffers from this contact, must, as we have seen, be exactly equal to the opposite measure of heat abstracted by it in dissolving its corresponding share of moisture. Wherefore, at the temperature of the wet ball, atmospheric air would take up moisture amounting to the 20,000th part of its weight for each degree marked by the hygrometer. Thus, supposing the hygrometer to mark fifty degrees, the air would then require humidity equal to the 320th part of its weight for saturation at its reduced temperature. When the papered ball of the hygrometer is frozen, the degrees of this instrument must have their value increased by one seventh, so that each of them will now correspond to an absorption of moisture equal to the 17,000th part of the weight of the air.

But the value of those degrees becomes augmented in a much higher proportion if the hygrometer be immersed in hydrogen gas. This very dilute medium appears to have about eight times the capacity of common air, and the quantity of heat which under similar circumstances it will deposite on the evaporating surface must likewise, from the same principle of mutual balance, be eight times greater, and consequently, each hygrometric degree will indicate an absorption of moisture equal in weight to the 750th part of the solvent. The energy of hydrogen gas is therefore scarcely less remarkable in dissolving moisture than in containing heat. Confined with a powerful absorbent substance, whilst common air marks eighty degrees of dryness, hydrogen gas will indicate seventy. This gas must, in similar circumstances, therefore, hold in solution seven times as much moisture as the atmospheric medium.

To discover the precise law by which equal additions of heat augment the dryness of air, or its power to retain moisture, is a problem of great delicacy and importance. Two different modes were employed in that investigation, but which led to the same results. The one was, in a large close room, to bring an hygrometer, conjoined with a thermometer, successively near to a stove intensely heated, and to note the simultaneous indications of both instruments; or to employ two nice thermometers, placed beside each other, and having their bulbs covered respectively with dry and with wet cambric. By taking the mean of numerous observations, and interpolating the intermediate quantities, the law of aqueous solution in air was laboriously traced. But the other method of investigation appeared better adapted for the higher temperatures. A thin hollow ball of tin four inches in diameter, and having a very small neck, was neatly covered with linen, and being filled with water nearly boiling, and a thermometer inserted, it was hung likewise in a spacious close room, and the rate of its cooling carefully marked. The experiment was next repeated by suspending it to the end of a fine beam, and wetting with a hair pencil the surface of the linen, till brought in exact equipoise to some given weight in the opposite scale; ten grains being now taken out, the humid ball was allowed to rest against the point of a tapered glass tube, and the interval of time, with the corresponding diminution of temperature, observed, when it rose again to the position of equilibrium. The same operation was successively renewed; but, as the rapidity of the evaporation declined, five, and afterwards two, grains only were, at each trial, withdrawn from the scale. From such a series of facts, it was easy to estimate the quantities of moisture which the same air will dissolve at different temperatures, and also the corresponding measures of heat expended in the process of solution.

By connecting the range of observations, it would appear that air has its dryness doubled at each rise of temperature, answering to 15 centesimal degrees. Thus, at the freezing point, air is capable of holding a portion of moisture represented by 100 degrees of the hygrometer; at the temperature of 15 centigrade, it could contain 200 such parts; at that of 30, it might dissolve 400; and, at 45 on the same scale, 800. Or, if we reckon by Fahrenheit's divisions, air absolutely humid holds, at the limit of congelation, the hundred and sixtieth part of its weight of moisture; at the temperature of 59 degrees, the eightieth part; at that of 86 degrees, the fortieth part; at that of 113 degrees, the twentieth part; and at that of 140 degrees, the tenth part. While the temperature, therefore, advances uniformly in arithmetical progression, the dissolving power which this communicates to the air mounts with the accelerating rapidity of a geometrical series.

It hence follows, that, whatever be the actual condition of a mass of air, there must always exist some temperature at which it would become perfectly damp, as M. Leroi had first advanced. Nor is it difficult, from what has been already stated, to determine this *dewing point* in any given case. Thus, suppose the hygrometer to mark 52, while its wet ball has a temperature of 20 centesimal degrees, or 68 by Fahrenheit; the dissolving power of air at this temperature being 252, its distance from absolute humidity will therefore be 200, which is the measure of solution answering to 15 centesimal degrees, or 59 by Fahrenheit. The same air would consequently, at the depressed temperature of 59 degrees, shrink into a state of absolute saturation; and if cooled lower, it would even deposite a portion of its combined moisture, losing the eightieth part of its weight at the verge of freezing.

Annexed is a small table of the solvent power of air, from the temperature of 15 centesimal degrees below zero to 44 above it, or from — 5° of Fahrenheit's scale to 111°.2.

*Quantities of Moisture dissolved in Atmospheric Air at different Temperatures by the Centesimal Scale.*

| Temp. | Moist. | Temp. | Moist. | Temp. | Moist. | Temp. | Moist. |
|---|---|---|---|---|---|---|---|
| ° | | ° | | ° | | ° | |
| — 15 | 50.0 | 0 | 100.0 | 15 | 200.0 | 30 | 400.0 |
| — 14 | 52.4 | 1 | 104.7 | 16 | 209.5 | 31 | 418.9 |
| — 13 | 54.9 | 2 | 109.7 | 17 | 219.4 | 32 | 438.7 |
| — 12 | 57.4 | 3 | 114.9 | 18 | 229.7 | 33 | 459.5 |
| — 11 | 60.1 | 4 | 120.3 | 19 | 240.6 | 34 | 481.2 |
| — 10 | 63.0 | 5 | 126.0 | 20 | 252.0 | 35 | 504.0 |
| — 9 | 66.0 | 6 | 132.0 | 21 | 263.9 | 36 | 527.8 |
| — 8 | 69.1 | 7 | 138.2 | 22 | 276.4 | 37 | 552.8 |
| — 7 | 72.4 | 8 | 144.7 | 23 | 289.5 | 38 | 578.9 |
| — 6 | 75.8 | 9 | 151.6 | 24 | 303.1 | 39 | 606.3 |
| — 5 | 79.4 | 10 | 158.7 | 25 | 317.5 | 40 | 635.0 |
| — 4 | 83.1 | 11 | 166.2 | 26 | 332.5 | 41 | 665.0 |
| — 3 | 87.1 | 12 | 174.1 | 27 | 348.2 | 42 | 696.4 |
| — 2 | 91.2 | 13 | 182.3 | 28 | 364.7 | 43 | 729.4 |
| — 1 | 95.5 | 14 | 191.0 | 29 | 381.9 | 44 | 763.9 |

These temperatures, and the corresponding quantities of moisture dissolved, may be represented by the abscissæ and ordinates of the logarithmic curve, as in Plate VI. fig. 8. where some of the principal terms are marked.

The influence of warmth in augmenting the dryness of the air, or its disposition to absorb moisture, affords also the most satisfactory explication of the singular fact already noticed. If two equal surfaces of water be exposed in the same situation, the one in a shallow and the other in a deep vessel of metal or porcelain, the latter is always found, after a certain interval of time, to have suffered, contrary to what we might expect, more waste by evaporation than the former. Amidst all the changes that happen in the condition of the ambient medium, the shallow pan must necessarily receive, more completely than the deeper vessel, the chilling impressions of evaporation, since it exposes a smaller extent of dry surface to be partly heated up again by the contact of the air. The larger mass being, therefore, kept invariably warmer than the other, must in consequence support a more copious exhalation.

From the principles which have been explained, it likewise results, that the hygrometer does not indicate the actual dryness of the air, but only the dryness which it retains after being reduced to the temperature of the humid ball. The real condition of the medium, however, could easily be determined, from the gradations already ascertained in the power of solution. Suppose, for example, that the hygrometer should mark 42 degrees, while the thermometer stands at 16 centigrade; the moist surface has therefore the temperature of 11.8 centigrade, at which the dissolving energy is less by 37 degrees than at 16 centigrade; and hence the total dryness of the air, at its former temperature, amounted to 79 degrees. The following table will greatly facilitate such reductions. It is computed for as wide a range of dryness and temperature as will probably occur in any climate.

*Correction of the Hygrometer, and Position of the Point of Saturation at different Centesimal Temperatures.*

| | — 15° | | — 14° | | — 13° | | |
|---|---|---|---|---|---|---|---|
| Hyg. | Dryness. | Point Sat. | Dryness. | Point Sat. | Dryness. | Point Sat. | Hyg. |
| 10 | 12 | — 21.0 | 12 | — 19.5 | 12 | — 18.3 | 10 |
| 20 | 24 | — 29.2 | 24 | — 27.6 | 25 | — 26.1 | 20 |
| 30 | 36 | — 42.7 | 36 | — 39.8 | 37 | — 37.2 | 30 |

*Correction of Hygrometer, &c.—continued.*

| | — 12° | | — 11° | | — 10° | | |
|---|---|---|---|---|---|---|---|
| Hyg. | Dry-ness. | Point Sat. | Dry-ness. | Point Sat. | Dry-ness. | Point Sat. | Hyg. |
| 10 | 13 | — 17.8 | 13 | — 16.3 | 13 | — 15.0 | 10 |
| 20 | 25 | — 24.7 | 25 | — 23.2 | 26 | — 21.5 | 20 |
| 30 | 37 | — 35.0 | 38 | — 32.9 | 38 | — 30.1 | 30 |
| | — 9° | | — 8° | | — 7° | | |
| 10 | 13 | — 13.8 | 13 | — 12.5 | 13 | — 11.4 | 10 |
| 20 | 26 | — 19.8 | 26 | — 18.2 | 26 | — 16.8 | 20 |
| 30 | 39 | — 28.3 | 39 | — 26.1 | 39 | — 24.4 | 30 |
| | — 6° | | — 5° | | — 4° | | |
| 10 | 13 | — 10.0 | 14 | — 9.3 | 14 | — 7.9 | 10 |
| 20 | 27 | — 15.4 | 27 | — 14.0 | 27 | — 12.6 | 20 |
| 30 | 40 | — 22.2 | 40 | — 20.3 | 41 | — 18.8 | 30 |
| | | | 53 | — 29.2 | 54 | — 26.9 | 40 |
| | — 3° | | — 2° | | — 1° | | |
| 10 | 14 | — 6.8 | 14 | — 5.6 | 14 | — 4.5 | 10 |
| 20 | 28 | — 11.3 | 28 | — 9.9 | 28 | — 8.6 | 20 |
| 30 | 41 | — 16.9 | 42 | — 15.2 | 42 | — 13.7 | 30 |
| 40 | 55 | — 24.6 | 55 | — 22.2 | 56 | — 20.4 | 40 |
| | — 0° | | 1° | | 2° | | |
| 10 | 15 | — 3.4 | 15 | — 2.3 | 15 | — 1.2 | 10 |
| 20 | 29 | — 7.4 | 29 | — 6.1 | 30 | — 4.8 | 20 |
| 30 | 43 | — 12.2 | 43 | — 10.6 | 44 | — 9.1 | 30 |
| 40 | 57 | — 18.2 | 58 | — 16.3 | 59 | — 14.5 | 40 |

*Correction of Hygrometer, &c.—continued.*

| | 3° | | 4° | | 5° | | |
|---|---|---|---|---|---|---|---|
| Hyg. | Dry-ness. | Point Sat. | Dry-ness. | Point Sat. | Dry-ness. | Point Sat. | Hyg. |
| 10 | 15 | 0.0 | 15 | 1.0 | 16 | 2.1 | 10 |
| 20 | 30 | — 3.6 | 31 | — 2.3 | 31 | — 1.1 | 20 |
| 30 | 45 | — 7.7 | 45 | — 6.3 | 46 | — 4.9 | 30 |
| 40 | 59 | — 12.7 | 60 | — 11.0 | 61 | — 9.4 | 40 |
| | | | | | 76 | — 15.0 | 50 |
| | 6° | | 7° | | 8° | | |
| 10 | 16 | 3.1 | 16 | 4.3 | 16 | 5.4 | 10 |
| 20 | 32 | 0.0 | 32 | 1.3 | 33 | 2.4 | 20 |
| 30 | 47 | — 3.6 | 48 | — 2.2 | 49 | — 0.9 | 30 |
| 40 | 62 | — 7.8 | 63 | — 6.3 | 64 | — 4.7 | 40 |
| 50 | 77 | — 13.0 | 79 | — 11.1 | 80 | — 9.4 | 50 |
| | 9° | | 10° | | 11° | | |
| 10 | 17 | 6.5 | 17 | 7.5 | 18 | 8.6 | 10 |
| 20 | 33 | 3.6 | 34 | 4.8 | 35 | 5.9 | 20 |
| 30 | 50 | 0.4 | 50 | 1.9 | 52 | 3.0 | 30 |
| 40 | 66 | — 3.3 | 67 | — 1.8 | 68 | — 0.4 | 40 |
| 50 | 81 | — 7.6 | 83 | — 5.9 | 84 | — 4.3 | 50 |
| | | | 98 | — 11.0 | 100 | — 9.0 | 60 |
| | 12° | | 13° | | 14° | | |
| 10 | 18 | 9.6 | 18 | 10.7 | 19 | 11.8 | 10 |
| 20 | 35 | 7.1 | 36 | 8.2 | 37 | 9.4 | 20 |
| 30 | 53 | 4.3 | 54 | 5.5 | 55 | 6.7 | 30 |
| 40 | 69 | 1.0 | 71 | 2.4 | 72 | 3.7 | 40 |
| 50 | 86 | — 2.7 | 88 | — 1.1 | 89 | 0.4 | 50 |
| 60 | 102 | — 7.1 | 104 | — 5.4 | 106 | — 3.7 | 60 |

*Correction of Hygrometer, &c.—Continued.*

| | 15° | | 16° | | 17° | | |
|---|---|---|---|---|---|---|---|
| Hyg. | Dry-ness. | Point Sat. | Dry-ness. | Point Sat. | Dry-ness. | Point Sat. | Hyg. |
| 10 | 19 | 12.9 | 20 | 13.9 | 20 | 15.0 | 10 |
| 20 | 38 | 10.5 | 39 | 11.6 | 39 | 12.7 | 20 |
| 30 | 56 | 7.9 | 57 | 9.1 | 58 | 10.3 | 30 |
| 40 | 74 | 5.0 | 75 | 6.4 | 77 | 7.6 | 40 |
| 50 | 91 | 1.8 | 93 | 3.3 | 95 | 4.7 | 50 |
| 60 | 108 | — 1.8 | 111 | — 0.2 | 113 | 1.3 | 60 |
| 70 | 125 | — 6.3 | 128 | — 4.5 | 131 | — 2.6 | 70 |
| 80 | 142 | — 11.8 | 145 | — 9.5 | 148 | — 7.2 | 80 |
| | 18° | | 19° | | 20° | | |
| 10 | 20 | 16.0 | 21 | 17.0 | 21 | 18.1 | 10 |
| 20 | 40 | 13.8 | 41 | 14.9 | 42 | 16.0 | 20 |
| 30 | 60 | 11.5 | 61 | 12.7 | 63 | 13.8 | 30 |
| 40 | 79 | 8.9 | 81 | 10.2 | 83 | 11.4 | 40 |
| 50 | 97 | 6.0 | 100 | 7.4 | 102 | 8.8 | 50 |
| 60 | 116 | 2.8 | 118 | 4.4 | 121 | 6.1 | 60 |
| 70 | 134 | — 0.8 | 136 | 0.9 | 140 | 2.5 | 70 |
| 80 | 151 | — 5.2 | 154 | — 3.2 | 158 | — 1.3 | 80 |
| | | | | | 176 | — 5.9 | 90 |
| | | | | | 193 | — 11.5 | 100 |
| | 21° | | 22° | | 23° | | |
| 10 | 22 | 19.1 | 22 | 20.1 | 23 | 21.2 | 10 |
| 20 | 43 | 17.1 | 44 | 18.2 | 46 | 19.3 | 20 |
| 30 | 64 | 15.0 | 66 | 16.1 | 68 | 17.3 | 30 |
| 40 | 84 | 12.6 | 87 | 13.8 | 89 | 15.0 | 40 |
| 50 | 104 | 10.1 | 107 | 11.4 | 110 | 12.7 | 50 |
| 60 | 124 | 7.4 | 127 | 8.7 | 130 | 10.2 | 60 |
| 70 | 143 | 4.1 | 146 | 5.7 | 150 | 7.2 | 70 |
| 80 | 162 | 0.5 | 165 | 2.3 | 169 | 4.0 | 80 |
| 90 | 180 | — 3.8 | 184 | — 1.8 | 188 | 0.2 | 90 |
| 100 | 198 | — 9.0 | 202 | — 6.5 | 207 | — 4.3 | 100 |

*Correction of Hygrometer, &c.—Continued.*

| | 24° | | 25° | | 26° | | |
|---|---|---|---|---|---|---|---|
| Hyg. | Dry-ness. | Point Sat. | Dry-ness. | Point Sat. | Dry-ness. | Point Sat. | Hyg. |
| 10 | 24 | 22.2 | 24 | 23.3 | 25 | 24.3 | 10 |
| 20 | 47 | 20.4 | 48 | 21.4 | 49 | 22.5 | 20 |
| 30 | 69 | 18.4 | 71 | 19.5 | 73 | 20.6 | 30 |
| 40 | 91 | 16.2 | 94 | 17.4 | 96 | 18.6 | 40 |
| 50 | 112 | 13.9 | 116 | 15.2 | 119 | 16.5 | 50 |
| 60 | 133 | 11.5 | 137 | 12.8 | 141 | 14.1 | 60 |
| 70 | 154 | 8.7 | 158 | 10.1 | 162 | 11.6 | 70 |
| 80 | 174 | 5.6 | 178 | 7.2 | 183 | 8.7 | 80 |
| 90 | 193 | 2.0 | 198 | 3.9 | 203 | 5.6 | 90 |
| 100 | 212 | — 2.0 | 218 | 0.0 | 223 | 2.0 | 100 |
| | | | 237 | — 4.5 | 242 | — 2.3 | 110 |
| | | | 255 | — 10.2 | 261 | — 7.4 | 120 |

| | 27° | | 28° | | 29° | | |
|---|---|---|---|---|---|---|---|
| 10 | 26 | 25.3 | 26 | 26.4 | 27 | 27.4 | 10 |
| 20 | 51 | 23.6 | 52 | 24.7 | 54 | 25.7 | 20 |
| 30 | 75 | 21.7 | 77 | 22.8 | 79 | 23.9 | 30 |
| 40 | 99 | 19.8 | 102 | 21.0 | 104 | 22.1 | 40 |
| 50 | 122 | 17.7 | 125 | 18.9 | 129 | 20.1 | 50 |
| 60 | 144 | 15.4 | 148 | 16.7 | 152 | 18.1 | 60 |
| 70 | 166 | 13.0 | 171 | 14.3 | 175 | 15.7 | 70 |
| 80 | 188 | 10.3 | 193 | 11.7 | 198 | 13.2 | 80 |
| 90 | 209 | 7.2 | 214 | 8.9 | 220 | 10.4 | 90 |
| 100 | 229 | 3.8 | 235 | 5.6 | 241 | 7.4 | 100 |
| 110 | 249 | — 0.1 | 255 | 2.0 | 262 | 3.9 | 110 |
| 120 | 268 | — 4.8 | 275 | — 2.4 | 282 | — 0.1 | 120 |

| | 30° | | 31° | | 32° | | |
|---|---|---|---|---|---|---|---|
| 10 | 28 | 28.4 | 29 | 28.4 | 30 | 30.5 | 10 |
| 20 | 55 | 26.8 | 57 | 26.8 | 59 | 28.9 | 20 |
| 30 | 82 | 25.0 | 84 | 25.0 | 87 | 27.2 | 30 |
| 40 | 108 | 23.2 | 111 | 23.2 | 114 | 25.5 | 40 |

*Correction of Hygrometer, &c.—Continued.*

| | 30° | | 31° | | 32° | | |
|---|---|---|---|---|---|---|---|
| Hyg. | Dryness. | Point Sat. | Dryness. | Point Sat. | Dryness. | Point Sat. | Hyg. |
| 50 | 133 | 21.3 | 136 | 22.5 | 141 | 23.7 | 50 |
| 60 | 157 | 19.3 | 161 | 20.5 | 166 | 21.7 | 60 |
| 70 | 181 | 17.0 | 186 | 18.3 | 191 | 19.6 | 70 |
| 80 | 204 | 14.6 | 209 | 16.0 | 216 | 17.4 | 80 |
| 90 | 226 | 12.0 | 233 | 13.5 | 239 | 15.0 | 90 |
| 100 | 248 | 9.1 | 255 | 10.7 | 262 | 12.3 | 100 |
| 110 | 269 | 5.8 | 277 | 7.6 | 285 | 9.3 | 110 |
| 120 | 290 | 2.0 | 298 | 4.1 | 307 | 6.0 | 120 |
| | 33° | | 34° | | 35° | | |
| 10 | 31 | 31.5 | 32 | 32.5 | 33 | 33.6 | 10 |
| 20 | 61 | 30.0 | 63 | 31.0 | 65 | 32.0 | 20 |
| 30 | 90 | 28.3 | 92 | 29.4 | 95 | 30.5 | 30 |
| 40 | 118 | 26.6 | 121 | 27.7 | 125 | 28.9 | 40 |
| 50 | 145 | 24.9 | 149 | 26.0 | 154 | 27.1 | 50 |
| 60 | 171 | 22.9 | 177 | 24.1 | 182 | 25.3 | 60 |
| 70 | 197 | 20.9 | 203 | 22.1 | 209 | 23.4 | 70 |
| 80 | 222 | 18.7 | 229 | 20.0 | 236 | 21.3 | 80 |
| 90 | 246 | 16.4 | 254 | 17.8 | 262 | 19.2 | 90 |
| 100 | 270 | 13.8 | 278 | 15.3 | 287 | 16.8 | 100 |
| 110 | 293 | 11.0 | 302 | 12.6 | 311 | 14.2 | 110 |
| 120 | 316 | 7.9 | 325 | 9.7 | 335 | 11.4 | 120 |
| | | | | | 358 | 8.3 | 130 |
| | | | | | 380 | 4.7 | 140 |
| | 36° | | 37° | | 38° | | |
| 10 | 34 | 34.6 | 35 | 35.6 | 36 | 36.6 | 10 |
| 20 | 67 | 33.1 | 69 | 34.1 | 71 | 35.2 | 20 |
| 30 | 98 | 31.5 | 102 | 32.6 | 105 | 33.7 | 30 |
| 40 | 129 | 30.0 | 133 | 31.0 | 138 | 32.1 | 40 |
| 50 | 159 | 28.2 | 164 | 29.4 | 169 | 30.5 | 50 |
| 60 | 188 | 26.5 | 194 | 27.7 | 200 | 28.8 | 60 |

*Correction of Hygrometer, &c.—Continued.*

| | 36° | | 37° | | 38° | | |
|---|---|---|---|---|---|---|---|
| Hyg. | Dry-ness. | Point Sat. | Dry-ness. | Point Sat. | Dry-ness. | Point Sat. | Hyg. |
| 70 | 216 | 24.6 | 223 | 25.8 | 230 | 27.0 | 70 |
| 80 | 243 | 22.6 | 251 | 23.9 | 259 | 25.2 | 80 |
| 90 | 270 | 20.5 | 278 | 21.9 | 287 | 23.2 | 90 |
| 100 | 295 | 18.3 | 305 | 19.7 | 314 | 21.1 | 100 |
| 110 | 320 | 15.8 | 330 | 17.3 | 341 | 18.8 | 110 |
| 120 | 345 | 13.1 | 355 | 14.7 | 366 | 16.3 | 120 |
| 130 | 368 | 10.1 | 380 | 11.9 | 391 | 13.6 | 130 |
| 140 | 391 | 6.7 | 403 | 8.7 | 416 | 10.6 | 140 |
| | 39° | | 40° | | 41° | | |
| 10 | 37 | 37.6 | 39 | 38.6 | 40 | 39.7 | 10 |
| 20 | 72 | 36.2 | 76 | 37.2 | 79 | 38.3 | 20 |
| 30 | 108 | 34.7 | 112 | 35.8 | 116 | 36.9 | 30 |
| 40 | 142 | 33.2 | 147 | 34.3 | 152 | 35.4 | 40 |
| 50 | 175 | 31.6 | 181 | 32.7 | 187 | 33.8 | 50 |
| 60 | 207 | 30.0 | 214 | 31.1 | 221 | 32.3 | 60 |
| 70 | 238 | 28.2 | 246 | 29.4 | 254 | 30.6 | 70 |
| 80 | 267 | 26.4 | 276 | 27.6 | 286 | 28.9 | 80 |
| 90 | 296 | 24.5 | 306 | 25.8 | 316 | 27.1 | 90 |
| 100 | 324 | 22.4 | 335 | 23.7 | 346 | 25.1 | 100 |
| 110 | 352 | 20.2 | 363 | 21.6 | 375 | 23.0 | 110 |
| 120 | 378 | 17.9 | 390 | 19.4 | 403 | 20.8 | 120 |
| 130 | 404 | 15.3 | 417 | 16.9 | 430 | 18.5 | 130 |
| 140 | 429 | 12.4 | 443 | 14.2 | 457 | 15.9 | 140 |
| | | | 468 | 11.2 | 483 | 13.0 | 150 |
| | | | 492 | 7.8 | 508 | 9.8 | 160 |
| | 42° | | 43° | | 44° | | |
| 10 | 41 | 40.7 | 43 | 41.7 | 44 | 42.7 | 10 |
| 20 | 81 | 39.3 | 84 | 40.3 | 87 | 41.3 | 20 |
| 30 | 120 | 37.9 | 125 | 39.0 | 129 | 40.0 | 30 |
| 40 | 158 | 36.5 | 163 | 37.6 | 169 | 38.7 | 40 |

*Correction of Hygrometer, &c.—Continued.*

| | 42° | | 43° | | 44° | | |
|---|---|---|---|---|---|---|---|
| Hyg. | Dry-ness. | Point Sat. | Dry-ness. | Point Sat. | Dry-ness. | Point Sat. | Hyg. |
| 50 | 194 | 34.9 | 201 | 36.0 | 208 | 37.1 | 50 |
| 60 | 229 | 33.4 | 237 | 34.5 | 245 | 35.6 | 60 |
| 70 | 262 | 31.8 | 272 | 32.9 | 281 | 34.1 | 70 |
| 80 | 295 | 30.1 | 305 | 31.3 | 316 | 32.4 | 80 |
| 90 | 327 | 28.4 | 338 | 29.6 | 350 | 30.7 | 90 |
| 100 | 358 | 26.4 | 370 | 27.7 | 383 | 29.0 | 100 |
| 110 | 388 | 24.4 | 401 | 25.8 | 414 | 27.1 | 110 |
| 120 | 416 | 22.2 | 430 | 23.7 | 445 | 25.1 | 120 |
| 130 | 444 | 20.0 | 459 | 21.5 | 475 | 23.0 | 130 |
| 140 | 472 | 17.5 | 487 | 19.1 | 504 | 20.7 | 140 |
| 150 | 498 | 14.8 | 515 | 16.5 | 532 | 18.2 | 150 |
| 160 | 524 | 11.8 | 541 | 13.7 | 559 | 15.5 | 160 |

We may compute, that a cubic mass of air 40 inches every way, or a little more than the standard of French measures, and of the ordinary density, weighs 20,000 grains. The table now given exhibits, therefore, in grains, the weight of moisture which a metrical cube of air is capable of holding at different temperatures. Thus, at 20 degrees, which corresponds to 68 degrees of Fahrenheit, this body of air could retain 252 grains of humidity. But if a larger scale be preferred, the same numbers will express, in pounds troy, the quantity of water required to saturate a perfectly dry mass of air constituting a cube of twenty yards in every dimension.

It is remarkable how small a portion of the aqueous element is at any time suspended in the atmosphere. Reckoning the mean temperature over the surface of the globe to be nineteen centesimal degrees, the air could only hold 240.6 parts of humidity for 20,000 times its whole weight; but this weight is nearly the same as that of a column of water of 400 inches in altitude, and hence, if the atmosphere, from a state of absolute dampness, were to pass into that of extreme dryness, and discharge the whole of its watery store, it would form a sheet of 4.812 inches, or somewhat less than five inches in depth. To furnish a sufficient supply of rain, the air must therefore undergo very frequent changes from dryness to humidity in the course of the year.

But it was requisite to subject theory to the test of accurate experiment. For this purpose a globe or glass balloon was procured, of very large dimensions, containing nearly 4000 cubic inches, terminated by a neck of about three inches wide, having its mouth ground flat. The balloon was supported from the floor by a light circular stand or rim, and a round piece of plate glass perforated through the centre by a hole of about the twentieth part of an inch in diameter, through which passed a slender silver wire suspended from the end of a fine beam placed on the table. This wire was fastened to the top of the scale of a delicate hygrometer, from the lower part of which hung a bit of wet paper nearly three inches in diameter. The whole was balanced by a counterpoise in the opposite scale, so that the instrument occupied the middle of the balloon. As the moisture gradually evaporated from the wet ball of the hygrometer, and still more from the larger surface of the paper attached below it, a loss of weight became visible, while the ascent of the coloured liquid in its tube indicated the corresponding diminution of the dryness of the included air. The progress of humefaction was observed from fifty to ten degrees of the hygrometer, after each successive grain of water, amounting in all to five, had exhaled and dispersed itself through the medium. Having rectified the hygrometric degrees according to the principle already explained, it was easy to compare them with the weight of the whole mass of air contained within the balloon. The conclusions were perhaps as satisfactory as such a nice and fugacious inquiry will admit; and if they be not absolutely correct, they must, at least, approximate very nearly to the truth.

The table given above is in strictness applicable to air only of the ordinary density. Since this fluid has its capacity for heat enlarged by rarefaction, the same depression of temperature must intimate a proportional augmentation of dryness. Thus, for air at the elevation of three miles and a half, and consequently twice as rare as at the surface, it would be requisite to add the sixteenth part to the numbers in the first column. For the lower altitudes, the correction will be to multiply those numbers by triple the height in feet, and cut off six decimal places. Thus, suppose, while the thermometer stood at twenty-eight centesimal degrees, that the hygrometer marked 110° on the plains of Mexico, at the elevation of 8000 feet above the level of the sea; then $3 \times 8000 \times 255 = 6120000$, which, being divided by a million, gives 6 to be added to 255, increasing the actual dryness to 261°. In most cases, therefore, this modification may be neglected.

But, in estimating the distance of the point of saturation in rarefied air, a greater correction will be required. The solvent power of that medium is extended about fifty hygrometric degrees each time it has its rarefaction doubled. Hence it may be calculated that our atmosphere would, at the same temperature, become a degree drier for every 360 feet of ascent. Thus, on the preceding supposition, the air on the plain of Mexico would have its distance from the point of saturation enlarged $22\frac{1}{2}$ degrees, its whole range being thus $283\frac{1}{2}$ degrees.

If the papered ball of an hygrometer be suffered to become dry, the instrument, even in that state, will mark, though for a short time only, the different condition of the media into which it is transported. Thus, the air of a room being supposed to have fifty degrees of dryness, on carrying the quiescent hygrometer into another apartment of seventy degrees, the column of liquor will fall near twenty degrees, from the renewed evaporation of that portion of moisture which had still adhered to the coats of paper. But if the same instrument were carried into an apartment of only thirty degrees of dryness, the coloured liquor would actually rise near twenty degrees above the beginning of the scale, the paper now attracting the excess of humidity from the air. This vapour, in combining with it, passes into the state of water, and therefore evolves a corresponding share of heat. The equilibrium, however, unless the coats of paper have a considerable thickness, is again restored in a very few minutes.

Those changes are most readily perceived on immersing the quiescent hygrometer alternately in two receivers containing air drier and damper than that of the room. If a pyroscope, having both its balls covered with gold-beater's skin, be treated in the same way, it will indicate an effect, though momentary indeed, of a similar kind; for, in air which is drier, the pellicle of the naked ball will throw off its moisture more freely than that of the gilt ball; and in damper air it will, on the contrary, imbibe the surplus humidity with greater eagerness, thus losing some portion of heat in the one process, and gaining a minute accession in the other. The quantity of moisture concerned in producing such fleeting alterations, may not exceed the thousandth part of a grain.

If a large receiver, having a delicate hygrometer suspended within it, be placed on a brass plate, and over a metal cup containing some water, the included air will, from the solution of the moisture, become gradually damper, and this progressive change is marked by the instrument. Yet the mass of air will never reach its term of absolute humidity, and before the hygrometer points at five degrees, the inside of the receiver appears covered with dew. While the humefying process, therefore, still goes on, the close attraction of the glass continually robs the contiguous air of a portion of its moisture, so that a kind of perpetual distillation is maintained through the aerial medium, the vapour successively formed being again condensed on the vitreous surface. But if, instead of the receiver, there be substituted a vessel formed of polished metal, the confined air will pass through every possible degree of humidity, and the hygrometer will, after some interval, arrive at the beginning of its scale.

The contrasted properties of a vitreous and a metallic surface, in attracting and repelling moisture, may be shown still more easily. In clear calm weather, let a drinking-glass and a silver cup be placed empty near the ground, on the approach of evening, and, as the dampness begins to prevail, the glass will become insensibly obscured, and next wetted with profuse dew, before the metal has yet betrayed any traces of humidity. The effect is, indeed, augmented by the cold pulses darted from the sky, which act more powerfully on the glass than on the metal.

The hygrometer is an instrument of the greatest utility, not only in meteorological observations, but in aiding domestic economy, in regulating many processes of art, and in directing the purchase and selection of various articles of produce. It will detect, for instance, the dampness of an apartment, and discover the condition of a magazine, of an hospital, or of a sick-ward. Most warehouses require to be kept at a certain point of dryness, which is higher or lower according to the purposes for which they are designed. The printing of linen and cotton is carried on in very dry rooms; but the operations of spinning and weaving succeed best in air which rather inclines to dampness.

The manufacturer is at present entirely guided by observing the effects produced by stoves, and hence the goods are often shrivelled, or otherwise injured, before he can discern any alteration in the state of the medium. Wool and corn have their weight augmented sometimes as much as 10 or even 15 per cent. by the presence of moisture. But the condition of these commodities could be nicely and readily examined, by heaping them over a small wired cage, within which an hygrometer is placed.

V. The ATMOMETER.*—This instrument is an useful auxiliary, and might with some attention serve as a substitute of the hygrometer. It does not mark the mere dryness of the air, but it measures the quantity of moisture exhaled from a humid surface in a given time. The atmometer consists of a thin ball of porous earthenware, two or three inches in diameter, with a small neck, to which is firmly cemented a long and rather wide glass tube, bearing divisions, each of them corresponding to an internal annular section, equal to a film of liquid that would cover the outer surface of the ball to the thickness of the thousandth part of an inch. (Plate IV. fig. 1.) The divisions are marked by portions of quicksilver introduced, ascertained by a simple calculation, and they are numbered downwards to the extent of 100 or 200; to the top of the tube is fitted a brass cap, having a collar of leather, and which, after the cavity has been filled with distilled water, is screwed tight. The outside of the ball being now wiped dry, the instrument is suspended out of doors, exposed to the free access of the air. In this state of action the humidity transudes through the porous substance just as fast as it evaporates from the exter-

* From ἀτμὸς, *exhalation* or *vapour*, and μέτρον, *a measure*.

nal surface, and this waste is measured by the corresponding descent of the water in the stem.

If the atmometer had its ball perfectly screened from the agitation of wind, its indications would be proportional to the dryness of the air at the lowered temperature of the humid surface; and the quantity of evaporation every hour, as expressed in thousandth parts of an inch, would, when multiplied by twenty, give the hygrometric measure. For example, in this climate, the mean dryness in winter being reckoned 15°, and in summer 40°, the daily exhalation from a sheltered spot must in winter form a thickness of .018, and amount in summer to .048 decimal parts of an inch. Suppose a pool for the supply of a navigable canal exposed a surface equal to ten English acres, and that the atmometer sank eighty parts during the lapse of twenty-four hours; the quantity of water exhaled in that time would be $\frac{80}{12000} \times 660 \times 66 \times 10$, or 2904 cubic feet, which corresponds to the weight of 81 tons.

The dissipation of moisture is much accelerated by the agency of sweeping winds, the effect being sometimes augmented five or even ten times. In general, this augmentation is proportional, as in the case of cooling, to the swiftness of the wind, the action of still air itself being reckoned equal to that produced by a celerity of eight miles each hour. Hence the velocity of wind is easily computed, from a comparison of the indications of a hygrometer with an atmometer, or of a sheltered with those of an exposed atmometer. Thus, suppose the hygrometer to mark 40 degrees, or the column of water in a sheltered atmometer to subside at the rate of two divisions every hour, while in one exposed to the current the descent is twelve divisions; then as two is to ten, the superadded effect of the wind, so is eight to forty miles, the distance through which it has travelled in that time.

The atmometer is an instrument evidently of extensive application, and of great utility in practice. To ascertain with accuracy and readiness the quantity of evaporation from any surface in a given time, it is an important acquisition, not only in meteorology, but in agriculture, and the various arts and manufactures. The rate of exhalation from the surface of the ground is scarcely of less consequence than the fall of rain, and a knowledge of it might often direct the farmer advantageously in his operations. On the rapid dispersion of moisture depends the efficacy of drying-houses, which are too frequently constructed most unskilfully, or on very mistaken principles.

It is obvious, that though the atmometer should be exposed to the free air, it must be sheltered from rain, which, by wetting the ball, would derange the proper action of the instrument. This could easily be done, by fixing a small canopy over it; or, in the case of drifting showers, to have a sort of shelved open screen, like Venetian blinds, turned by the wind. The only objection to this atmometer is, that it cannot be used during intense frost, since the expansion of the included water, by a sudden congelation, might burst the ball and even the tube. But the instrument could still be made to act in another way: Let it be emptied, and a certain portion of the water, measured in the stem, be spread over the outside of the ball, by successive layers, to form a coat of ice. The time is to be noted when the whole of this crust has disappeared; or if any portion should remain, it may be deducted from the whole, and thus the hourly quantity of evaporation ascertained.

VI. PHOTOMETER.*—This instrument, which was contrived to indicate the power of illumination, by the slight

* From φως, *light*, and μετρον, *a measure*.

elevation of temperature which it occasions, has been shortly noticed in the article CLIMATE. It consists of a differential thermometer, having one of its balls diaphanous, and the other coated with china ink, or rather blown of deep-black enamel. (Plate V. fig. 11. and 12.) The rays which fall on the clear ball pass through it, without suffering obstruction; but those which strike the dark ball are stopt and absorbed at its surface, where, assuming a latent form, they act as heat. This heat will continue to accumulate, till its farther increase comes to be counterbalanced by an opposite dispersion, caused by the rise of temperature which the ball has come to acquire. At the point of equilibrium, therefore, the constant accessions of heat derived from the action of the incident light, are exactly equalled by the corresponding portions of it again abstracted in the subsequent process of cooling. But, in still air, the rate of cooling is, within moderate limits, proportioned to the excess of the temperature of the heated surface above that of the surrounding medium. Hence the space through which the coloured liquid sinks in the stem will measure the momentary impressions of light, or its actual intensity. To prevent any extraneous agitation of the air from accelerating the discharge of heat from the black ball, and thereby diminishing the quantity of aggregate effect, the instrument is always sheltered, and more especially out of doors, by a thin glass case. The addition of this translucid case is quite indispensable. It not only precludes all irregular action, but maintains, around the sentient part of the instrument, an atmosphere of perpetual calm. Under the same force of incident light, the temperature of the black ball must still rise to the same height above that of its encircling medium. The case will evidently have some influence to confine the heat actually received, and hence to warm up the internal air. Wherefore, corresponding to this excess, the black ball will acquire a farther elevation of temperature; but the clear ball, being immersed in the same fluid, must experience a similar effect, which will exactly counterbalance the former. The difference of temperature between the opposite balls thus continues unaltered; and neither has the size or the shape of the case, nor the variable state of the exterior atmosphere with respect to rest or agitation, any sensible influence to derange or modify the results exhibited by this delicate instrument.

The photometer has, like the hygrometer, two general forms; the *stationary*, represented by fig. 9, and the *portable*, delineated in fig. 12. But they are both of them easily transported from one place to another. Their glass cases can be screwed off, and the former instrument, being cemented into a small slip of brass, which slides with a spring into the bottom, may be packed separately, if required, while the latter is protected by an external wooden case, in which it is carried in the pocket as safely as a pencil. This case, if screwed below, serves also as a handle to hold the photometer in a vertical position out of doors.

The photometer, placed in open air, exhibits distinctly the progress of illumination from the morning's dawn to the full vigour of noon, and thence its gradual decline till evening has spread her mantle; it marks the growth of light from the winter solstice to the height of summer, and its subsequent decay through the dusky shades of autumn; and it enables us to compare, with numerical accuracy, the brightness of different countries, the brilliant sky of Italy, for instance, with the murky atmosphere of Holland.

In this climate, the direct impression of the sun about midsummer amounts to about 90 degrees; but it regularly declines as his rays become more oblique. The greatest force of the solar beams with us in the depth of winter reaches only to 25 degrees. At the altitude of 17 degrees, it is already reduced to one half; and at 3 degrees above the horizon, the whole effect exceeds not one millesimal degree.

The quantity of indirect light reflected from the sky, though extremely fluctuating in our climate, is often very considerable. It may be estimated at 30 or 40 degrees in summer, and 10 or 15 in winter. This secondary light is most powerful when the sky is overspread with thin fleecy clouds; it is feeblest, either when the rays are obstructed by a mass of congregated vapours, or when the atmosphere is clear and of a deep azure tint. On the lofty summits of the Alps or Andes, the photometer, screened from the sun, and only exposed to the dark hue of the broad expanse, would indicate a very small effect. During the solar eclipse which took place on the 7th of September 1820, the sky being completely overclouded, it showed, both before and after the passage of the moon's disc, only 12 degrees of light; but when the obscuration was the greatest, it marked not more than a single degree.

The delicacy of this instrument renders it a valuable auxiliary in various scientific inquiries. It ascertains the diminution which the rays of light suffer in reflection, and during their passage through different transparent substances. By combining it with the transferrer of an air-pump, it likewise detects the comparative powers for conducting heat of the several gases, whether in their ordinary state, or when variously attenuated or condensed. Hence we learn, that air expanded 256 times conducts heat nearly twice as slow, or in the ratio of 7 to 13; but that hydrogen gas transfers it with more than redoubled celerity, or in the ratio of 9 to 4. But the photometer will measure also the conducting powers of different liquids. It is only wanted to remove the case, and plunge the instrument erect in a wide metallic vessel, bright on the outside, but blackened within, containing the liquid to be examined, and exposed to the sun's rays. In this way, it was found that water, which conveys heat about 30 times faster than air of the mean temperature of our climate, transfuses it with still greater rapidity if warmed to a higher pitch. If the photometer, enclosed within its case, be immersed under the surface of water, the impression of the light will be much stronger than when the balls were encircled by the actual contact of this liquid. Yet will the effect be less than if the case had been surrounded externally by a body of air instead of water, which, by its powerful action in drawing off the accumulated heat, hastens the transmission of it through the internal medium, and reduces the elevation of the temperature of the black ball to nearly one third part.

From observations made with this instrument, we likewise discover that, in the clearest and most serene sky, one half only of the sun's light, sloping at an angle of 25°, will reach the ground; and that at an angle of 15°, the proportion is reduced to one third; but with an obliquity of 5°, the length of track being then extended ten times, one twentieth part only of the whole incident light can reach the surface. When the sun has approached within a degree of the horizon, and his rays now traverse a track of air equal in weight to a column of about 905 feet of water, no more than the 212th part of them can penetrate to the ground.

The photometer discovers the relative density of different artificial lights, and even contrasts their force of illumination with that of the solar rays. It may be mentioned as a curious inference, that the light emitted from the sun is 12,000 times more powerful than the flame of a wax-candle; or that, if a portion of the luminous solar matter, rather less than an inch in diameter, were transported to our planet, it would throw forth a blaze of light equal to the effect of 12,000 candles.

To compare the illumination of candles or lamps, and of the flame of coal or oil gas, the best form of the photometer is that of Plate V. fig. 9, guarded both in front and behind by a pair of thin spreading plates of mica, set parallel at the mutual interval of about half an inch. It may be sufficient to notice at present, that the flame of coal-gas has more than triple the brilliancy of that of a wax-candle.

A photometer of the branched form is easily adapted to measure the diminution which light suffers in penetrating through a body of water. The scale may then be shortened, and the balls enlarged. A bottom of lead is turned to receive the instrument, with its case, which are cemented to it. Thus loaded, the photometer is suspended vertically by cross silk threads, to which a cord of some definite length is attached, terminating by a small bladder. The sky being clear, and the sun shining bright, the instrument is, by the help of a long pole, stretched from the side of a boat, held a few minutes suspended about four inches below the surface of the water, and then drawn up, and the number of degrees marked. When the direct action of the solar rays amounted to 90 photometric degrees, their enfeebled influence on the instrument, while thus encompassed externally by a dense chilling body of water, was commonly found to be reduced to 32 degrees. From this point, therefore, the subsequent diminution, occasioned by the descent of the instrument, was computed. The photometer, being let down, was left to float near a quarter of an hour at the depth of perhaps three or six fathoms. On drawing it up, the diminished action of the light, occasioned by the length of oblique passage, was at once perceived.

From experiments performed in this way last summer, it follows, that half of the incident light which might pass through a field of air of the ordinary density and $15\frac{1}{2}$ miles of extent, would penetrate only to the perpendicular depth of 15 feet in the clearest sea-water, which is, therefore, 5400 times less diaphanous than the atmospheric medium. The light is hence diminished four times for every five fathoms of vertical descent; and, consequently, the 64th part only could reach to the depth of 15 fathoms. Supposing the bottom, then, to consist of a clear white sand, the portion of light reflected, and sent back to the surface, would be attenuated more than 64 × 64, or 4096 times, and would therefore hardly be perceptible to the most acute eye. But the water of shallow lakes, though not apparently turbid, betrays a still greater opacity, insomuch that the perpendicular light was diminished one half in descending only through the space of six feet in Loch Leven, or even two feet in a fine artificial sheet of water at Raith, near Kirkaldy. These results, however, are to be considered as mere approximations, the state of the weather having been very unfavourable for such experiments.

It would be easy, by a small modification, to adapt the photometer as a diaphanometer, for measuring the comparative transparency of different collections of water. The black and the clear ball might be blown larger than usual, and the instrument covered with two thin parallel cases of glass, separated by an interval of about three eighths of an inch. The transparency of a lake, or of the sea, would be inversely as the length of passage traversed by the light, when it had suffered a proportional diminution of intensity.

VII. ÆTHRIOSCOPE.*—Such is the name of another very delicate modification of the differential thermometer, intend-

* From αἴθριος, which signifies at once *clear*, *dry*, and *cold*.

ed to measure those frigorific impressions which are showered incessantly from the distant sky. The history of this invention, and of its progressive improvement, has been already given under the article CLIMATE. For about two years few good observations were added, owing to the prevalence of very cloudy and windy weather. The chief object, then, was to render the æthrioscope more portable, in the hope of obtaining, through it, some correct information regarding the state of the atmosphere in other quarters of the globe. The upper ball is now scarcely half an inch in diameter; but to compensate for this diminution, the lower ball has a diameter of about four fifths of an inch. The tube, which exceeds not four inches in length, has its bore contracted, a little above its junction, to the very short cylindrical cavity that holds the coloured liquor. This simple contrivance, augmenting greatly the capillary action, prevents the descent of the column into the ball from any sudden change of temperature, while it only retards the motion of the fluid, without affecting the accuracy of its play. Plate VII. fig. 1. represents the instrument in this abridged form, and fig. 2. shows the way of packing it, the bottom being merely screwed to the top of the case. The only precaution needed is, not to shake the æthrioscope, or invert it; and as it takes very little room, it may easily be carried by the traveller in his pocket.

In ordinary cases, the hot or cold pulses propagated through the air only assist the energy of the transfer of the different portions of the fluid in promoting an equilibrium of temperature. But the æthrioscope proves that those pulses are incessantly forwarding such a balance, even while the mere transfer and commixture of the medium would not contribute to the effect. In the article CLIMATE, it was shown that the rapid interchange which takes place between the higher and lower strata of the atmosphere maintains an equal distribution in the *quantity* and not in the *intensity*, of heat. Since air has its capacity for heat increased by rarefaction, it must, with the same igneous infusion, indicate a proportionally depressed temperature. But this inequality of temperature, resulting from the internal commotion produced by the sun's rays acting more powerfully near the surface of the earth, is partly corrected by the influence of the cold or hot pulses which are at all times darted, and in every direction, unless obstructed or absorbed by the interposition of the clouds. While the cold pulses from the upper strata of the atmosphere are constantly chilling the lower strata, the warm pulses again from below are exerted in warming the higher regions. In most cases this mutual influence, indeed, is comparatively feeble; but if the rays of the sun were withdrawn for any considerable time, a great progress would be made by such a mutual interchange of the pulsations towards an equality of temperature through the mass of atmosphere. The lowest strata would become unusually colder, whilst the highest regions would grow warmer, and sparkle with augmented clearness and lustre. Such are some of the effects of the long protracted nights within the arctic circle.

Much yet remains to be explored in the higher strata of our atmosphere. If the differential thermometer, included within the æthrioscope, had its position reversed, that instrument would become adapted to measure the hot pulses which are no doubt shot incessantly upwards with various obliquity from the warmer beds incumbent over the surface of the earth. It would be most interesting to obtain the reports of both the erect and the pendant opposite æthrioscope, when carried up in the car of a balloon to the elevation of four miles. In that region of mid-air, we might expect the hot and the cold pulses, as they crossed in opposite directions, to act with nearly equal energy. The measures of these effects, compared with the simultaneous indications of the photometer, could not fail to dispel much obscurity, and to open new views of the disposition of the elements, and of the economy of nature.

VIII. CYANOMETER.*—This instrument was contrived by M. de Saussure, to measure the variable intensity of the cærulean hue which the sky assumes in different climates and elevations, according to the progress of the day or the advance of the season. It consists of fifty-three slips of paper of about a quarter of an inch broad, stained with the successive shades of blue, from the palest sapphire to the deepest azure, which are pasted around the circumference of a circle of pasteboard of about four inches in diameter. The colours were obtained from fine Prussian blue, diluting it with white chalk, or darkening it with a mixture of ivory black. He likewise compared those coloured spaces with the pure tints of a solution of copper in ammonia, which resemble most the soft transparent hues of the atmosphere. To represent the effect of clouds, and diffuse aqueous vapours, he dropped into that liquid a portion of very fine divided argillaceous earth, precipitated by ammonia from a solution of alum.

In observing with the cyanometer, it should be held out of doors, between the eye and the part of the heavens which is to be compared, and, with a little practice, the corresponding tint is easily distinguished.

In this way, Saussure found, that the deepest blue of the zenith on the summit of Mont Blanc, at his station on the Col du Géant, at Chamouni, and at Geneva, corresponded respectively to the shades denoted *39*, *37*, *34*, and *26½*. From morning till noon, the colour of the vertical sky dark-

* From κυανος, *cærulean* or *sky-blue*, and μετρον, *a measure.*

ened, but became lighter again as the evening advanced; and this transition was wider and more rapid in great elevations. On the Col du Géant, the tint of the horizontal air at sunrise was 5; it deepened to 11½ at noon, but again relapsed to 5 towards night. On the 15th of July, which was a very clear day, the atmosphere at the horizon had the 11th shade; at the altitude of 10°, the 20th; at that of 20°, the 31st; at that of 30°, the 34th; at that of 40°, the 37th; and thence with any sensible variation to the zenith. Baron Humboldt, in his voyage from Corunna to Cumana, found the tints of the sky to vary, by the cyanometer, from 13 to 24, and again to 16, while the colour of the ocean fluctuated between 34 and 44.

The misfortune is, that we cannot annex any very distinct ideas to these numbers. We are not informed even of the proportions of the ingredients of the series of colours. The manner of composition likewise will modify the colorific effect; and most of the pigments, and especially the Prussian blue, not only want uniformity of tone, but are subject to great alteration. It would be quite impossible to paint with any water colours two cyanometers that should continue to agree, after being exposed for some time to the action of the air and the sun.

If an accurate method could be devised to discriminate colours, and mark their different tints with a sort of numerical precision, it would prove a valuable acquisition to philosophy and the arts. This was first attempted by the famous painter Leonardi da Vinci. Zahn proposed, in 1702, to accomplish it, by the graduating mixture of primary colours dispersed over the surface of a triangle; but he reckoned five of those colours, including *black* and *white,* with *red, yellow,* and *blue.* The celebrated Professor Mayer of Göttingen, after various trials, simplified the procedure, in a posthumous work, published by Lichtenberg in 1775. Having distinguished each side of an equilateral triangle into 13 equal parts, he subdivided the whole space into 91 small triangles, which he painted with the successive mixtures of *vermilion, ultra-marine,* and *bright orpiment.* Lambert assumed three colours, *carmine, Prussian blue,* and *gamboge,* to cover a triangular base, upon which he erected a coloured pyramid, having *white* planted at its apex. But Dr Thomas Young, whose authority in those matters has deservedly great weight, prefers the simple triangle, and adopts *red, green,* and *violet,* for the primary colours. Their binary combinations are *yellow,* formed by mixing red with green; *crimson,* consisting of red and violet; and *blue,* produced by blending green with violet. The difficulty, however, is to regulate the intensity of the compounds; nor can the powders be safely mixed except in a dry state, lest some chemical action should be introduced which might alter their tints. But the colours thus combined must evidently want the freshness and brilliancy of those which nature paints, or which the prism reveals.

Air, like water, is no doubt, by its constitution, a coloured fluid. The former is naturally blue, as the latter is green; but these colours acquire intensity only from the depth of the transparent mass. A small body of limpid water has the appearance of crystal, but, in proportion as it accumulates, it assumes all the successive shades, till it rivals the tints of the emerald and the beryl. This gradation is distinctly seen in the profound lakes of Switzerland, whose lustre is never stained by any vegetable infusion. The same series of colours emerges on receding from our shores and approaching the vast abyss of the Atlantic Ocean. At first, the water on the shelving banks is merely translucid; at the depth of ten fathoms it appears greenish; and the tint, by degrees, becomes more intense, till it passes into a full green at the depth of fifty fathoms; but beyond soundings it darkens almost into azure.

In like manner, the blue shade of the air becomes more intense in proportion to the length of the track of light. This we perceive in viewing distant objects, whose colours are always tinted by the deepening hues of the interjacent range of atmosphere. The remotest hills seem lost in a cærulean vesture. The mixture of aqueous vapours only diffuses a mist, which tarnishes rather than dilutes the fine blue.

It must be observed, that no substance can disclose its inherent colour, but by a sort of internal secretion or dissection of the rays of light. The mere reflexion from the surface of a solid body could never betray its tints; for, when rendered most perfect by polish, it would only, like a mirror, send back unchanged the incident beams. To detect the subjacent colour, it is necessary that the particles of light should at least penetrate under the surface, and, after suffering a sort of chemical separation, should be again emitted. In transparent substances, whether solid or fluid, the penetration is greater, but the mode of evolving the native colours must be still the same. The atmosphere, besides dispersing internally the blue rays, likewise reflects in various proportions the white light unaltered. This fact is established by some experiments of *polarization*, which show that such simple reflexions are the most copious from the portion of the sky which is 90 degrees from the sun, and regularly decline on either side to the opposite points, where they cease altogether.

The white or compound beam of light suffering, in its passage through the air, a continual defalcation of the blue rays, must, as it advances, assume the complementary colour, or the tints of the remaining portions of the spectrum, and therefore merge successively into yellow, orange, red, and crimson. Such, accordingly, are the graduating colours of the solar rays, as they approach to their extreme obliquity. Near sun-setting, the shadow of a pencil along a blank card appears a bright azure on a lilac ground. When a diffuse attenuated vapour reflects the incident light unaltered, the western sky, as the sun declines from his altitude, glows with the successive shades of yellow and orange, which deepen finally into a blush red. These colours again may, under certain circumstances, come to be blended with the natural blue of the atmosphere. Hence the explication of a curious phenomenon, which rarely occurs in this climate, the existence of *green clouds*. This happens in the mornings and evenings, when a thin cloud is illuminated at once by the yellow rays of the sun, and the bright azure of the upper sky, these contrasted colours producing a *green* by their mixture. For the same reason, sometimes a portion of the bright sky appears, in the finer climates, tinted with violet. This was remarked by Humboldt in his voyage to America, and we have had occasion to observe the same at Avignon. It was no doubt occasioned by the reddish rays of the declining sun dyeing the intense blue of the higher atmosphere.

The easiest and readiest way of ascertaining the tints of different portions of the sky, is perhaps to employ a sharp wedge of blue glass, of which the base and the parallel sides are painted black, and cased with thin brass, and the slanting sides are ground to true planes, and highly polished. To these angular surfaces two slides might be adapted, having each a broad slit, or intermediate opening, to permit the entrance and transmission of white light. Such rays having a greater length of passage to traverse, according to their distance from the top of the wedge, must emerge with a proportional intensity of blue. The scale would hence be determined by dividing the slides into ten or twenty equal parts, which might probably be sufficient.

To examine the orange and crimson tints which gild the east in the morning, or suffuse the western sky on the approach of evening, it would be necessary to combine a series of the complementary or accidental colours. A wedge of glass, stained of a gold-red or deep orange, might answer the purpose; or perhaps a nearer approximation would be obtained by joining two reversed wedges, one of an angle of 8°, and crimson-coloured, and another having only 3°, but of a yellow body.

These instruments might be reduced to a very convenient size, not exceeding four inches in length; but they would require to be formed out of the same mass of glass, and exactly after the same pattern. With some skill in the execution, they could be made to unite elegance and correctness.

IX. Anemometer.*—Various attempts have been made to construct an instrument that should readily indicate the force and velocity of the wind. One method was to employ a very small model of a wind-mill, and either to reckon its revolutions, or to estimate its power by the application of a weight to a conical barrel or axis. But a more direct and accurate procedure consisted in measuring the impulse of wind against a vertical plane, as intimated by the contraction of a spiral spring. All these instruments, however, act with such extreme irregularity, as scarcely ever to furnish any definite results. They are, besides, racked by incessant motion, and soon put out of order.

We may notice, however, a material improvement made in this construction of the machine by Mr Waddel of the Trinity-House, Leith, who, amidst other objects of useful experimental inquiry, has long directed his ingenuity to ascertain the force and velocity of the wind. A circular plate, of four inches diameter, is opposed to the blast; but instead

* From ἄνεμος, *the wind*, and μέτρον, *a measure*.

of pushing against a spring, it presses against a fine cylindrical bag, of about an inch long, and the third part of an inch diameter, filled with quicksilver, and joined tight to a vertical tube of glass, a foot or fifteen inches high, but having a bore only the twentieth part of an inch wide. The compression of the bag, caused by the impulse of the wind upon the plate, squeezes the quicksilver up into the tube, carrying with it a small steel mark, which slightly adheres to the sides of the bore. The height of the mercurial column, diminished in the ratio of the surface of the plate to the section of the bag, must evidently give the measure of the force of the wind. But, in the actual exposure of this anemometer, the quicksilver oscillates excessively, so that the extreme effects only are indicated. The instrument, however, is very sensible, and may continue to act for a long period without being impaired.

The direct action of the wind in supporting a column of water, appears to furnish the best and simplest kind of anemometer. This principle was first employed, in 1731, by Pitot, the French engineer, in his recurved tube for estimating the force of the current of a river; and, forty years afterwards, it was applied by Dr Lind to measure the impulse of a stream of air. With some modifications to correct, or at least to diminish, the oscillations of the liquid, this instrument is rendered quite manageable. The tube may consist of two pieces, each about a foot in height, having bores of the fiftieth and the fifth parts of an inch, the narrow piece being swelled out into a cylinder, perhaps an inch wide and two inches long, near the end where it is joined hermetically to the other piece. The top of the narrow tube is bent horizontally, and cemented into the centre of a vertical circle of plate-glass, of about three inches in diameter; or, instead of this plane, a hollow segment of a sphere of the same expansion, but including only 30° or 40°, is substituted. The top of the wide tube is likewise bent horizontal, and drawn to a point at the same height as the minute central orifice of the cavity, and bent in the opposite direction. A portion of nut-oil, tinged by the alkanet root, had been previously introduced into the cylindrical cistern. On turning the small plate or basin to front the wind, a condensation, corresponding to its force, is immediately produced on the opposing surface, a small portion of air enters the orifice, and continues to press upon the oil, till this rises to form an equiponderant column in the wide tube. As the air can with difficulty penetrate through the very narrow bore, the irregular action of the blast is, in a great measure, corrected, and the oil moves rather tardily.

A scale is adapted, bearing two sorts of divisions, the one indicating the impulse, and the other the velocity, of the wind. Reckoning the weight of the atmosphere equivalent to a column of oil of 400 inches in altitude, this space is subdivided into 10,000 equal portions, each degree thus corresponding to the twenty-fifth part of an inch. It would hence be easy to show, that the pressure of the wind upon every square foot of surface is expressed in pounds avoirdupois, by dividing by five the number of degrees through which the oil ascends.

But we may place an adjacent line of subdivisions, that shall mark the velocity of the wind in miles each hour. Since air will rush into a vacuum at the rate of 1350 feet in a second, it would, under a predominating pressure of the 100th part of an atmosphere, or at 100 degrees, flow with a celerity ten times less, or 135 feet in a second, which corresponds to 92 miles in an hour. Wherefore, 25 degrees of the scale of impulse would be marked by a velocity of 46 miles, and $6\frac{1}{4}$ degrees by that of 23 miles, in an hour. The subordinate divisions could hence be easily formed.

Such are the velocities which theory would assign to the

different altitudes of the columns supported by the force of the wind. But the actual resistance of fluids, owing chiefly to their detention at the obstructing surface, generally exceeds the result of calculation. In the case of water and air, the ratio of excess appears from experiment to be nearly that of eight to five. We may therefore modify the velocities after this proportion above stated. The relations of celerity and impulse will stand thus :—

| Celerity in Miles per Hour. | Impulse in 10,000th Parts of the Weight of the Atmosphere. |
|---|---|
| 10 | 1°.9 |
| 20 | 7 .6 |
| 30 | 17 .1 |
| 40 | 30 .4 |
| 50 | 47 .5 |
| 60 | 68 .4 |
| 70 | 93 |
| 80 | 121 .5 |

This anemometer, being furnished with a vane to make it always face the wind, might also, by an index, point out the direction. Nor is it absolutely requisite that the instrument should be exposed out of doors. The funnel, with its vane, may have a socket of bell-metal, nicely fitted to the top of a long perforated brass tube, which descends from the roof of the house, and terminates below in the recurved tube and its double scale. The impressions of the wind would thus be conveyed with great regularity and undiminished effect to the surface of the oil. Still, however, it would be impossible to avoid entirely the oscillations of the liquid column. Even the steadiest wind will be found to blow with a reciprocating force, now swelling and again relaxing.

and, at certain short intervals, concentrating all its vehemence.

X. Ombrometer,* or Rain-Gauge.—A very simple instrument, contrived to indicate the quantity, or rather the depth, of the rain which falls upon any spot. It is likewise named *Hyetometer,* and has been sometimes called by the barbarous compound *Pluviameter.* It is composed generally of a circular bason of tinned iron, soldered to the top of a vertical cylinder, which is contracted in some given proportion, and closed below. A small float is introduced, bearing a slender rod, distinguished by the corresponding divisions. In the most ordinary construction, the bason being a foot wide, the attached cylinder has $3\frac{4}{5}$ inches in diameter, and its section is consequently ten times smaller. The inches on the rod are hence marked only tenth parts.

This method of measuring the fall of rain is evidently not susceptible of much accuracy; and it would require the gauge to be very frequently visited, on account of the loss of the water by continual evaporation. The more correct ombrometers have their bason made of brass, and turned to a fine sharp edge; the rain, as it falls, runs through a small orifice into the vertical cylinder, which has only about the fourth part of the diameter, and communicates by means of a cock, with another perpendicular tube still narrower, and consisting of glass having a scale affixed. The divisions of this scale are determined from the proportion of the joint sections of the cylinder and tube to the horizontal surface of the bason. In making an observation, the cock, being turned, lets the collected water rise to the same level in the glass tube, and thus indicate its quantity, which, by another ope-

---

* From ὄμβρος, *rain,* and μετρον, *a measure.*

ration, is now drawn off. In the time of frost or snow, it becomes necessary to warm the instrument gently, and make the water flow.

We may suspect that the measure of the rain, hail, or snowy flakes received by the ombrometer, is not exactly proportioned to the extent of surface which it presents; for, while torrents pour down from the heavens, an eddy plays about the rim of the bason, deranging the regularity of the discharge. A bason of several feet in diameter would perhaps be preferable, or the platform of a roof could be adopted, if it were sufficiently sloped to allow the rain to collect quickly.

But the most perplexing circumstance affecting the ombrometer is, that it has been found to indicate very different quantities of rain as falling upon the very same spot, according to the different elevation at which it was placed. In general, less rain is collected in high than in low situations, even though the difference of altitude should be inconsiderable. Thus it was discovered, that, in the space of a year, while 12.1 inches only fell on the top of Westminster Abbey, 18.1 inches were collected on the roof of a house sixteen feet lower, and even 22.6 inches of rain at the ground. Similar observations have been made at the summit and near the base of hills of no great elevation. In such situations, we can hardly suppose the clouds to stretch down to the surface, or to augment the lower portion of rain. We must hence refer the copious fall near the ground to some other cause. Most of the rain which falls proceeds from drifting showers of short duration. The current moves more slowly along the surface, and allows the drops to descend as fast as they are formed. But being forced to mount a swelling eminence, and thus compressed into a narrower stream, it hurries the mass of vapour along with it, and does not suffer the free or full discharge on the summit. On both sides of the hill, an ombrometer placed near the bottom indicates always a greater fall of rain than on the exposed top.

The observations furnished by this instrument are hence liable to considerable inaccuracy, unless made in an open champaign country. Thus, a register kept at Keswick gives 67½ inches, which evidently exceeds greatly the annual fall of rain in that district; the quantity at Carlisle, not twenty-five miles distant, being only 20 inches. Again, the measures of rain, being 33 and 34½ inches in the open country about Manchester and Liverpool, are found to amount to 45 and 60 inches at Lancaster and Kendal, which approach the flanks of a mountainous range. In general, twice as much rain falls on the western as on the eastern side of our island; and the average annual quantity may be reckoned at 30 inches, or it would form, if all collected, a sheet of water of that depth. According to this estimate, the whole discharge from the clouds in the course of a year, on every square mile of the surface of Great Britain, would, at a medium, be 1,944,643, or nearly two millions of tons. This gives about three thousand tons of water for each English acre.

XI. Electrometer,* which detects the electrical state of the lower atmosphere. The best instrument of this kind undoubtedly is Bennet's, consisting of two slips of thick gold leaf, suspended from a knob within a small cylinder of glass, which is surmounted by a cap of brass. This may be connected with an insulated rod or wire, extending a few feet beyond the window.

The electrometer indicates the condition of the air only in its immediate vicinity. But when thunder-storms pre-

* From ἤλεκτρον, *amber*, and μέτρον, *a measure.*

vail, the atmosphere becomes affected to a very considerable extent. Yet the indications of the electrometer are often capricious and evanescent. Whenever clouds are suddenly formed, or melt away, whether the air changes to dryness or humidity, the electrical equilibrium is disturbed. The observations made with the electrometer are hence of much less importance than was once expected, and have been gradually falling into neglect.

XII. Drosometer.*—An instrument so called was proposed by Weidler, a German professor, in 1727, to measure the quantity of dew which gathers on the surface of a body which has been exposed to the open air during the night. It consisted of a bent balance, which marked in grains the preponderance which a piece of glass of certain dimensions, laid horizontally in one of the scales, had acquired from the settling and adhesion of the globules of moisture.

The main objection to a drosometer of such a construction is, that it would require to be protected from the action of the wind; and being thus screened, it could not receive the whole of the dew which might otherwise have been deposited. The steel beam, too, from continual exposure to the weather, would soon lose its polish, and become unfit for any accurate performance. Besides, it is in general easier to measure than to weigh a portion of liquid.

A simpler and more convenient drosometer could be formed on the principle of the ombrometer, or rain-guage. Suppose a glass funnel, of about three inches diameter, whose interior surface is very smooth, and slopes towards the centre at an angle of 15 or 20 degrees, to be joined hermetically to a long tube, sealed at the lower end, and hav-

---

* From δροσος, *dew*, and μετρον, *a measure.*

ing an equable bore not exceeding the quarter of an inch, with an attached scale, divided into portions corresponding to the thousandth parts of an inch on the external aperture. The only difficulty is to make the dew which gathers during the night to run down the sides of the funnel into the tube. To facilitate this descent, a coat of deliqueate salt of tartar may be spread with a hair pencil over the shallow surface, and renewed as often as occasion requires. The dew, instead of settling in minute detached globules, would then be attracted by the alkaline lye, which thus, becoming dilute, would gradually flow into the narrow cavity of the tube. It would be easy at any time to make an allowance for the very small portion of liquid alkali blended with the dew.

Such is the complete apparatus required for keeping a meteorological register. But those instruments are not all of equal importance. The barometer, the thermometer, and the hygrometer, may be considered as indispensable. Next to them deserve to be ranked the photometer and æthrioscope, which disclose the more recondite condition of the atmosphere. The atmometer, the ombrometer, and the anemometer, besides, are of consequence, from the practical results which they furnish.

The value of any meteorological register, however, depends on the accuracy with which it is kept. The observations should be made in a place rather elevated, and exposed freely on all sides to the aspect of the sky; and they should be repeated either at equal intervals, during day and night, or at least at those hours which represent most nearly the mean state of the atmosphere. These requisites are seldom attained, and very few registers of the weather, accordingly, are entitled to much confidence.

It cannot be expected that registers of the weather will possess much value, so long as they are kept merely as objects of curiosity. Like astronomical observations, as now conducted, they should no longer be left to the chance of individual pursuit. They would require to be unremittingly prosecuted in all variety of situations, and at the public expense. Proper sets of meteorological instruments should be placed, not only in the regular observatories, but sent to the different forts and light-houses, both at home and at our principal foreign stations. They might also be distributed among the ships employed in discovery, or engaged on distant voyages. The cost of providing those instruments would be comparatively trifling, and the charge incurred by keeping registers on a regular and digested plan might shrink to nothing in the scale of national expenditure.

The state of the barometer alone is kept with tolerable accuracy, because that instrument, being little influenced by collateral circumstances, marks nearly the same impressions over a wide extent of surface. The thermometer, again, is seldom observed at the proper hours, or in a situation sufficiently detached from the buildings and solid walls.

It is customary, for the sake of convenience, to note the thermometer in the morning, at the height of the day, and again in the evening. But these three observations must evidently furnish results below the medium temperature of the whole twenty-four hours, since the accumulated warmth is reckoned but once, while the freshness partaking of the night is counted twice. It would be nearer the truth to assume the middle point between the maximum and minimum, though even this cannot be deemed absolutely correct, because the heat neither mounts nor declines in an uniform progression. The hottest time of the day is generally about two o'clock, and the coldest just before sunrise. This hour of extreme descent is consequently very variable; and it would be difficult to fix the times suited for observing, unless they were more multiplied. But even fewer observations may sometimes be made to serve the purpose. In this climate, the daily average may be reckoned from eight o'clock in the morning, and the month of October is found to have nearly the mean temperature of the whole year.

The observations usually made with the hygroscopes of Deluc or Saussure cannot be regarded as affording any definite indication of the dryness of the atmosphere. It would essentially contribute to the advancement of meteorological science, if the hygrometer which we have described were introduced into general practice. This adoption cannot be very distant.

Some of the monks, in the religious houses scattered over the Continent, might find an agreeable and useful occupation in recording the state of the atmosphere. Many of these establishments are scattered in lofty and romantic situations; and several of them, destined by their founders for the charitable accommodation of travellers, occupy the summits of the most elevated and inaccessible mountains. Accurate registers kept in such towering spots are peculiarly interesting.

Light-houses would, from their usual position, be well fitted for observing the force and direction of the wind, and the swell and relapse of the tide. The navigators who traverse the ocean in every latitude, might, besides keeping meteorological soundings, record the variation of the needle, and examine the intensity of magnetic attraction.

To promote the science of meteorology, it would be most expedient that the various learned associations planted in different parts of the globe, should institute inquiries into the state and internal motions of the higher strata of our atmosphere. As the ultimate results could not fail to prove advantageous to the public, the several governments might be expected to defray the moderate expense incurred in carrying this plan into effect. Light small balloons might from time to time be launched towards the most elevated regions, to detect by their flight the existence and direction of currents which now escape observation. Barometers, thermometers, hygrometers, and perhaps æthrioscopes, in compact forms, and which should register themselves, might be sent up in the car. Observers, furnished with accurate and complete instruments, could likewise be despatched occasionally to the intermediate heights in large balloons. By classing the various meteorological journals, and combining those ulterior facts, some new lights could not fail to be struck out, which would gradually reveal that simple harmony which no doubt pervades all the apparent complication of the universal frame. Till we obtain such insight, we must content ourselves with the best explications of the phenomena of the atmosphere which our imperfect and limited knowledge will admit. We shall, therefore, treat in succession of the origin of winds; the generation of clouds and fogs; and their precipitation in the form of rain, snow, or hail. Other collateral objects will be discussed as they present themselves to view.

I. Wind.—It is a curious circumstance, that in all languages the ordinary name of *air* refers to its mobility, and merely signifies *to blow*. This impulse alone of the fluid appears to have awakened our sensations; and had the atmosphere continued perfectly still, we might for ages have remained ignorant of the very existence of the fluid which we breathe.

The main cause of *wind*, or the flow of air, is undoubtedly the variable distribution of heat through the atmosphere, which incessantly affects the local density, and disturbs the equilibrium of the mass. The presence of the sun affects the surface of the terraqueous globe, which again warms and dilates the lower strata of atmosphere. The calorific action of the solar beams is greatly diminished by their obliquity; it rapidly accumulates on the land, but becomes attenuated and diffused in the waters of the ocean.

The alternation of day and night, and the annual revolution of the seasons, are hence the perpetual sources of winds. If the surface of the globe, however, had been wholly covered by the ocean, and not disparted by land and seas, those winds must have been scarcely perceptible. The daily illumination of the sun does not warm the ground to the depth of an inch; but the same quantity of illumination penetrates, though with a decreasing intensity, many fathoms into water, spreading and dividing its influence. We may reckon the tenth part of the incident light to be intercepted by a superficial stratum of the thickness of one foot; and it will hence follow, that the solar beams communicate every day a hundred times less heat to the surface of a body of water than to an expansion of level ground. The subsequent influence, again, of those contrasted surfaces in warming the incumbent air must be proportionally different, though slightly modified by the portion of light reflected from the water. In a general view, the diurnal variation of temperature in the atmosphere may be considered as limited to the lowest stratum, not exceeding 2000 feet in height. Such a body of air will intercept commonly the fiftieth part of all the light which traverses it. We may hence conclude, that the change of temperature in the air, caused by the succession of day and night, is, on the whole, about thirty times less above a spacious lake than over the surrounding land.

The current which rushes from all sides towards any heated portion of the atmosphere, is easily explained and computed from the diminution of pressure which rarefaction produces at that place. The celerity of the flow is precisely the same as that of the efflux from a small aperture under the pressure of a column equivalent to that diminution, or to the difference between the weight of the warm air, and of an equal volume of the exterior fluid. Thus, suppose a chimney 20 feet high were heated up 50 centesimal degrees, or 90 on Fahrenheit's scale; the air in the flue being therefore expanded one fifth part, or four feet, would be driven upwards by the pressure of a column compensating this difference. The velocity of the discharge would hence be $8\sqrt{4} = 16$ feet every second, or at the rate of about eleven miles in the hour. If a fire be kindled in an open field, it is evident that the rush of air must proceed from all sides. At the spot itself, therefore, the opposite currents will produce a counterbalance, and no dominant wind can prevail. But if the warm air should cover a very wide extent of surface, its influence may be felt at a great distance, and the several converging winds may have space to blow without any mutual interference.

These views afford a complete explication of the phenomena of sea and land breezes, which are occasionally met with in every latitude, but are constantly observed near the shores of the continent, and of the larger islands within the tropics. In those sultry regions, as the day advances, a refreshing wind blows from the sea, and is succeeded by an opposite current from the interior of the land on the approach of evening, and during a great part of the night. In open seas, and especially near the equator, the thermometer scarcely varies a degree, and very seldom two degrees, by Fahrenheit's scale, in the whole course of a day. But on the land, the change of temperature between the night and day, in similar situations, will rise often higher than 76 centesimal degrees, or 126 of Fahrenheit. If we, therefore, conceive a stratum of air 2000 feet in altitude, heated only to the mean difference, or three centesimal degrees, it would receive an expansion of 24 feet; whence the velocity of the wind produced would be $8\sqrt{24} = 39$ feet every second, or at the rate of 26 miles in the hour. This is a very moderate estimate; but the celerity of the current must, no doubt, be diminished, from the retardation which it suffers in proportion to the length of track over which it has to sweep.

During the night, the lower atmosphere is colder on land than at sea, owing partly to the descent of the more elevated and colder portions of air which chill the surface of the ground, and partly to those cold pulses which are incessantly darted from every point of the azure sky. If we reckon the reduced temperature of the land only a centesimal degree and a half below the standard of the adjacent ocean, this would give 12 feet for the contraction of the vertical column of air, and, consequently, a stream would flow towards the sea with a celerity of $8\sqrt{12} = 28$ feet per second, or very nearly 20 miles every hour. In general, the land breeze may be considered not so powerful as what blows from the sea.

The ordinary appearances are clearly and graphically described by that very intelligent and enterprising navigator, Captain Dampier.

"These sea breezes do commonly rise in the morning about nine o'clock, sometimes sooner, sometimes later. They first approach the shore so gently, as if they were afraid to come near it; and ofttimes they make some faint breathings, and, as if not willing to offend, they make a halt, and seem ready to retire. I have waited many a time, both ashore to receive the pleasure, and at sea to take the benefit of it.

"It comes in a fine, small, black curl upon the water, when as all the sea between it and the shore, not yet reached by it, is as smooth and even as glass in comparison. In half an hour's time after it has reached the shore, it fans pretty briskly, and so increaseth gradually till twelve o'clock; then it is commonly strongest, and lasts so till two or three a very brisk gale. About twelve at noon, it also veers off to sea two or three points, or more, in very fair weather. After three o'clock it begins to die away again, and gradually withdraws its force till all is spent, and about five o'clock, sooner or later, according as the weather is, it is lulled asleep, and comes no more till the next morning.

"These winds are as constantly expected as the day, in their proper latitudes, and seldom fail but in the wet season. On all coasts of the main, whether in the East or West Indies, or Guinea, they rise in the morning, and withdraw towards the evening; yet capes and headlands have the greatest benefit of them, where they are highest, rise earlier, and blow later.

"Land breezes are as remarkable as any winds that I have yet treated of: they are quite contrary to the sea breezes; for those blow right from the shore, but the sea breeze right in upon the shore; and as the sea breezes do blow in the day, and rest in the night, so, on the contrary, these do blow in the night, and rest in the day, and so they do alternately succeed each other. For when the sea breezes have performed their offices of the day, by breathing on their respective coasts, they in the evening do either withdraw from the coast, or lie down to rest. Then the land winds, whose office is to breathe in the night, moved by the same order of divine impulse, do rouse out of their private recesses, and gently fan the air till the next morning; and then their task ends, and they leave the stage.

"There can be no proper time set when they do begin in the evening, or when they retire in the morning, for they do not keep to an hour; but they commonly spring up between six and twelve in the evening, and last till six, eight, or ten in the morning. They both come and go away again earlier or later, according to the weather, the season of the year, or some accidental cause from the land; for on some coasts they do rise earlier, blow fresher, and remain later, than on other coasts, as I shall show hereafter.

"These winds blow off to sea, a greater or less distance, according as the coast lies more or less exposed to the sea winds; for in some places we find them brisk three or four leagues off shore, in other places not so many miles, and in some places they scarce peep without the rocks, or, if they do sometimes in very fair weather make a sally out a mile or two, they are not lasting, but suddenly vanish away, though yet there are every night as fresh land winds ashore at those places as in any other part of the world.

"Indeed, these winds are an extraordinary blessing to those that use the sea in any part of the world within the tropics; for as the constant trade-winds do blow, there could be no sailing in these seas; but, by the help of the sea and land breezes, ships will sail 200 or 300 leagues, as particularly from Jamaica to the Lagune of Trist, in the Bay of Campeachy, and then back again, all against the trade wind.

"The seamen that sail in sloops or other small vessels in the West Indies, do know very well when they shall meet a brisk land wind, by the fogs that hang over the land before night; for it is a certain sign of a good land wind to see a thick fog lie still and quiet, like smoke over the land, not stirring any way; and we look out for such signs when we are plying to windward. For if we see no fog over the land, the land wind will be but faint and short that night. These signs are to be observed chiefly in fair weather; for in the wet season fogs do hang over the land all the day, and it may be neither land wind nor sea breeze stirring. If in the afternoon, also, in fair weather, we see a tornado over the land, it commonly sends us forth a fresh land wind.

"These land winds are very cold, and though the sea breezes are always much stronger, yet these are colder by far. The sea breezes, indeed, are very comfortable and refreshing; for the hottest time in all the day is about nine, ten, or eleven o'clock in the morning, in the interval between both breezes; for then it is commonly calm, and then people pant for breath, especially if it is late before the sea breeze comes, but afterwards the breeze allays the heat. However, in the evening again, after the sea breeze is spent, it is very hot till the land wind springs up, which is sometimes not till twelve o'clock or after."—(*Voyages*, vol. ii.)

The TRADE WINDS, which, within the tropics, at all times constantly blow from the east, but somewhat vary their force, and decline a little to the north or the south, according to the latitude and the season, are the most remarkable of all the aerial currents, and of signal importance in navigation. These steady breezes favoured the voyage of Columbus, and conducted him to the discovery of the Mexican archipelago. The same powerful stream afterwards drew the Portuguese from their southern course, and carried them to the shores of the Brazils. Since the character and extent of those winds have become perfectly known, the navigator reckons safely on their aid, and shapes his voyage in such a way as to reduce its performance almost to a calculation.

The cause of the trade winds, however, is not obvious, or very easily traced. Various attempts have been made to explain the phenomenon, yet seldom on any solid or accurate principles. It would form an interesting discussion to examine the different hypotheses advanced; but we can afford room to notice, very briefly, the more considerable only of those opinions.

Descartes and his followers imputed the trade winds to the inertia of the atmosphere, which they conceived to prevent this fluid from acquiring the full rotation of the earth, especially near the equator. The air being thus left behind as the globe rolled from the west, would have an apparent motion in the contrary direction, and seem to blow from the east. But it may be urged, that as passengers almost insensibly gain the celerity of the ship which carries them, so every portion of the incumbent atmosphere, though more loosely adherent to the terraqueous surface, must soon acquire the peculiar motion corresponding to the parallel of latitude. Nor would the inequality of such combined movements in the air at all disturb the order and arrangement of its general mass.

Dr Halley gave a different explication of the origin of trade winds, which seems very plausible, and has long been deemed quite satisfactory. This able philosopher and experienced navigator supposed, that the spot where the sun's vertical rays exert their utmost heating energy, being in the lapse of a day successively transferred from east to west round the circumference of the globe, must, as a centre of confluence, draw in its train a current of air. The current thus formed would result from the excess of the streaming from the east above that from the west; and it would therefore advance with a tardy pace, following at a distance the powerful energy of the sun. The same easterly wind might incline towards the north or the south, according as the great luminary appears to approach to the northern or the southern tropic.

But it should be observed, that the torrid zone stretches mostly over the ocean, and includes only a narrow portion of land. The heat excited in succession through that liquid track, by the diurnal passage of the sun, is hence extremely small, and hardly sufficient to produce the aspiration of the gentlest air. Nor could even this feeble current have a decided and constant direction. It would only tend towards the heated part of the surface of the ocean. In the morning, it would breathe from the west; about noon, it would become neutral, and die away; and, in the evening, it would again spring up, and flow from the east. Near midnight, this current would sink into a perfect calm. The hypothesis will, therefore, not bear any strict examination. It is neither adequate to the production of such effects, nor accordant with the actual phenomena of the trade winds. It casts a false glare over the subject, without elucidating its real bearings.

The first who succeeded in taking a correct view of the question was George Hadley, in a short paper inserted in the Philosophical Transactions for 1735. By combining in some measure the idea of Descartes with the opinion of Halley, he produced a clear and simple account of the cause of trade winds, which appears entirely consistent, and free from every objection. Though the daily variation of temperature be very inconsiderable within the tropics, yet the annual accumulation of heat renders the equatorial regions much warmer than the higher latitudes, and consequently maintains a perpetual current of air from either side. If those aerial motions were not modified by the figure and rotation of the globe, there would always be two opposite winds blowing directly from the north and from the south to the equator. But the stream which perhaps originates at the northern tropic, in advancing to the equator, must seem gradually to deflect towards the west, in consequence of the increasing velocity with which the successive parallels of latitude are carried eastwards. During the time this current takes to perform its journey, it is apparently transported to the west, through a space equal to the excess of the arc described by the equator above the corresponding arc traced by the tropic. The current from the southern tropic is equally bent towards the west. When both of them meet at the equator, their opposite impulsions from the north and the south are extinguished, and they flow directly west in a single united stream, and with accumulated force. The apparent motions of the different streamlets which from both hemispheres conspire to constitute the trade wind, is represented in Plate VII. fig. 9.

But it is not enough to connect the general facts; a complete theory should harmonize in all the subordinate details. An easy calculation, accordingly, is conducted to those precise results which are commensurate and exactly congruous with the actual phenomena. The trade wind may be reckoned to begin about the latitude of 25 degrees. At this parallel, the mean temperature is four centesimal degrees colder than immediately under the equator, which difference of heat may graduate through the atmosphere to the altitude of 10,000 feet. Wherefore, the expansion of the air at the equator, which draws to it a meridional wind, will amount to a column of 100 feet. The velocity of the current hence produced must be $8 \sqrt{100}$, or 80 feet every second, which corresponds to fifty-four miles in the hour. But each point on the parallel of 24° is carried eastwards by its revolution about the earth's axis seven miles faster every hour than on the parallel of 25°. Consequently, when the wind arrives at the parallel of 24°, it will seem to have acquired a tendency of seven miles an hour to the west. As it reaches the successive parallels of 23°, 22°, 21°, &c. it will gain continual, though decreasing, additions to its apparent westerly course, which at the equator will have augmented to 104 miles in the hour.

In this calculation, we have made no deduction for the resistance which the streams of air must experience in sweeping over the surface of the globe, because no experiments have been made to ascertain the effect of such retardation. It is no doubt less on the ocean than on the land, and must evidently be diminished in proportion to the depth of the mass of fluid which is borne along. Still, however, this obstruction, joined to this impediment of internal motion, must be very considerable ; and we may safely reduce the numbers before stated to one-third, which would give eighteen miles an hour for the celerity of the primary meridional wind, and thirty-five miles for that of the oriental or trade wind, resulting from the influence of the figure and rotation of the earth.

Our northern hemisphere, presenting to the action of the solar beams a larger surface of land than the southern, is, on the whole, rather warmer. Hence the parallel of greatest heat runs not exactly through the equator, but about three degrees farther north. This circle is therefore strictly the mean path of the aggregate easterly streams of air.

But though the hottest part within the torrid zone, taking the average of a whole year, occupies the parallel of three degrees north latitude, it must, to a certain extent, shift its position with the seasons. In the summer months the sun shines twice vertical upon the tropic of Cancer, and consequently raises the temperature of the northern half of the zone. During winter, again, this effect is transferred to the southern half of the torrid region. In the progress of summer, therefore, the trade wind gradually bends about a point towards the north ; but as winter advances, it declines as much to the south.

Such is the character of that general wind which encircles the globe, flowing with slight deviation constantly from the east, and spreading over a zone of more than 50° in breadth. It sweeps the Atlantic Ocean from the coast of Africa to Brazil, and the Pacific from Panama to the

Philippine Isles and New Holland, and again the Indian Seas partially from Sumatra to Zanguebar.

The trade-wind undergoes an essential modification, however, where the continent stretches into the torrid zone. The sun acting more powerfully upon the land than upon the surface of the sea, the accumulated warmth is much greater. and shifts with the revolution of the seasons on either side of the equator. The centre of heat approaches in summer to the northern, and in winter to the southern tropic. Instead of the great eastern stream, those regions have two opposite periodic winds alternating towards the north and the south, and called the *monsoons*. When these winds advance to the equator, they conjoin an apparent easterly velocity; but when they recede from the equator, they carry their excess of velocity from the west. A diagonal motion results from the combined tendencies. In the Arabian and Indian Seas, on the north side of the equator, the monsoon blows north-west during the summer months, from April to October; and in the opposite direction, or south-east, during the winter. But on the south side of the equator, near Java and Sumatra, the course of the monsoon is north-east in summer and south-west in winter.

The primary winds, which blow from the parallels of 25° or 30° to the equator, must evidently give rise to opposite currents that flow in the higher atmosphere towards the poles. These streams, after they have travelled beyond the tropics, may descend to the surface, transporting the celerity of equatorial rotation. They will appear, therefore, to blow from the western quarter, with the excess of their previous velocity above that of the parallel which they reach. Hence a westerly breeze, of considerable force and regularity, prevails in either hemisphere above the latitude of 30°. The same winds cross the Atlantic from Newfoundland to Cornwall, and traverse the Southern Ocean from the Plata to the Cape of Good Hope, and thence to New Holland. Any wind which blows from the quarter inclining to the south of the west comes really from the equatorial region, and is therefore relatively warm. Such is the disposition of our westerly winds, which commonly prevail for nine months in the year.

On the same principle, a wind which blows directly from the arctic pole, and impregnated with intense cold, must, in consequence of the rotation of the globe, appear to arrive from some point to the north of the east. In passing through the first degree of latitude, it will suffer a deflection of eighteen miles in the hour towards the west; in a short space, therefore, it will seem to flow with impetuous force, and almost directly from the east. Hence our easterly and north-easterly winds have a polar origin, and are always bitterly cold.

Local winds could be explained, if the different circumstances which affect them were distinctly known. The latitude and temperature of the place, its relative position, the figure and contour of the surrounding country, would all enter into the calculation. We shall content ourselves with a concise notice of some peculiar winds. The *Bize* is a cold piercing wind which blows from the ridge of the Jura, and the frozen summits of the Pyrenees. The *Sirocco* is a hot, moist, and relaxing wind, which visits Naples and the south of Italy from the opposite shores of the Mediterranean. The *Harmattan* seems to be a cold and dry wind, of a very parching quality, which is frequent in Africa and some of the eastern countries. The *Samiel,* or *Simoom,* is a burning pestilential blast, extremely arid, which springs up at times in the vast deserts of Arabia, and rushes with tremendous fury, involving whole pillars of sand.

II. Clouds and Rain.—Their formation and dissolution produce all the varied train of the meteorological phenomena. The humidity suspended in the atmosphere is derived by exhalation partly from the land, but ultimately from the vast expanse of the ocean. A surface of lake, of pasture, corn fields or forest, supports a continual evaporation, augmented only by the dryness of the air, and the rapidity of its successive contacts. Even ploughed land will supply nearly as much moisture to the exhaling fluid as an equal sheet of water. It is only when the ground has become quite parched that it obstinately retains its latent store.

If the whole of the waters which fall from the heavens were to return again, the evaporation from the ground might be sufficient alone to maintain the perpetual circulation. But more than one third of all the rains and melted snows are carried by the rivers into the ocean, which must hence restore this continued waste. The commerce of land and sea is thus a necessary part of the economy of nature.

The air, in exhaling its watery store, is rendered quite damp; but it may afterwards become dry on being transported to a warmer situation. Such is the case of the sea-breeze, particularly in summer. It arrives on the shore cold and moist; but as it advances into the interior of the continent it grows milder and drier. The same principle accounts for the disposition of different winds in respect to humidity. At Colombo, in the island of Ceylon, as we gather from some remarks of Dr Davy, the north-east monsoon, with a temperature of only 68 on Fahrenheit's scale, has yet a dryness of 75 hygrometric degrees; but the opposite monsoon, from the south-west, though at 82 by the thermometer, is so damp as to indicate scarcely 30 degrees. The cold wind, coming from the north, was rendered warmer and drier in its progress; while the hot wind, flowing from the equator, was somewhat chilled and made damper as it approached Ceylon.

Since air in mounting upwards has its capacity for heat enlarged, and becomes colder, it will hence likewise grow proportionally damper. But a continual intercourse being maintained between the lower and the higher atmosphere, the middle region must, from its chillness, be soon charged with moisture. If this tendency were to act, therefore, without control, the heavens would have been shrouded with perpetual clouds and darkness, and never could the cheering rays of the sun have visited the surface of the earth. A principle of conservation happily occurs to restrain, and finally to overpower, the effect of cold, in disposing air to part with its moisture. By expansion, this fluid is made capable of holding, at the same temperature, a larger share of humidity. Each portion of air, in rising vertically, grows, from the predominance of cold, constantly damper; but after having reached a certain altitude, it again becomes gradually drier, from the influence of its wide dilatation. Every time the air has its volume doubled, it acquires an additional dryness corresponding to fifty hygrometric degrees. Hence one degree would be the effect of the rarefaction of only the 72d part. This small variation again answers to a depressed temperature of $1\frac{4}{9}$ on the centesimal scale, which, near the surface, will occasion an increase of humidity equal to the actual range of the solvent power of the air divided by 31.4. Suppose the thermometer to mark 15° centesimal at the ground, the air would, for each ascent of about 390 feet, be $\frac{200}{31.4}$, or 8 hygrometric degrees damper, which would be reduced to 7 by the influence of dilatation. Had the temperature at the surface been as low as — 25°, which answers to a solvent power of $31\frac{1}{2}$°, the opposite agencies of cold and rarefaction would evidently have produced a perfect balance, and the same dryness would have continued to a moderate height.

It would be easy to show that $d$, expressing the density of the air at any altitude, and $h$ the corresponding indication of the hygrometer, $\frac{h}{62.8}\left(\frac{1}{d}+d\right)$ will denote the increment of humidity occasioned by depressed temperature, while the corresponding decrement resulting from expansion is one degree. Hence at the pole the position of the *maximum* humidity in the atmosphere must occur at an elevation of 13,300 feet; where the density is .6, the temperature would stand at 26.7°, and the hygrometric range only 29°. Under the equator, that limit would attain a much greater altitude, and yet not rise so far above the curve of perpetual congelation. It is probable, however, that the canopy of clouds descends considerably lower, being warmed by the hot pulses darted from the ground and the inferior strata of the atmosphere. We shall not err much if we estimate the position of extreme humidity at the height of two miles at the pole, and four miles and a half under the equator, or a mile and a half beyond the limit of congelation. This range is represented in Plate VII. fig. 9. running nearly parallel to the curve of perpetual congelation, but bending nearer in approaching the equatorial parts. It marks the mean height of the clouds in different latitudes, and intimates the shading into the fine etherial expanse.

The moisture deposited by a body of air in minute globules, which remain suspended, or subside slowly in the atmosphere, constitutes a *Cloud*. When it comes near us, whether it hovers on the tops of the hills, or spreads over the valleys, it receives the name of a *Fog*. The cold occasioned by the ascent or transfer of air may be sufficient to form thin clouds, but a more powerful and extended energy is required for the production of *Rain*. The subject has from the earliest times engaged the attention of philosophers, who have made numerous unavailing attempts to explain it. At length the very ingenious Dr James Hutton subjected the problem to a correct analysis, and succeeded in deducing a most satisfactory solution. His fine theory of Rain, which first appeared in the Transactions of the Royal Society of Edinburgh for the year 1787, constitutes an epoch in meteorological science. Its merits, however, have been slowly perceived by the public, because the author, full of his original conception, satisfied himself with merely sketching the general outline. But it was not enough that the operation of the principle advanced should always cause rain; it was farther requisite that the results arising from its application should quite accord with the actual phenomena. We shall, therefore, endeavour to render this theory more definite and more complete.

Air in cooling becomes ready, we have seen, to part with its moisture. But how is it cooled in the free atmosphere, unless by the contact or commixture of a colder portion of the same fluid? Now, the portion of the air which is chilled must in an equal degree warm the other. If, in consequence of this mutual change of condition, the former be disposed to resign its moisture, the latter is more inclined to retain it; and, consequently, if such opposite effects were balanced, there could, on the whole, be no precipitation of humidity whatever. The separation of moisture, on the mixing of two masses of damp air at different temperatures, would therefore prove, that the dissolving power of air suffers more diminution from losing part of the combined heat, than it acquires augmentation from gaining an equal measure of it; and, consequently, this power must, under equal accessions of heat, increase more slowly at first than it does afterwards, thus advancing always with accumulated celerity.

The quantity of moisture which air can hold, thus increases in a much faster ratio than its temperature. This great principle in the economy of nature was traced by Dr Hutton from indirect experience. It is the simplest of the a celerating kind, and perfectly agrees with the law of solution which the hygrometer has established. Suppose equal bulks of air in a state of saturation, and at the different temperatures of 15 and 45 centesimal degrees, were intermixed, the compound arising from such union will evidently have the mean temperature of 30°. But since at these temperatures the one portion held 200 parts of humidity, and the other 800, the aggregate must contain 1000 parts, or either half of it 500; at the mean or resulting temperature, however, this portion could only suspend 400 parts of humidity, and, consequently, the difference, or 100 parts, amounting to the two hundredth part of the whole weight of air, must be precipitated from the compound mass.

As another illustration, let air of 15° be mixed with air at the temperature of 35°, in three different proportions, all at the point of saturation; one part being combined with three parts, two with two, and three with one. The temperatures arising from the commixture would be 20°, 25°, and 30°; the corresponding parts of moisture precipitated from the mass being derived from the intermediate proportions of 200 and 504, are, 352, 317.5 or 34.5; 276, 252 or 24; 352, 317.5 or 34.5; and 428, 400 or 28. These depositions are represented in Plate VI. fig. 8. by the several intervals between the logarithmic curve and the oblique line which connects the summits of the ordinates of 15° and 35.

In these examples we have assumed the portions of differently heated air to be quite charged with moisture before mixing; but it is only required that they should approach to the point of humidity. The effect, however, of simple commixture would, in most cases, be very small. To explain the actual phenomena, we must have recourse to the mutual operation of a chill and of a warm current, driving swiftly in opposite directions, and continually mixing and changing their conterminous surfaces. By this rapidity, their volume of the fluid is brought into contact in a given time. Suppose, for instance, the one current to have a temperature of 50, and the other that of 70 degrees, by Fahrenheit's scale; the blendiug surfaces will, therefore, assume the mean temperature of 60°. Consequently, the two streams throw together 200 and 334.2 parts of moisture, making 567.1 parts for the compound, which, at its actual temperature, can hold only 258.6 parts; the difference, or 8.6 parts, forms the measure of precipitation, corresponding to the 2325th of the whole weight of the commixed air. It would thus require a column of air 30 miles in length to furnish, over a given spot, and in the space of an hour, a deposite of moisture equal to the height of an inch. If the sum of the opposite velocities amounted to 60 miles an hour, and the intermingling influence extended but to a quarter of an inch at the grazing surfaces, there would still, on this supposition, be produced in the same time a fall of rain reaching to half an inch in altitude.

These quantities come within the limits of probability, and agree sufficiently with experience and observation. But, in the higher temperatures, though the difference of the heat between the opposite strata of air should remain the same, the measure of aqueous precipitation is greatly increased. Thus, while the mixing of equal masses of air, at the temperatures of 40° and 60°, is only 6.6, that from a like mixture at 80° and 100° amounts to 19°. This result is entirely conformable to observation, for showers are most copious during hot weather and in the tropical climates.

The quantity of moisture precipitated from the atmosphere thus depends on a variety of circumstances; on the previous dampness of the commixed portions of the fluid, their difference of heat, the elevation of their mean temperature, and the extent of the combination which takes place. When this deposition is slow, the very minute aqueous globules remain suspended, and form clouds; but if it be rapid and copious, those particles conglomerate, and produce—according to the state of the medium with regard to heat—rain, hail, or snow.

The profuse precipitation of humidity is caused by a rapid commixture of opposite strata. The action of swift contending currents in the atmosphere brings quickly into mutual contact vast fields of air over a given spot. The separation of moisture is hence proportionally copious. In temperate weather this deposition forms rain; but, in the cold season, the aqueous globules, freezing in the mid-air into icy spiculæ, which collect in their slow descent, become converted into flakes of snow. Hail is formed under different circumstances, and generally in sudden alternations of the fine season, the drops of rain being congealed during their fall, by passing through a lower stratum of dry and cold air.

The drops of rain vary in their size perhaps from the twenty-fifth to a quarter of an inch in diameter. In parting from the clouds, they precipitate their descent, till the increasing resistance opposed by the air becomes equal to their weight, when they continue to fall with an uniform velocity. This acquired or *terminal* velocity is therefore in the sub-duplicate ratio of the diameters of the aqueous globules. A thunder-shower hence pours down much faster than a drizzling rain. In general, if $d$ express the diameter of a drop in parts of an inch, the terminal velocity, according to theory, will be denoted by $78\sqrt{d}$, or, if the usual correction be made for the discrepancy in fluids, it will be $67\sqrt{d}$. Thus a drop of the twenty-fifth part of an inch, in falling through the air, would only gain a celerity of 11½ feet, while one of a quarter of an inch would acquire a celerity of 33½ feet. A flake of snow, being perhaps nine times more expanded than water, would descend thrice as slow. But hailstones are often of considerable dimensions, exceeding sometimes an inch in length. They may hence fall with a velocity of 70 feet each second, or at the rate of about 50 miles in the hour. Striking the ground with such impetuous force, it is easy to conceive the extensive injury which a hail-shower may occasion in the hotter climates. The destructive power of those missiles in stripping and tearing the fruits and foliage increases, besides, in a faster ratio than the momentum, and may be estimated by the square of their velocity multiplied into mass. This fatal energy is hence as the fourth power of the diameter of the hailstone.

II. Optical Phenomena.—It remains for us to explain the general optical appearances of the sky. When the rays of the sun strike upon a cloud, they are copiously reflected, but partly absorbed by the minute suspended globules. In working their progress through the mass of vapour, they suffer a great diminution from the multiplied acts of absorption. The quantity of light thus finally detained depends on the density of the cloud, and its thickness. But the portion which penetrates through the nebulous medium is always much less than what traverses an equal body of air. In extreme cases, perhaps, the solar beams will suffer greater defalcation by repeated repercussions within a congregated cloud, than from passing through fifty times the same extent of a clear aerial expanse. Hence such clouds always appear dark and black, by their scanty transmitted light. Whiteness, being produced by the copious emission of intermingled rays, can belong only to very thin clouds. The depth of shade indicates the mass of floating vapour.

Owing to the excessive minuteness of the aqueous globules, the particles of light are only reflected or absorbed at their external surface, without entering them. But when they collect into large drops, the luminous pencil which strikes at a certain angle converges by refraction to a point of the posterior surface, and, after suffering one or more interior reflections, it emerges dissected into its primitive colours. Hence the glorious vision of the rainbow, which was reduced to mathematical calculation by Descartes, but only received its complete explication from the optical discoveries of Newton. The phenomena occur whenever the sun shines upon the falling drops of rain behind the spectator, the coloured arch being a portion of a circle whose centre is a point in the sky directly opposite to the sun. The primary or interior bow is formed by a single reflection, and lies 45° beyond that centre; but the secondary or superior bow, produced by a double reflection, appears with inverted tints at the distance of 56°. A ternary bow may exist, but being so extremely faint from the repeated reflections, it is scarcely ever perceived. It hence follows, that rainbows are only visible when the altitude of the sun is below 45° and 56°. In summer, accordingly, they are not seen in this climate about the middle of the day. For the same reason, they generally appear less than a semicircle; but viewed from the top of a spire, or any lofty pinnacle, they embrace nearly the whole circumference. Lunar rainbows may be frequently observed, only the faintness of their colours makes them far less conspicuous.

The coloured rings or *halos* which are often seen surrounding the moon and sun, are evidently occasioned by very thin vapour diffused through the atmosphere. They are supposed chiefly to encircle the moon; but, in this climate, hardly a day passes with light fleckered clouds, when at least portions of halos may not be perceived near the sun. It is only necessary to remove the glare of light which makes the delicate colours appear white. Thus, if we examine the reflection from a smooth surface of water, we shall perceive that the sun gilds the fleecy clouds with segments of beautifully coloured rings. This effect is still more distinctly seen if the rays from a hazy or mottled sky be received upon a sheet of white paper, held before a small hole in the window-shutter of a dark room. But even when the sun shines from an azure firmament, circles of the richest tints may be produced by experiment. Holding a hot poker below, and a little before the small hole of the shutter, throw a few drops of water upon it, and the sun will be painted upon the paper like the glowing radiations of the *passion-flower*. The appearance is exactly similar to what the traveller, in awakening from a short slumber, perceives, in a winter's morning, on opening his wearied eyes to a burning candle—concentric rings of violet, green, yellow, and red.

The explication formerly given of the cause of halos, even that proposed by Newton himself, is inadmissible; since it would confine them, like the rainbows, to certain definite limits, whereas they appear with every possible degree of extension. Our earliest inquiries led us to refer the origin of halos to the *deflection* of light, or that property of the rays to bend and divide as they pass near the edge of a body. Thus the light admitted through a very narrow slit in a card, or a bit of tinfoil, spreads into bright coloured fringes. The finer also is the slit, the broader are the fringes. A similar appearance is obtained by looking at the elongated flame of a candle through the delicate fibres of a feather, or even through the streaks of grease rubbed by the finger along a piece of glass. But if a very small round hole be substituted for the slit, the fringes will change into coloured rings.

Thus, if a piece of tinfoil, punctured with the point of a needle, be held close to the eye, the sun will appear through it surrounded by a halo very near his disc, but spreading more in proportion as the hole is contracted. That ingenious artist Mr Troughton constructed for us a slide of brass, and afterwards another of platina, perforated with a series of the finest conical holes, which were measured by his delicate micrometer. The purpose was to compare the angle subtended by the coloured ring with the diameter of the perforation, it being inferred that an aqueous globule of the same dimension might, by the exterior deflection of the solar rays, produce a similar halo. But our variable sky is very seldom fit for any refined optical experiment, and many delays happening to intervene, we could arrive at no very precise or certain result. We may state, however, as at least an approximation, that the globules of the diffuse vapour which occasions the appearance of coloured circles about the sun and moon, vary from the 5000th to the 50,000th part of an inch in diameter. When the halo approaches nearest to the luminous body, the largest globules are floating, and therefore the atmosphere is surcharged with humidity. Hence the justness of the vulgar remark, that a dense halo close to the moon portends rain.

Nearly the same theory has been struck out by Dr Thomas Young, to whose profound ingenuity and most extensive information we are glad to bear honourable testimony. By a skilful application of the principle of deflected light, he has likewise constructed the *Eriometer,* a curious instrument for measuring the size of the fibres of wool and other filamentous substances. But we cannot at present enter into the details.

The same acute philosopher has given perhaps the only true account of the origin of the *parhelia* or *mock suns,* which are frequently seen in the arctic regions during certain dispositions of the atmosphere. This gorgeous appearance of intersecting luminous arches, studded with opposite and transverse images of the sun, he ascribes to the combined reflections of the rays from the natural facets of the snowy *spiculæ* floating abundantly in the air.

Another most remarkable optical deception occurs in a peculiar state of the atmosphere on the verge of the horizon in various countries, and especially in the warmer climates, whether on the level plains, or on the margin of rivers or lakes, and near the sea-shore. In such situations the remote objects often appear with extraordinary elevation, and in double or inverted images. This singular phenomenon is obviously caused by the irregular refractions which the rays of light occasionally suffer by passing through the different strata of the lower atmosphere. When the effect is confined to the apparent elevation of an object, our seamen call it *looming;* but if inverted images be formed, the French and Italians give to this play of vision the appellations of *mirage* and *fata morgana*. The shipping and range of buildings on the shore of Naples have from Messina sometimes appeared floating inverted in the air. In the autumn of 1798 the coast of France was distinctly seen raised above the sea, from the beach of Hastings; the appearance lasted about an hour, and then sunk beneath the horizon. In the following year, Professor Vince watched the phenomena at Ramsgate with a large telescope. On the afternoon of the 1st of August he first descried the sails of a ship, but as she came fuller in view, he perceived another inverted image just below the surface of the water. Plate VII. fig. 7. represents the successive appearances; A is the ship when first observed, C after she had approached, and B the inverted image. Fig. 8. exhibits a variation of the effect; *a* marks the ship entering the horizon, C and B the double image of the ship when near, a portion of the sea lying between its opposite traces. To produce the appearances now represented, it was quite requisite that the rays of light, in traversing the lower strata of the air, should describe curves, whose final tangents point in the visual directions of the objects. These curves or trajectories are delineated in fig. 4, where E marks the place of the observer, *zb* the horizon, *b* the hull and *a* the mast of the ship; the lowest image is formed by the curves *bs*E and *ar*E; above this the curves *bv*E and *aw*E give an inverted image, and the curves *bn*E and *am*E exhibit the highest erect image.

In ordinary cases, a ray of light, in crossing different layers of the atmosphere, describes a trajectory, which is nearly the portion of a circle, having six times the diameter of the earth. The rate of inflection is proportional to the tangent of obliquity, and to the difference of the refractive power of the conterminous strata, which may be assumed as the same as their difference of density. This variation in a stratum of air twenty-eight feet thick is only the thousandth part of the whole refraction, and is hence equal to the effect of the quarter of a centesimal degree. But since a much greater difference of heat often occurs within that limit, the incurvation of the trajectory must be proportionally increased. Humidity, by dilating the air, will produce a similar effect, though in a much inferior degree, unless in very high temperatures.

Dr W. H. Wollaston, whose acuteness in devising philosophical experiments was unrivalled, exhibited on a small scale the various appearances occasioned by irregular atmospheric attraction. His paper inserted in the Philosophical Transactions for the year 1800, contains a very clear exposition of the phenomena. On looking lengthwise over the side of a red-hot poker at a distant object, besides the ordinary image, another inverted one was seen within the edge of the streaming air, and a larger erect one still nearer the poker, as represented in fig. 5. The deviation was here only half a degree; but along a red-hot bar of iron, the separation of the images was increased to a degree and a quarter. On directing the eye over the surface of a green board, which had been heated by exposure to the sun, a double image was distinctly seen. Another board, merely wetted with water, betrayed a similar disposition, though it was very difficult to distinguish the irregular image, its elevation being only 3′. But on moistening the surface of the board with alcohol, the images were easily seen at a separation of 15′. Still more conspicuous was the appearance on spreading a little ether over a plate of glass, for the irregular image showed itself $7\frac{1}{2}°$ raised.

It was easy to imitate and examine the natural phenomena by means of a square parallelopiped composed of plate glass. Having filled the half of this with cold water, let hot water be gently poured over it, and the cover placed upon the top. (Plate VII. fig. 6.) On standing a few feet behind the parallelopiped, and looking at distant objects through the stratum where the cold and hot water have become blended, the double and inverted images will for a short time be clearly perceived. But a more durable effect is obtained by filling one third of the glass case, as represented in fig. 6, with syrup or a strong solution of white sugar, another third with distilled water, and the rest with pure alcohol. A mutual penetration slowly takes place between the conterminous surfaces of the alcohol A and the water B, and between this and the syrup C; and after the lapse of perhaps two or three days, the blending spaces become sufficiently broad for observation. The appearances then are nearly such as those that have been sketched in the plate.

This mode of experiment is at once simple and convincing. The theory which Dr Wollaston draws from it is equally ingenious, but not so demonstratively established as its author presumes. It rests chiefly on the supposition, that the stratum occupied by the penetration of two fluids of different densities, graduates in refractive intensity more slowly at its boundaries, and most rapidly in the middle. Or, to borrow the elucidation of geometry, if the successive densities be denoted by ordinates, their summits will form a line of double curvature, having consequently a point of contrary flexure. This principle may be sound, though we can perceive no cogent reason why the curve of intermediate refraction should be convex or concave, or should combine both sorts of incurvation. But admitting this double curvature to be a constituent law, it will very satisfactorily explain the phenomena. The refraction will be the same as if the light had traversed a uniform medium bounded by a surface of that reverted curvature, the convex portion diverging, and the concave portion converging, the parallel rays. The different effects are exhibited in fig. 3.; the distant object O is viewed in its true position by the pencil that arrives at O; it is seen inverted at I, by the rays which enter at *m*; the rays which fall on *r* make it appear again erect at A. This explication is certainly very plausible, and may probably approximate to the truth. But though much has been already achieved, the subject of irregular refraction needs to be revised, and still more closely investigated.

For some further illustrations of the subject of this article, see BAROMETER, CLIMATE, and COLD.

FINIS.

---

H. & J. PILLANS, Printers, Edinburgh.

BAROMETER. PL. 1.

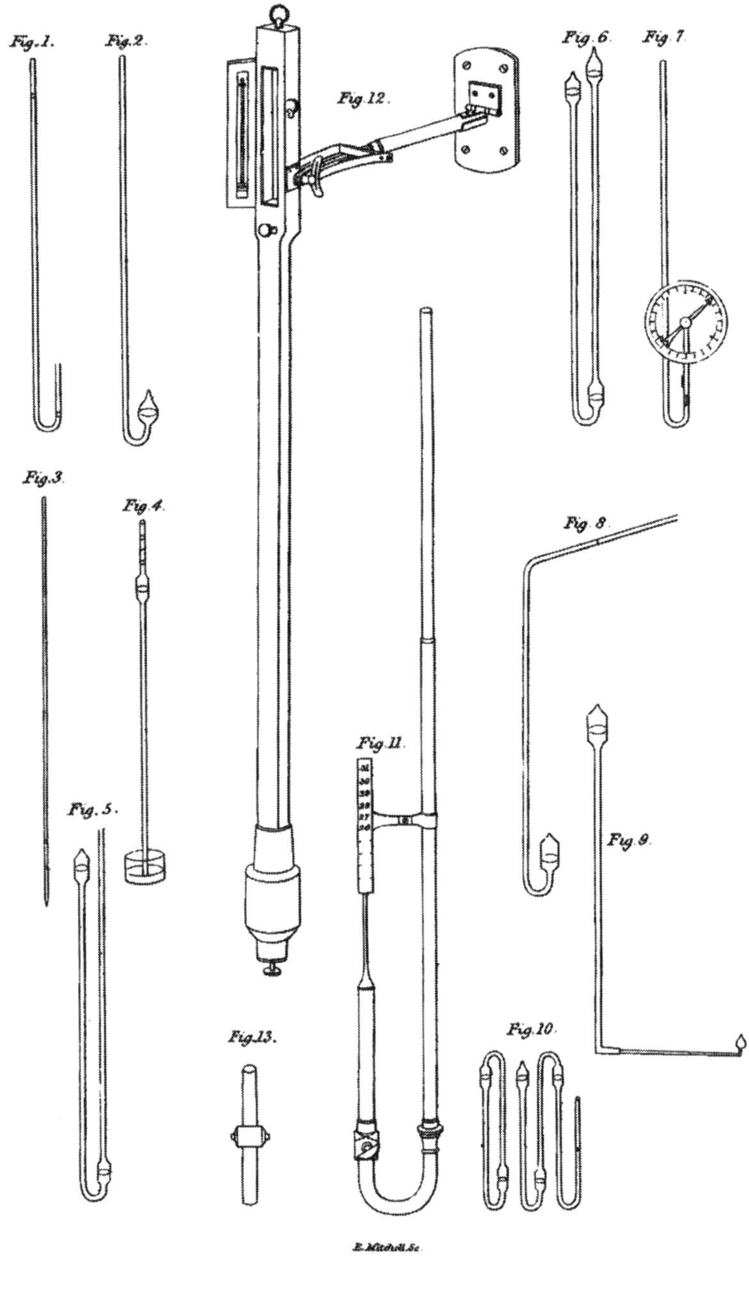


E. Mitchell, Sc.

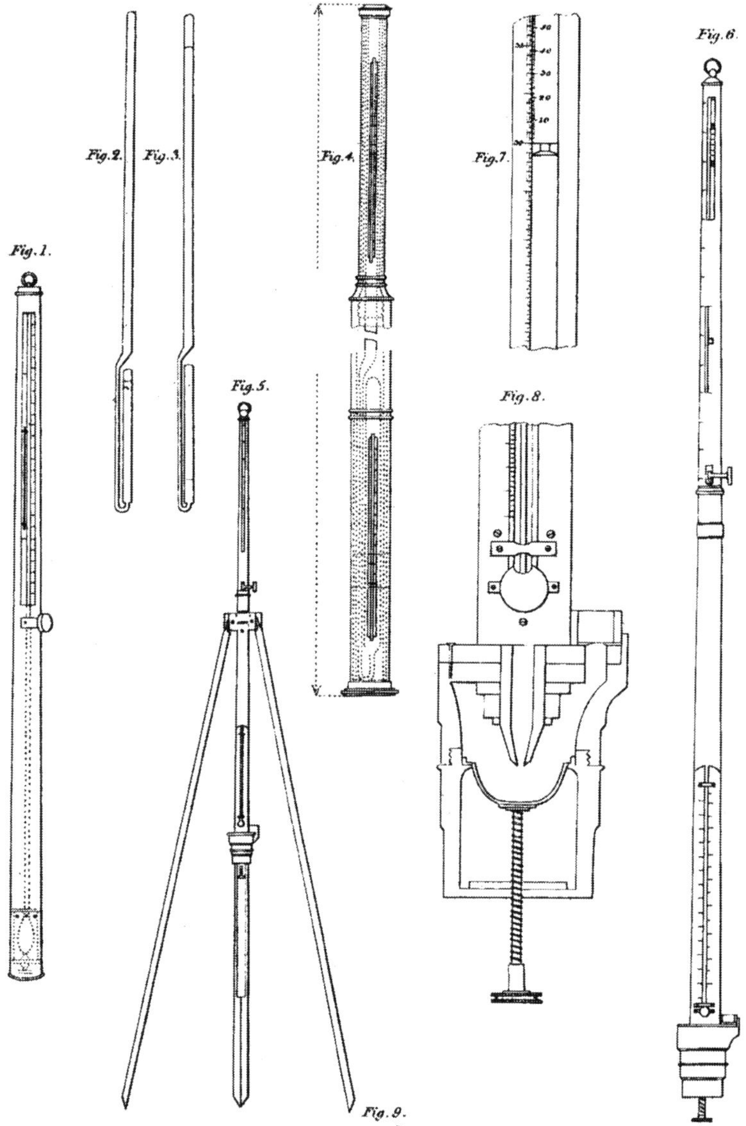


E. Mitchell, Sc.

## COLD. PL. III.

PROFESSOR LESLIE'S APPARATUS FOR COOLING AND FREEZING.

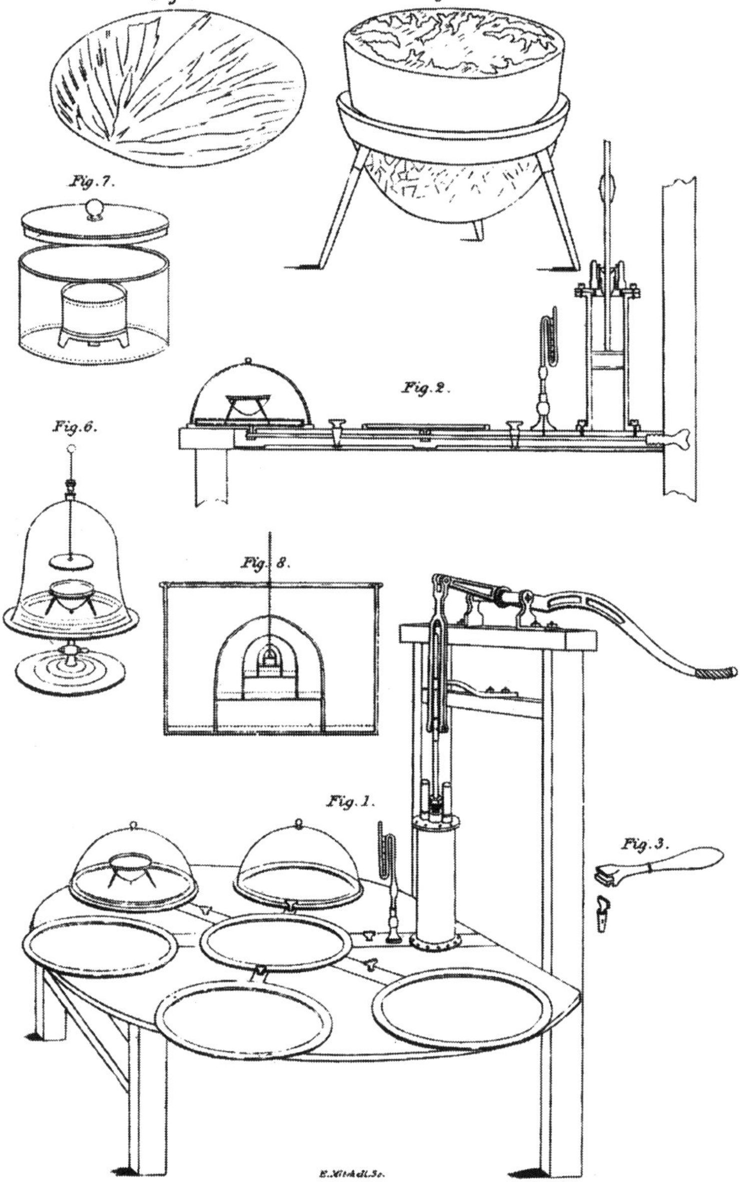

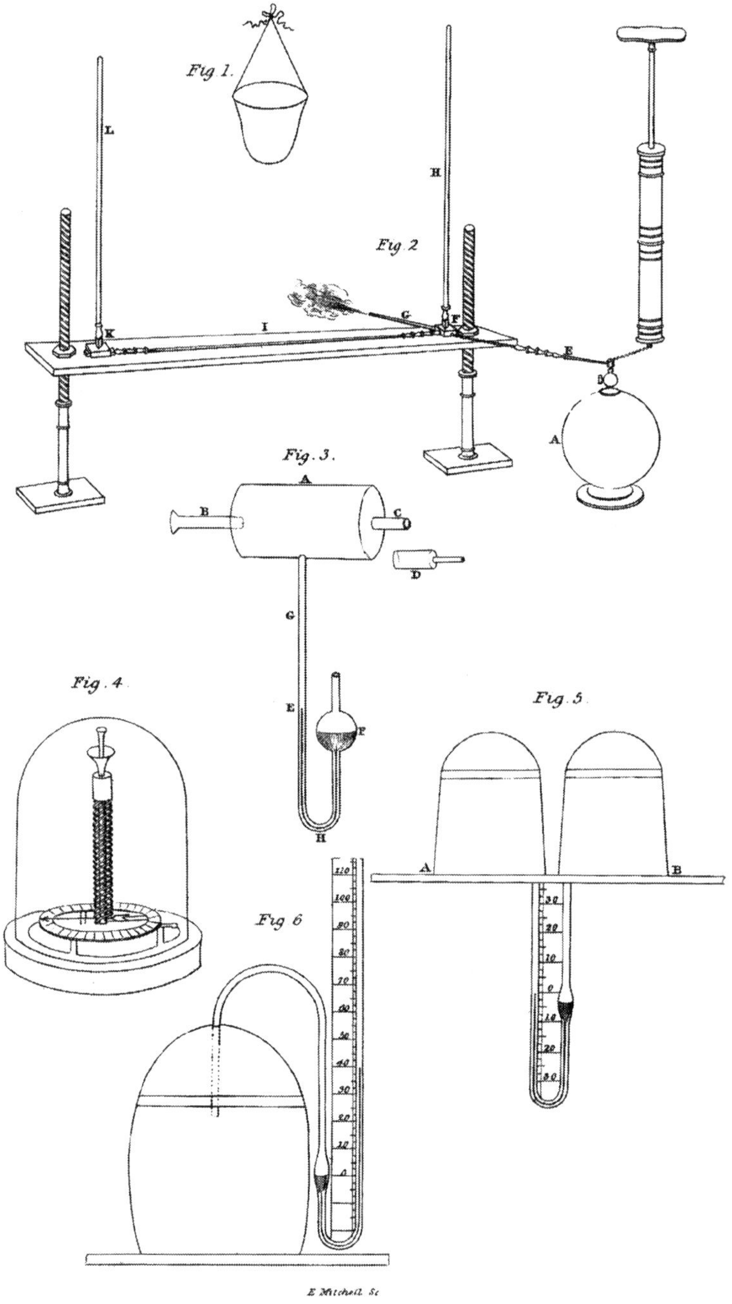


E Mitchell Sc

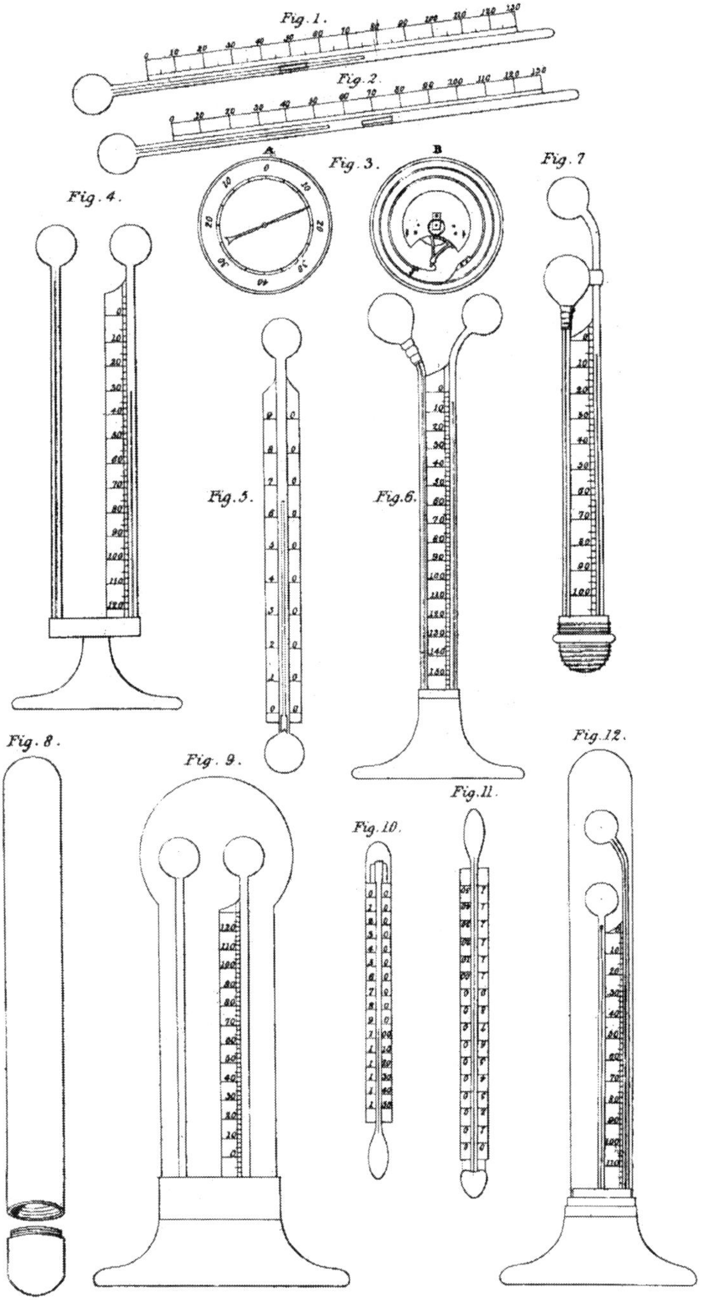


E.Mitchell, sc.

Fig. 2.

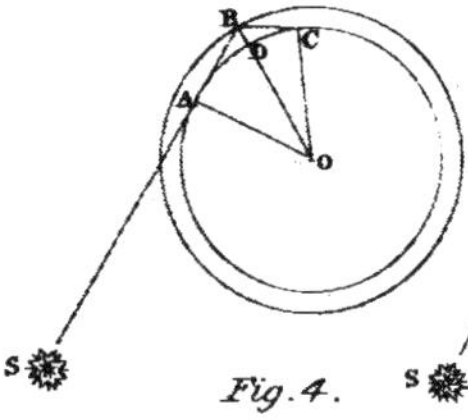


Fig. 3.

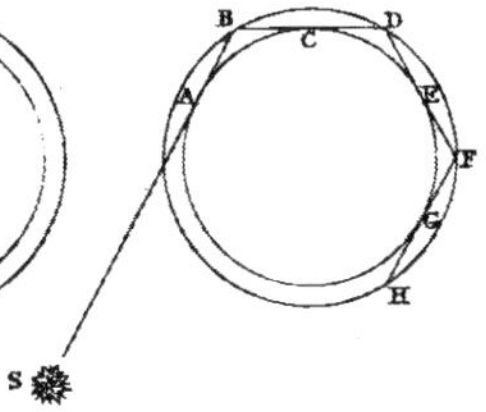


Fig. 1.

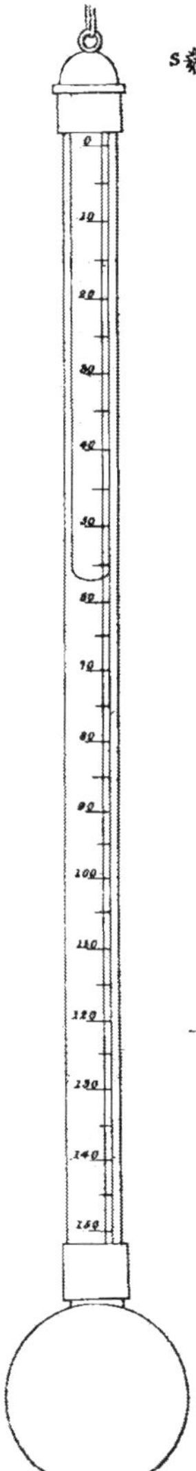

Fig. 4.

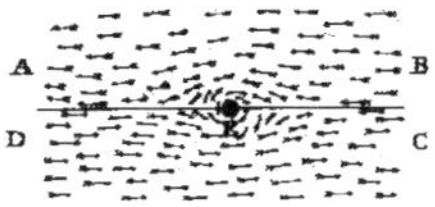


Fig. 7.

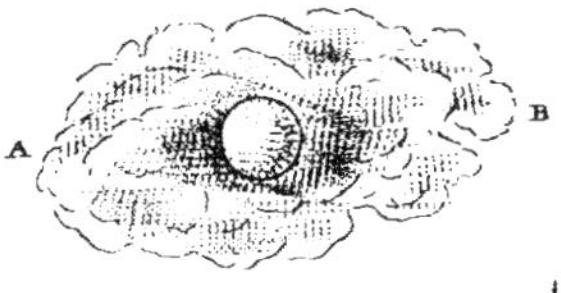


Fig. 5.

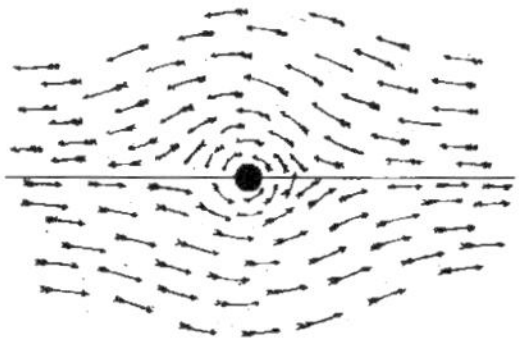

Fig. 8.

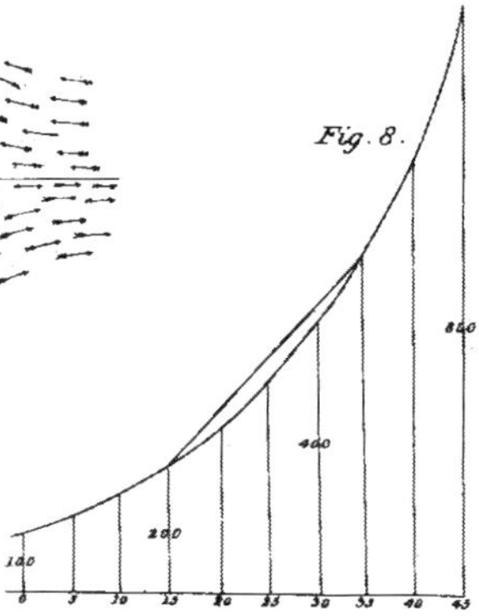


Fig. 6.

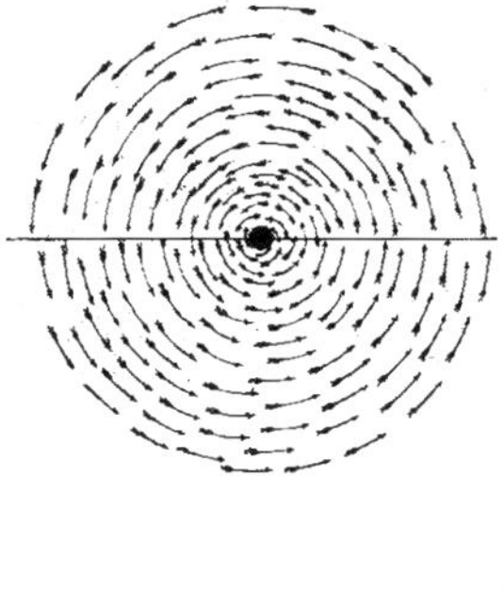

Fig. 9.

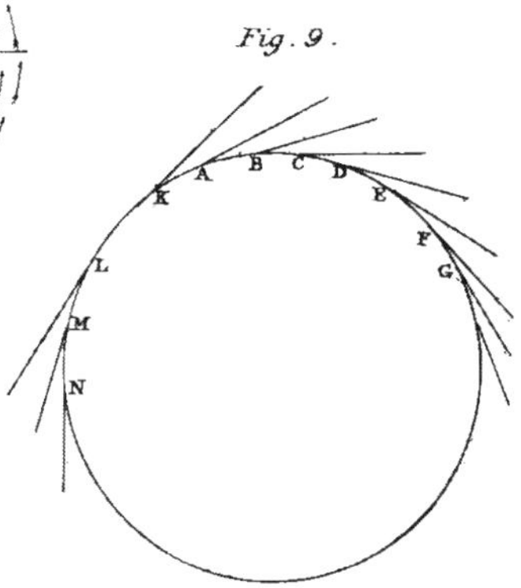


E. Mitchell, Sc.

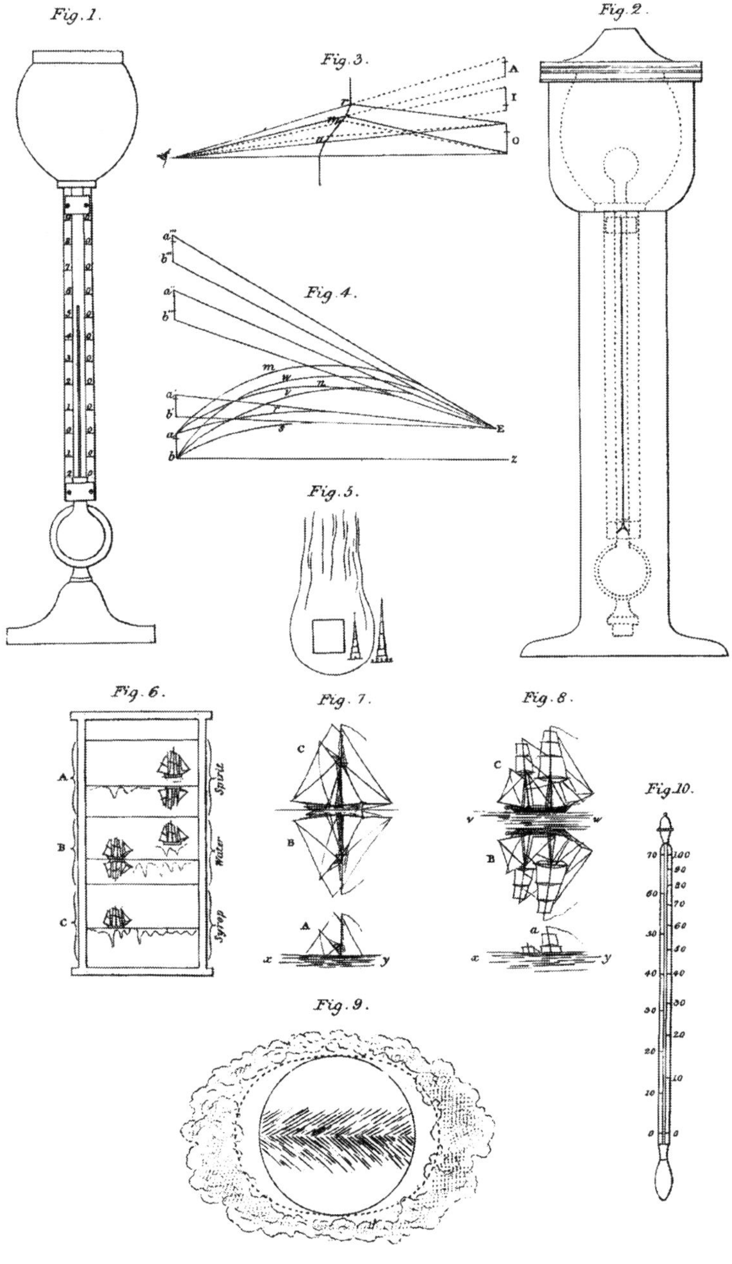


E. Mitchell, Sc.

Printed in Dunstable, United Kingdom

91224059R10317